CELL DEATH

Apoptosis and Other Means to an End

SECOND EDITION

OTHER TITLES FROM COLD SPRING HARBOR LABORATORY PRESS

Cell Death Techniques: A Laboratory Manual

Cell Survival and Cell Death

The p53 Protein: From Cell Regulation to Cancer

Navigating Metabolism

Cancer Evolution

MYC and the Pathway to Cancer

Cell Polarity

Signal Transduction: Principles, Pathways, and Processes

Mammalian Development: Networks, Switches, and Morphogenetic Processes

CELL DEATH

Apoptosis and Other Means to an End

SECOND EDITION

Douglas R. Green

St. Jude Children's Research Hospital

COLD SPRING HARBOR LABORATORY PRESS
Cold Spring Harbor, New York • www.cshlpress.org

Cell Death: Apoptosis and Other Means to an End, Second Edition

Printed in the United States of America

Publisher	John Inglis
Acquisition Editor	Richard Sever
Director of Editorial Development	Jan Argentine
Developmental Editor	David Hatton
Project Manager	Inez Sialiano
Permissions Coordinator	Carol Brown
Director of Publication Services	Linda Sussman
Production Editor	Kathleen Bubbeo
Production Manager	Denise Weiss

Front cover artwork: Painting by Max Ernst, French, born Germany, 1891–1976. *Halleluiah,* 1948. Oil on canvas 40×30 1/8 in. (101.6×76.5 cm). Bequest of Richard S. Zeisler, 2007.275. © 2016 Artists Rights Society (ARS), New York/ADAGP, Paris; The Art Institute of Chicago/Art Resource, NY.

Library of Congress Cataloging-in-Publication Data

Names: Green, Douglas R., author.
Title: Cell death : apoptosis and other means to an end / Douglas R. Green (St. Jude Children's Research Hospital).
Other titles: Means to an end
Description: Second edition. | Cold Spring Harbor, New York : Cold Spring Harbor Laboratory Press, [2018] | Includes bibliographical references and index.
Identifiers: LCCN 2018019767 (print) | LCCN 2018021403 (ebook) | ISBN 9781621822622 (ePub3) | ISBN 9781621822639 (Kindle-Mobi) | ISBN 9781621822134 (cloth) | ISBN 9781621822141 (pbk.)
Subjects: LCSH: Cell death. | Apoptosis.
Classification: LCC QH671 (ebook) | LCC QH671 .G74 2018 (print) | DDC 571.9/36--dc23
LC record available at https://lccn.loc.gov/2018019767

To Rona Mogil, who puts up with me, encourages me,
and inspires me, and to Maggie Green,
who makes me proud every day

Contents

Foreword from the First Edition

Biologists have been relatively slow to recognize and study the degradation processes that operate in cells, compared with the generative processes. Our understanding of protein degradation, for example, lagged well behind our comprehension of protein synthesis. And so has it been for cell death, where understanding followed many years behind the comprehension of cell division. Although it had long been recognized that cell death can be an important part of normal animal development and tissue homeostasis, it was only in 1972 that Kerr, Wyllie, and Currie drew a clear distinction between the conserved cytological features of these normal cell deaths and the very different features of acute pathological cell deaths. They coined the term apoptosis for the former type of cell death and, importantly, suggested that it might reflect the operation of a conserved intracellular death program, by which animal cells can actively kill themselves in a tidy and controlled way.

This important idea remained largely dormant for almost 20 years, and the study of apoptosis remained confined to a small group of aficionados working on diverse organisms. The big bang in the cell death field came from Horvitz and colleagues at the end of the 1980s and early 1990s with the genetic identification of the intracellular proteins that mediate and regulate apoptosis in the nematode *Caenorhabditis elegans* and, soon thereafter, the demonstration that related proteins operate in similar ways in other animals, including humans. In this way, it rapidly emerged that a family of cysteine proteases—the caspases—mediate the apoptotic death program and that a family of regulatory proteins—the BCL-2 proteins—either activate or repress the program. These spectacular findings indicated that apoptosis is a fundamental property of animal cells and that the proteins that mediate and regulate it have been largely conserved in evolution from worms to humans. The findings launched the subject into the cell biological stratosphere, where it remains to this day, having gone from neglect to hysteria in only a few years.

As the cell death field matured, it became increasingly clear that there are multiple ways of activating and repressing the apoptotic program from both inside and outside the cell. It also emerged that the molecular details can vary from organism to organism and that other nonapoptotic death programs can operate in animal cells.

This added complexity has created a pressing need for a comprehensive stock taking—a cool, clear, overview of cell death that cuts through the detail in a logical and engaging way while making it clear where controversy and mystery remain. The author of *Means to an End*, a highly respected leader in the field, has achieved all of this admirably. The writing is remarkably clear and is bolstered by simple, informative figures.

Whether you are a cell death expert or a neophyte, or even a retired cell biologist like me, you are likely to find the book informative, clarifying, and enjoyable. If you are a scientist just starting your career in the cell death field, it is unlikely that you will find a better place to identify important unsolved problems on which to work. If you are a drug developer, you will find an enlightened discussion of how one might design drugs to either encourage dangerous cells to kill themselves or discourage transiently injured cells from doing so. All you need to know about cell death is covered here, with panache, and all in fewer than 250 pages—a remarkable achievement.

MARTIN RAFF
London, July 2010

Preface from the First Edition

This book is not a text book, nor is it a monograph. It is not a history of the field of cell death, nor is it an exhaustive treatment. What it is *intended* to be is a starting point for those who are interested in cell death, especially in mammals, a subject about which I am admittedly passionate. The field has reached an interesting point at which I believe we have a reasonable understanding of the processes that are central to or consequences of cell death and its regulation. The book might be considered an argument that we have, indeed, reached this point.

I wrote this book primarily for students and informed individuals who would like to learn more about cell death. A college level understanding of cell biology and biochemistry is assumed, but I am not sure that it is absolutely essential. I have tried to define terms and processes that are necessary to comprehend the material. But there is certainly a limit: You must have a fairly clear idea of what a protein is, the parts of a cell, and cell signaling and transcription. That said, I have also tried to write the book so that it is of interest to those who work in the field (who will undoubtedly discover my mistakes, biases, and eccentricities).

This book is meant to be read from beginning to end. The order of the chapters is intentional, so that concepts developed in one chapter are built upon in chapters that follow. It is unlikely that any reader other than an expert will learn enough from any given chapter to gain anything close to a complete understanding, but if it is put into the context of the chapters that precede it, there should be many "aha" moments. The illustrations should be helpful, but they do not contain all the information that you need—that is in the text.

At the end of the book is a list called Additional Reading. This list serves two purposes. First, it provides sources of further information and alternative viewpoints. Second, many (but not all) of the landmark papers upon which I based the positions I have taken are included.

As the author, I am responsible for the content of the book, including the mistakes and misconceptions. But under no circumstances should it be thought that I had anything to do with the vast majority of the discoveries on which the concepts, ideas, notions, and musings are based. I admit that I do draw attention to some discoveries that were made in my own lab, by people I am proud of, and I have tried not to do this unduly.

The field of cell death owes a tremendous debt to many pioneers, and I have not given them suitable credit in the text for their phenomenal contributions. In particular, the field as we know it was the work of pathologists who recognized that there are discrete forms of cell death, scientists working in invertebrate models (especially nematodes and flies) who unraveled the genetic basis of cell death, and cancer biologists who identified molecules that control cell death. As I mention above, this book is not a history of the field. However, some of those pioneers who directly influenced my views and understanding of the field bear mention and thanks (in no particular order): Andrew Wyllie, John Kerr, John Cohen, Sten Orrenius, Stanley Korsmeyer, Yoshihide Tsujimoto, Jerry Adams, Martin Raff, Suzanne Cory, David Vaux, John Reed, Junying Yuan, Guido Kroemer, Andreas Strasser, Shigekazu Nagata, Peter Krammer, Michael Hengartner, Xiaodong Wang, David Wallach, Gerard Evan, Jurg Tschopp, Yigong Shi, Guy Salvesen, Tak Mak, Herman Steller, Hao Wu, Michael Karin, Richard Youle, Emad Alnemri, Vishva Dixit, Gerry Melino, and all of the others that I may have inadvertently missed (my apologies).

I must also mention some of those who made this book possible: Richard Sever, my editor and tormentor (just kidding), who cajoled me into writing this, and then stood steadfastly by me throughout; Tudor Moldoveanu, who generated many of the figures showing molecular structures; Guy Salvesen, Eric Baehrecke, and Carlos Lopez who read selected chapters and provided advice; Lisa Bouchier-Hayes and Melissa Parsons who read the entire book, corrected my errors, questioned some of my arguments, and helped enormously; and my family, Rona Mogil and Maggie Green, who put up with this project and with me.

I also would like to thank my parents, who inspired my love of science, and to whom this book is dedicated. When my sister and I were growing up, we spent our family time in the forest, learning to identify trees, birds, and animal tracks, and we thought that it was perfectly normal to veer off highways to collect wildflowers to plant in the yard. Although our parents are not scientists, we both became biologists, and my sister studies penguins on her base in the Antarctic. I chose to study cells, in part because I like to live where one can get a good martini.

I hope you enjoy reading this. Those of us who do science for a living know that often science is very hard work, but it can also be enormous fun. Most of all, I hope that the book encourages you to experience the latter.

Doug Green
Memphis, July 2010

Foreword

"Nobody never dies once created" is a phrase from a famous Japanese poem written by a samurai more than 800 years ago. This is also true for the cells that make up our body. Every day, billions of cells die inside us to maintain tissue homeostasis. Cells also die during animal development, under special conditions such as mechanical or chemical injuries, and from bacterial or viral infections. The first "programmed cell death" process was recognized in silkworm ecdysis by Lockshin and Williams in 1965. Seven years later, Kerr, Wyllie, and Currie discovered an active and programmed death process that was accompanied by cytoplasmic and nuclear condensation, followed by cell shedding from the epithelial layers or engulfment. This cell death process was called "apoptosis."

For years, only a limited number of scientists worked in this field. We find fewer than 100 papers published in 1989 that mention "programmed cell death" or "apoptosis" in their title or abstract. This state of the field changed dramatically in the early 1990s, when apoptosis was shown to be mediated by gene products. In 2000, the number of papers mentioning "programmed cell death" or "apoptosis" was more than 10,000, and last year it exceeded 30,000. As is often seen in rapidly developing fields, not all of the published data are consistent with each other, and some data are not reproducible. Thus, students and postdoctoral fellows who begin working on cell death can be overwhelmed by huge amounts of data that are not always consistent.

Dr. Douglas Green has been working on programmed cell death from its early days. Today he is one of the most prominent scientists in this field, and he is one of the most qualified persons to review it. When I read the first edition of this book, *Means to an End*, in 2010, I was deeply impressed by the clear and straightforward overview of the field, with simple but informative figures. The controversial and mysterious unresolved issues were well discussed, effectively directing the future of the field. Since then, the cell death field has witnessed many breakthroughs. The mechanisms of apoptosis, including the clearance of apoptotic cells, have been described in increasing detail. In addition, necrosis, which used to be regarded as an accidental passive cell death, was found to be mediated by gene products and is now categorized into two processes, necroptosis and pyroptosis.

As we can infer from the title of this book, this is no ordinary textbook on cell death. Although we can certainly learn all about cell death from the book, it is also a story that reflects the philosophy of the author, a leading scientist, which makes this book and the field even more interesting and attractive. I hope that many students and young scientists, and even established scientists in different fields, read this book and are inspired to join our society of cell death, which will never die!

SHIGEKAZU NAGATA
Osaka, Japan, April 2018

Preface

You are most likely holding in your hands an actual, physical book, written by a single author (i.e., me), about how cells die, why they die, what happens after they die, and the consequences of such processes for health and disease. At the time I am writing this, in the latter part of the second decade of the 21st century, books such as this one are not especially common. Academics in other disciplines write books (very often as solo authors), but science is more of a collaborative venture, as are the books written about scientific fields. More to the point, though, most of what we read in science is in the form of short publications, primary articles and reviews, that fly into our info-spheres, apparently at the speed of electrons. Books (and especially physical books) are a bit "retro." But there are a number of reasons I wanted to write a book about this field, cell death and survival, and then wanted to revise it in a second edition.

The field of cell death is certainly not "new." However, when I wrote the first edition of this book, I felt we had reached a point where our understanding of the processes involved in how cells die were sufficiently developed that we cell death researchers had a pretty good idea of how much of it works. As with any field, there remained differences of opinion regarding some of the finer aspects, but overall there was a lot of agreement. I thought that by putting this together, as a single author, I might be able to outline what amounts to a big picture view of these processes and describe them so that readers with some understanding of biology (perhaps at the college level) could follow all of it, and other readers with a more general interest in science could follow most of it. It was meant to be read in a few sittings, from beginning to end, with the hope that it would stimulate others to think about cell death and how it might relate to other fields of biomedical research. If some of it gets a bit "heavy," just skip ahead—most of it is pretty easy to grasp, and most of you (I hope) will find it smooth sailing.

In the several years since the publication of the first edition, there has been significant advancement of our understanding of processes involved in and surrounding cell death. Many of the mysteries raised as questions in that edition have been answered, as were questions we did not ask (or imagine!). It was time, I thought, to bring things up to date, and 2 years later, we have the edition you now hold. Surprisingly, I found that the view from 2011 was mostly correct from the current perspective, and we had refined rather than changed many of our ideas, while "filling in" some of the blanks. But our understanding continues to evolve, and some of what we have learned has dramatically shifted our views. I suspect that a few years from now much of what we think we know will continue to have been refined and may well undergo further shifts. We are far from "done." Cell death is a core biological process and, as such, is integrated with the whole of biology at every level. We are only beginning to apply our knowledge, and as we take our ideas to the proving grounds, we will continue to learn how little we actually "know." That is how science works, and a large part of the fun of it.

In the preface to the first edition, I noted the many pioneers (in no particular order) who have brought this field to its current state. Without repeating them here, I would add a number who have continued to propel the field (again, in no particular order): Seamus Martin, Eric Baerhecke, Feng Shao, Tony Letai, John Silke, Loren Walensky, Bill Kaiser, Andreas Linkermann, Marcus Conrad, Peter Czabotar, Peter Vandenabeele, Thirumala Devi-Kanneganti, Kim Newton, Andrew Oberst, Cristina Munoz-Pinedo, Scott Lowe, Jerry Chipuk, Liming Su, Karen Vousden, David Huang, Henning Walczak, Junying Yuan, Alexi Degterev, Joe Opferman, Mike Overholtzer, Pascal Meier, Kevin Ryan, Andy Villunger, Poul Sorenson, Marja Jaatela, Kodi Ravichandra, Francis Chan, Richard Krawaki, Marion MacFarlane, Manolis Pasparakas, Sudan He, Ana Garcia-Saez, Stephen Tait, Ed Mocarski, Lorenzo Galluzi, James Vince, Brent Stockwell, Janelle Ayers, Sidd Balachandran, Igor Brodsky, Kate Fitzgerald, Mo Kamfi, Peter Brosch, Domagoj Vucik, James Murphy, Scott Dixon, Valerian Kagan, Lisa Bouchier-Hayes, Russell Vance, Julie Magarian Blander, Dagmar Kulms, Philipp Jost, Tom Ferguson, Tom Vandenberg, Mathieu Bertrand, Atan Gross, and many others I have inadvertently omitted.

This book would not have been possible without a number of individuals. Richard Sever, of Cold Spring Harbor Laboratory Press, again encouraged me to do this revision and provided me with a "getaway" to draft the entire book. My project manager at CSHLP, Inez Sialiano, has been forever patient, and our independent developmental editor, David Hatton, was simply amazing. Huge thanks as well to Tudor Moldoveanu, who continued to provide lovely structures that went through seemingly endless tweaks without complaint. Thank you as well to the members of the Green Lab, who put up with my absences and distractions but probably appreciated that I was not ever present to interfere with their research, and who (together with past lab members) provide the rich intellectual environment in which such thought flourishes. This field does not progress independently, and I thank the very many researchers who have elucidated the processes described in these pages, and apologize for my mistakes, misinterpretations, and misrepresentations of their important studies. Finally, thank you to Rona Mogil, who has had to listen to my frustrations and put up with the disaster areas that are my work spaces (and closet!) while I worked on this project, and to Maggie Green, who buys me drinks and to whom Rona complains. This book is dedicated to them.

DOUG GREEN
May 2018

Introduction

Like all living things, cells die. Indeed, a great many cells in our bodies die throughout our lives, and their deaths are essential for our survival. They die by highly conserved mechanisms that might have their evolutionary origins more than a billion years ago. This book is about how that death happens, and how it contributes to physiological homeostasis and disease. The focus is on cell death in animals and, to a great extent, on only one form of cell death, called apoptosis. This is for two reasons. First, most cells that die in humans die by apoptosis. Second, it is the type of cell death about which we currently know most. So, although other types of cell death are covered here in some detail, most of our discussion concerns apoptosis.

The underlying mechanisms of apoptotic cell death are found throughout the animal kingdom, but probably nowhere else.[1] In all animals studied (including many of the phyla), the features of this type of cell death are the same—the dying cell effectively "packages" itself to be eaten and digested by healthy cells. Furthermore, many of the specific molecules involved in this process are conserved in animals. However, the specific molecular pathways, although similar, can have fundamental differences. Throughout most of this book, we focus on the molecular pathways of apoptosis (and cell death in general) that function in humans. This unabashedly anthropocentric (or, at least, "backbone-centric") view is our goal, with apologies in advance to those readers who consider themselves "fly people" or "worm people," and those with interests in other organisms.[2]

We will quickly step off into the deep end of the molecular pool by covering the biochemical mechanisms of cell death. For reasons that will become clear, a "bottom-

[1] There is a literature that explores cell death in other types of organisms, including plants and yeast, and it remains possible that within the cell death mechanisms in such organisms is a vestige of a far more ancient process than we currently suspect. However, the molecules involved in the process are, at best, very distantly related to those in animals, and the actual pathways remain to be elucidated.

[2] In fact, it is largely owing to those who study such organisms (especially nematodes and insects) that we know so much about the molecular mechanisms of apoptosis, and hence the apology (however flippant it appears) is meant with sincerity. Indeed, Robert Horvitz was awarded a Nobel Prize in Medicine and Physiology in 2002 for his pioneering studies on cell death in nematodes.

up" view of cell death by apoptosis makes up the first several chapters. But before we dive in, it might be useful first to say a word about how the chapters that follow are organized.

- Chapter 1 is essentially a synopsis, a quick take on the rest of the book. It is a chance to get our bearings and take a stab at the big picture, starting with why cells die and the three major types of cell death. It goes on to outline the molecular mechanisms covered in subsequent chapters.
- Chapters 2 and 3 concern caspases, the proteases that orchestrate apoptosis by cleaving substrates in the cell. Chapter 2 introduces the caspases and explores those substrates that have known roles in apoptosis. We also discuss caspases that are not involved in apoptosis, per se. Chapter 3 considers the biochemistry of activation of different types of caspases, as well as inhibitors of caspases and their roles.
- Chapters 4 and 5 cover the mitochondrial pathway of apoptosis, the major way in which apoptosis occurs, at least in the vertebrates. Chapter 4 discusses the events that occur once the mitochondrial outer membrane is permeabilized, and how this leads to caspase activation. It also introduces the caspase activation pathways of flies and nematode worms, together with ideas on how apoptosis may have evolved. Chapter 5 introduces the BCL-2 family of proteins, whose complex interactions link different signals for cell death to the permeabilization of the mitochondrial outer membrane.
- Chapter 6 considers another way apoptosis is engaged in vertebrates—by cell-surface death receptors—and how these specialized receptors engage a distinct pathway of caspase activation. This pathway can also link to the mitochondrial pathway to cause apoptosis.
- Chapter 7 looks at additional pathways of caspase activation. One is engaged by signals from infectious organisms and some inert substances and can result either in apoptosis or in a form of necrotic cell death called pyroptosis. It also triggers inflammatory responses. Another pathway of caspase activation involves the most highly conserved of the caspases and how it is activated, but its role in cell death is obscure.
- Chapter 8 explores the other major forms of cell death—necrosis and autophagic cell death. Although necrosis can occur passively, as a cell incurs more damage than it can repair, necrosis can also occur in a regulated manner. One form of regulated necrosis, called necroptosis, is covered in some detail. Autophagy, which normally functions as a survival mechanism, is overviewed, as are its roles in cell death.
- Chapter 9 covers what happens after a cell dies. Regardless of how it died, a dead cell is rapidly cleared from the body by phagocytosis. This is a complex process, involving many receptors and signals, and leads to "waste management" processes

in the engulfing cell. And once the dying cell is cleared, there are additional consequences, including effects on the immune system and proliferation of healthy cells.

- Chapter 10 provides examples of cell death in development, exploring how cell death functions in sculpting the embryo and in selecting functional cells. During development, cells are "specified" for cell death, and we discuss how this occurs and how these signals engage the cell death pathways.
- Chapter 11 introduces the idea that cancer is, in part, a disease of defective cell death. We discuss the mechanisms that are in place to prevent cancer and how these link to the machinery of apoptosis, as well as the roles cell death might play in promoting cancer and in cancer therapy.
- Finally, Chapter 12 explores the mechanisms of cell death as we understand them and how these are tested. These include formal models and their consequences for biology, as well as the practical applications of these mechanisms to the treatment of disease.

The last chapter is followed by a section called Additional Reading. These suggestions serve several purposes. First, these are materials from which readers can obtain further information and alternative viewpoints. Second, many (but not all) of the landmark papers are included here. The cell death processes discussed in the text are largely based on these papers.

The field of cell death is, perhaps paradoxically, vibrant, and it can be difficult to navigate the many concepts and controversies. It is hoped that this book will serve as a first step in the reader's exploration of the core biological processes of cell death and how it relates to health and disease. It is also hoped that some of the fun and excitement that infuses this area of research will rub off a bit as well.

CHAPTER 1

A Matter of Life and Death

CELL DEATH IS ESSENTIAL

Every second, something on the order of 1 million cells die in our bodies.[1] This is a good thing, because cell death is central to efficient homeostasis and adaptation to a changing environment. When, for some reason, it does not occur, the consequences can be catastrophic, manifesting as cancer, autoimmunity, or other maladies. Alternatively, if cell death occurs at the wrong time and place, this can also produce untoward effects, such as stroke, degeneration, heart attack, and many other injuries; thus, cell death is a major component of disease.

Cells die for a variety of reasons. For example, they can become physically or chemically stressed to the point that they cannot maintain their integrity. More often, however, such stresses engage an active cell death process before the stress becomes overwhelming; that is, the cell essentially commits suicide. Most, but not all, cell suicide occurs by a process we call apoptosis.

Another stress that triggers cell death is infection of the cell by an organism that would use the cell for its own parasitic ends (i.e., to make additional parasites). Such cell death can protect the body from further infection, unless the parasite can prevent this protective death until it has reproduced. In organisms with immune systems, lymphocytes and phagocytic cells can detect an infected cell and instruct the cell to die. This cell death can be apoptosis or another type of active cell death.

But it is not only in response to stress that cells die. In the course of normal development, cells can serve functions that become superfluous, and developmental cues (in the form of specific molecules) can signal such a cell to die. Technically, the death of a cell at a prescribed time in development is called programmed cell death, although this term is often applied to apoptosis in general, probably

[1] We do not know the exact number, but if we consider only one type of cell, the neutrophil, it is estimated that up to 1 million of these cells die per second. The turnover of mucosal epithelium (in the lung and gut) is also very high. One million per second is probably an underestimate, but it is in the ballpark.

incorrectly.[2] Another situation in which cell death occurs is when a cell acquires a mutation that allows it to lose its social inhibitions and proliferate as a cancer. Tumor-suppressor mechanisms that are built into our cells instruct them to undergo suicide rather than imperil the body at large.

TYPES OF CELL DEATH

There are three major types of cell death[3]: apoptosis, autophagic cell death, and necrosis. Apoptosis (sometimes called "type I cell death") means, literally, a "falling off."[4] The term was coined to invoke leaves falling from a tree, because cells that die by apoptosis in tissues usually do so in an apparently random manner.

When cells die by apoptosis, the contents of the dying cell remain contained in membranes. The plasma membrane contorts into "blebs," and often the cell fragments into smaller membrane-bound "apoptotic bodies" (this term is also applied to apoptotic cell "corpses" that do not break up). The nucleus undergoes characteristic changes, including condensation, and it, too, often breaks up. Another characteristic feature of apoptosis is the cleavage of DNA, usually into pieces that are multiples of nucleosomes that appear as a "ladder" on agarose gels. Other organelles do not undergo dramatic morphological changes. The dying cell shrinks, and, if it is adherent, it detaches from surrounding cells. Some of the morphological changes associated with apoptosis are shown in Figure 1.1.

Apoptotic bodies are rapidly removed by other healthy cells through phagocytosis ("eating by cells"). In a relatively short time, nothing obvious remains to indicate that a cell had ever been there. Generally, apoptosis is not associated with a subsequent inflammatory response (but it can be in some settings).

Autophagic cell death (type II cell death) differs from apoptosis and is less well understood. It involves a cellular mechanism called autophagy (literally, "self-eating") that is normally involved in sustaining cell metabolism under conditions of nutrient deprivation. Whether autophagy kills the cell by autophagic cell death is controversial, and most, but not all, studies suggest that autophagy accompanies such cell death in a last-ditch effort to keep the compromised cell alive. Often, this form of cell death

[2] The idea that apoptosis is "programmed" comes from early suggestions that activation of gene transcription is required for this form of cell death (which is sometimes, but not always, true, as we discuss in later chapters). However, a distinction between programmed cell death and apoptosis is worth making because not all apoptosis is "programmed" by genetic events, and not all developmentally programmed cell death occurs by apoptosis. But this may be a losing proposition because a great many people use the terms interchangeably. In this book, we avoid the use of "programmed cell death" unless we are discussing specific developmental events.

[3] This is based on recommendations made by the Nomenclature Committee on Cell Death.

[4] Although it is often related that the first use of the term comes from Homer, a careful perusal of that esteemed author's works by Mauro Degli-Esposti failed to turn up this word. Instead, the first use of the term appears to be in the works of Hippocrates, who used it in a medical context 2400 years ago, referring to the "falling off" of bones in gangrene. It was used as well some 550 years later by the celebrated physician Galen and extended to the "falling off" of scabs.

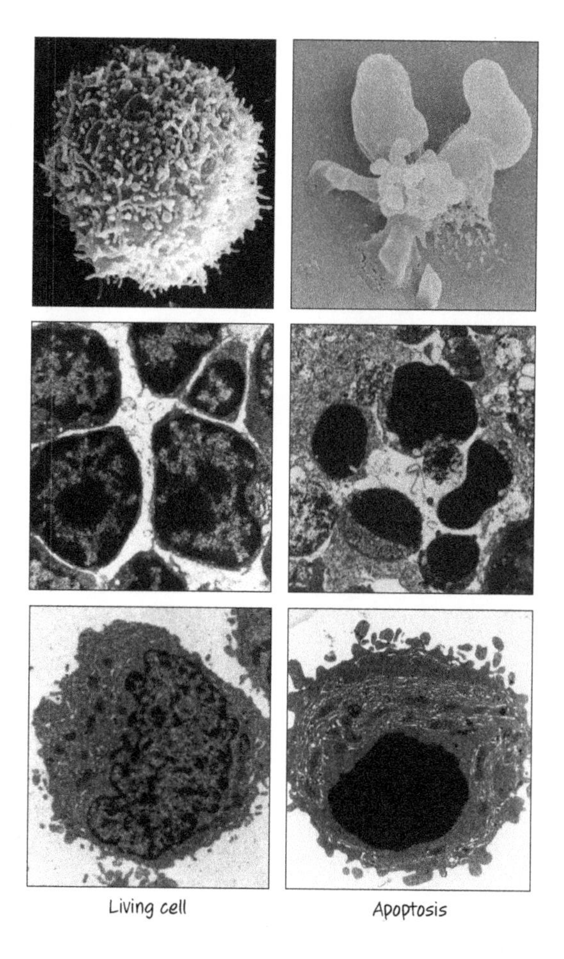

Figure 1.1. Apoptosis snapshots.

occurs when, for some reason, apoptosis is blocked. Autophagic cell death is characterized by the appearance of vacuoles in the cell and does not include extensive condensation of the nucleus (Fig. 1.2).

The third major type of cell death—necrosis (also known as type III cell death)—is a messy sort of death in which the cell swells, bursts, and decomposes. It can occur when the plasma membrane is ruptured or when energy levels drop so quickly that

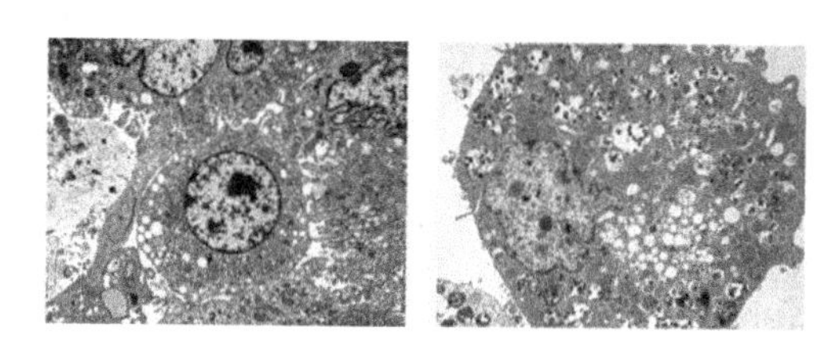

Figure 1.2. Autophagic cell death. (*Left*) The cell in the *center* is undergoing autophagic cell death. (*Right*) Another autophagic cell death.

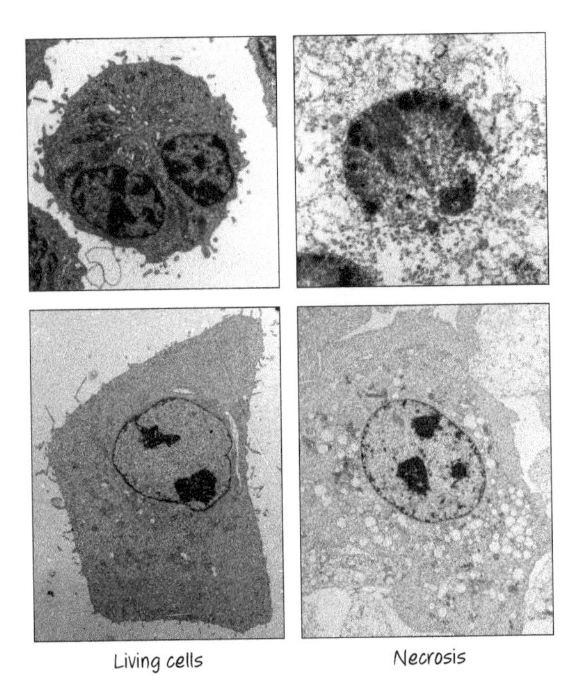

Figure 1.3. Necrosis.

the cells cannot sustain themselves (Fig. 1.3). However, there are forms of necrosis that are regulated by molecular processes; in other words, cells contain additional suicide pathways that result in necrosis rather than apoptosis.

There is another form of cell death worth mentioning, but it is very specialized. When skin is created, epidermal keratinocytes die by a process called cornification; the nucleus is degraded and the proteins of the cell are extensively cross-linked to make a hard, dead material that has a very important barrier role in our bodies. This form of cell death is distinct from those we consider in this book, and, because it is a very specialized process, we do not discuss it further.

There are many other types of cell death that have been described, with terms such as pyroptosis, necroptosis, ferroptosis, entosis, aponecrosis, paraptosis, mitotic catastrophe, etc. In general, these do not qualify as unique classes of cell death, and thus we generally include them under the three major types of cell death outlined above.

ACTIVE CELL DEATH: SUICIDE VERSUS SABOTAGE

Another way we can think about cell death is whether it is active or passive. In passive cell death, the cell is damaged to the point that it simply gives up the ghost, and the cell does not participate in the actual death event. Alternatively, in active cell death, cellular processes have a role in the coup de grâce.

We can further parse active cell death into mechanisms that resemble suicide, in which the cell kills itself (this happens in all three types of death discussed above), or sabotage, in which the disruption of an active cellular process (not otherwise involved in cell death) kills the cell. Removing railway ties does not "kill" a train on the track unless that train is actively moving.[5]

The distinction might be useful in thinking about many different forms of cell death. It is tempting to think of cellular suicide as something that evolved as a means to engage cell death in response to a signal, whereas sabotage lacks such selective value. For example, it is possible (but not proven) that disruption of some parts of the cell cycle, or of normal metabolic processes, can kill cells in a manner that depends on the disrupted process being active, but such an active cell death by sabotage would not have been selected by evolution for that purpose. However, ultimately this is speculation, and the lines between these distinctions might blur.

APOPTOSIS FROM THE "BOTTOM UP"

Probably the best example of active, cellular suicide is apoptosis. The process of apoptosis is perhaps most easily understood by viewing it in reverse, beginning at the extreme end of a cell's life and tracking back to the healthy cell and what caused it to die. The "packaging" of the dying cell and the signals it sends to other cells to clear it away are brought about by the cleavage of many hundreds of different proteins in the cell. The proteases (protein-cutting enzymes) responsible for this cleavage are called caspases, and, in particular, they are referred to as a subset called "executioner caspases."

Executioner caspases are already present in healthy cells, but in an inactive form. They become activated when they are themselves cut by other proteases, called "initiator caspases." These, too, are present in healthy cells, but their activation mechanism is different. Initiator caspases are inactive until two identical chains are brought together by adapter proteins to produce an active enzyme ("adapter" is a general term used to describe proteins that bind to and bring together other proteins; we use the term here to refer to those that perform this function for the initiator caspase chains).

Different adapter proteins are engaged by different apoptotic stimuli and define distinct apoptotic pathways (Fig. 1.4). When adapters promote the activation of initiator caspases, the latter cut and thereby activate the executioner caspases. The executioner caspases then cleave hundreds of substrates, and the cell undergoes cell death and the morphological changes associated with apoptosis. The different pathways

[5] This analogy relates to the origin of "sabotage," from "sabot," French for "wooden railway tie," and arguably the root of the term. Subsequently, it was related to sabotage of looms by Dutch weavers in the 15th century who would throw wooden shoes (also "sabot") to break the cogs, but the term did not appear until much later, when railway "sabotage" occurred.

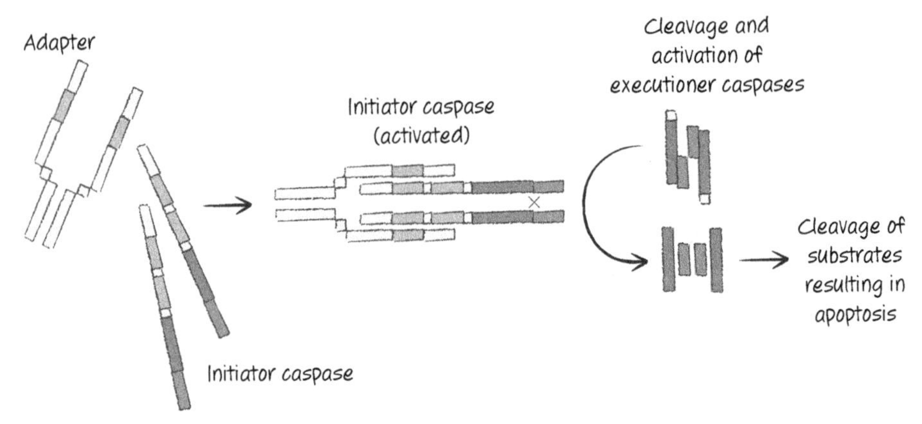

Figure 1.4. General caspase activation pathways.

converge on the same set of executioner caspases, which is why our bottom-up view is warranted and is the way in which we unravel these pathways in subsequent chapters of this book.

ROADS TO RUIN: THE PATHWAYS OF APOPTOSIS

Most apoptosis in vertebrates (at least) occurs by what is termed the "mitochondrial pathway." In this pathway, conditions that promote cell death engage a set of related proteins, called the BCL-2 family (Fig. 1.5), that control the integrity of the outer membranes of mitochondria in the cell. Proapoptotic BCL-2 effectors disrupt outer mitochondrial membranes, whereas antiapoptotic BCL-2 proteins prevent this disruption and thereby prevent apoptosis. A third set of BCL-2 proteins regulates the other two types.

If mitochondrial outer membrane permeabilization (MOMP) occurs, soluble proteins of the intermembrane space (between the outer and inner mitochondrial membranes) diffuse into the cytosol. These proteins include cytochrome *c*, which also has a central role in mitochondrial physiology.

When cytochrome *c* reaches the cytosol, it interacts with an adapter protein that is present there, causing the adapter to cluster (oligomerize) and bind to monomers of one of the initiator caspases. The initiator caspase is thus activated (Fig. 1.6) and, in turn, cleaves and thereby activates executioner caspases that themselves go on to cleave their substrates. Apoptosis ensues.

A second pathway for caspase activation and apoptosis involves specialized receptors on the cell surface, called death receptors. When the ligands for these receptors bind, the intracellular region of these receptors engages a specific adapter molecule (distinct from that of the mitochondrial pathway). This, in turn, binds to

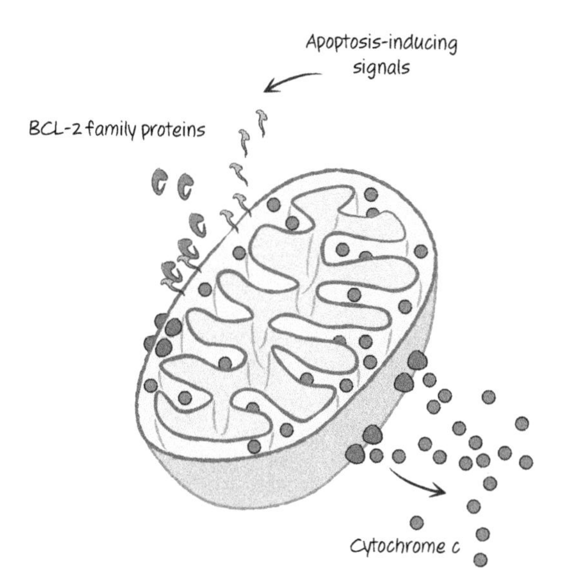

Figure 1.5. Simplified scheme of the first part of the mitochondrial pathway. BCL-2 proteins control mitochondrial outer membrane permeabilization (MOMP), releasing proteins that include cytochrome *c*.

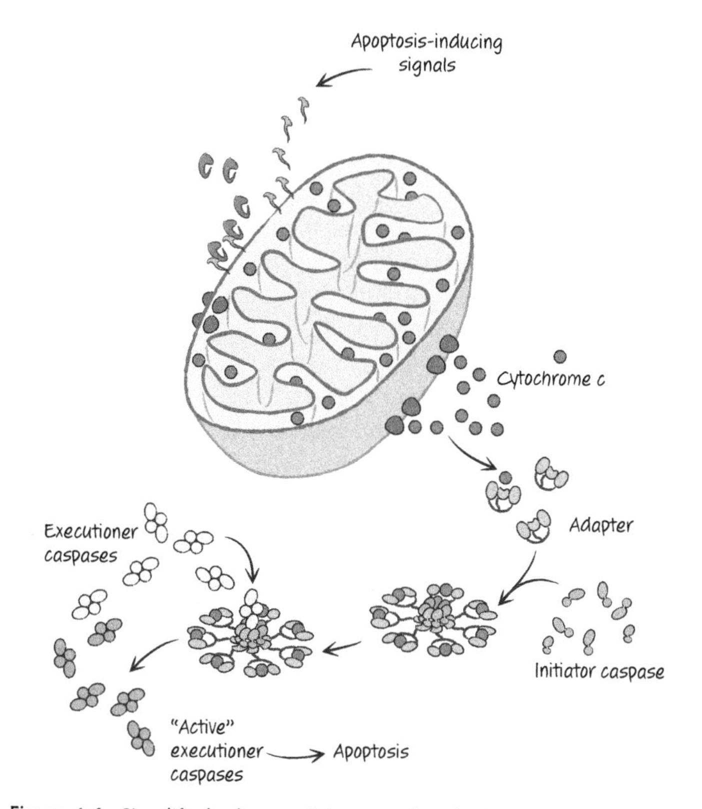

Figure 1.6. Simplified scheme of the mitochondrial pathway of apoptosis.

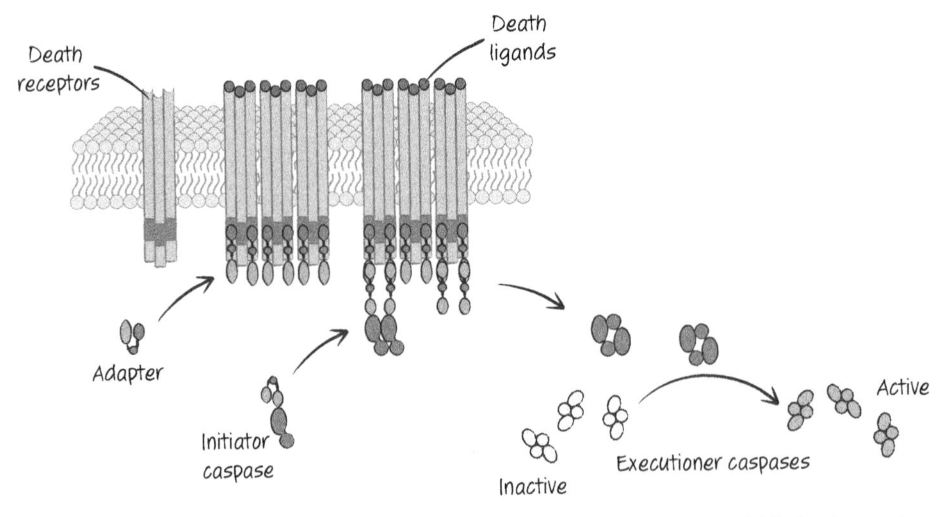

Figure 1.7. Simplified scheme of the death-receptor pathway of apoptosis, in which death receptors on the plasma membrane bind their ligands and, through several intermediaries, ultimately activate executioner caspases.

monomers of an initiator caspase (different from that of the mitochondrial pathway), activating it. The initiator caspase cleaves and thereby activates executioner caspases (the same as above), and apoptosis proceeds (Fig. 1.7).

As mentioned earlier, another type of signal that can induce cell death is infection of a cell. Our cells have sensors that detect such infection, and some of these also act to engage adapters (distinct from those of other pathways) for a caspase related to initiator caspases. Unlike the initiator caspases, however, this caspase (and another we will discuss in a later chapter) can also cleave a specific substrate that causes, not apoptosis, but a form of necrosis called pyroptosis. It also processes and allows the secretion of mediators that engage host defenses to fight the infection (Fig. 1.8).

The details of these apoptosis pathways are discussed in much more detail in the rest of this book, but the basics of the pathways are as described here. As we will see, apoptosis in other animals also follows these general schemes of adapter–caspase interactions.

THE "TOP" OF THE PATHWAY: WHAT INDUCES APOPTOSIS?

The pathways we have briefly considered above are triggered by distinct types of signals. Most events that stress a cell engage the mitochondrial pathway of apoptosis. These include loss of growth factors, agents that damage DNA, disruption of the cytoskeleton, extensive protein aggregation, and other forms of cell stress. This pathway is also often triggered by developmental signals and by tumor-suppression mechanisms. The death-receptor pathway is engaged by the specific ligands that bind to

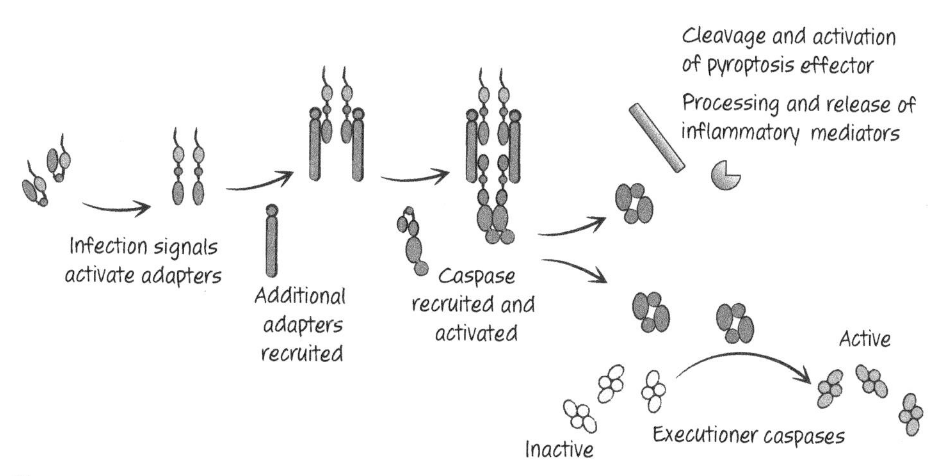

Figure 1.8. Simplified inflammasome, showing events leading to the twin events of activation of executioner caspases and the secretion of inflammatory mediators.

A GENERAL PROBLEM IN CELL DEATH RESEARCH

What death is might seem obvious. But, in practice, cell death turns out to be a bit tricky to define in a general way. If a cell loses its plasma membrane integrity, it is certainly dead, but, as we mentioned above, apoptotic cells are often engulfed before this event, and a cell that has been eaten and is being digested is dead, too. However, if we wait until that event to register this as a "cell death," we face a kinetic problem, because some cells will be long gone before others engage the pathway, and any count of dying cells could be an underestimate. Therefore, we might prefer to call a cell "dead" if it has activated caspases and is undergoing apoptosis. However, here we have additional problems because it often happens that cells that initiate apoptosis might not die (if, e.g., the extent of caspase activation is low) or, if caspases are experimentally blocked, death ensues nevertheless (although without the features of apoptosis). We might count a cell as "dead" if it loses the ability to proliferate or perform its functions. However, cells that lose the ability to proliferate can persist for years (indeed, this is very important for many of our cells), and functions can change. So, when is a cell "dead"?

In practice, the definition of "death" is often tied to the cells and systems under study. Therefore, surrogate markers of cell death are often used, and it is assumed that these represent cellular demise. Although extensive progress has been made using this approach, mistakes abound. Undoubtedly, possible mistakes in this book might themselves be the result of the erroneous interpretation of such surrogates as "death," and it is advisable, therefore, to keep this issue in mind.

the receptors. The inflammasome pathway is induced by signals resulting from infection, although some inert substances can act in the same way if taken up into the cell. Therefore, apoptosis is engaged by a wide range of conditions, and there can be extensive overlap with regard to which pathway can be triggered by a specific agent, depending on the type of cell. As we will also see, these pathways can also "cross talk," in that the death-receptor pathway or inflammasome pathway can engage the mitochondrial pathway as well.

DYING CELLS ARE SIGNALS

When a cell dies, other cells in the body rapidly remove the corpse.[6] Cells that undergo apoptosis produce signals, as a consequence of caspase activation, that attract cells capable of eating the dying cell and also other signals that cause removal of the apoptotic cell before the plasma membrane is disrupted. Necrotic cells, in contrast, release intracellular molecules that also lead to their clearance, but in a way that promotes inflammation of the tissue. Depending on how the cell died, our immune system can therefore be engaged to respond to whatever might have caused the cell death. Alternatively, it can be instructed not to respond.

It is evident that the ways in which dying cells affect the body are complex, and the consequences of cell death are varied—and when a cell is gone, it might not be forgotten.

[6] We seem to be violating our "bottom-up" view by discussing the consequences of cell death (admittedly, the "bottom") only after discussing the "top" of the pathway. But cells that die by any mechanism, not only apoptosis, can have an impact on physiology.

CHAPTER 2

Caspases and Their Substrates

KILLER PROTEASES

Apoptosis is orchestrated by a set of proteases, called caspases, which reside in an inactive form in nearly all of our cells. When activated, some caspases cleave hundreds, or perhaps thousands, of distinct target proteins. Such cleavage leads to all of the features that characterize this type of cell death, and apoptosis depends on the functions of caspases. In this chapter, we introduce these enzymes and the substrates that they cut to bring about apoptosis.

Caspases are endopeptidases—that is, they cut proteins internally rather than nibble away at the ends of the proteins. Unlike digestive proteases such as trypsin, caspases do not degrade their substrates but, rather, they clip them at discrete sites. In general, they cut specific sequences that end in aspartate residues and cut immediately after this amino acid (this is the "asp" in the term caspase). However, caspases do not cut after every aspartic acid that can be accessed in a protein—they have a preference for certain sequences.

In caspases, the active site of the enzyme includes a cysteine, which therefore classifies caspases as "cysteine proteases" (this is the source of the "c" in the word caspase; hence, a cysteine protease that cuts proteins after aspartic acid residues).

HOW CASPASES CUT PROTEINS

Cysteine proteases, also called thiol proteases, are one of the four major types of proteases characterized by their active sites (the others are serine, aspartyl, and zinc proteases). Although the focus here is on how caspases work, there are similarities among all proteases.

For a caspase to work, the interaction with its protein substrate must be "fast-on–fast-off." This involves the enzyme briefly holding the target peptide bond at the active site. In a functionally active caspase, a substrate specificity pocket is close to the active cysteine, which is itself near a histidine. This cysteine–histidine dyad is where the

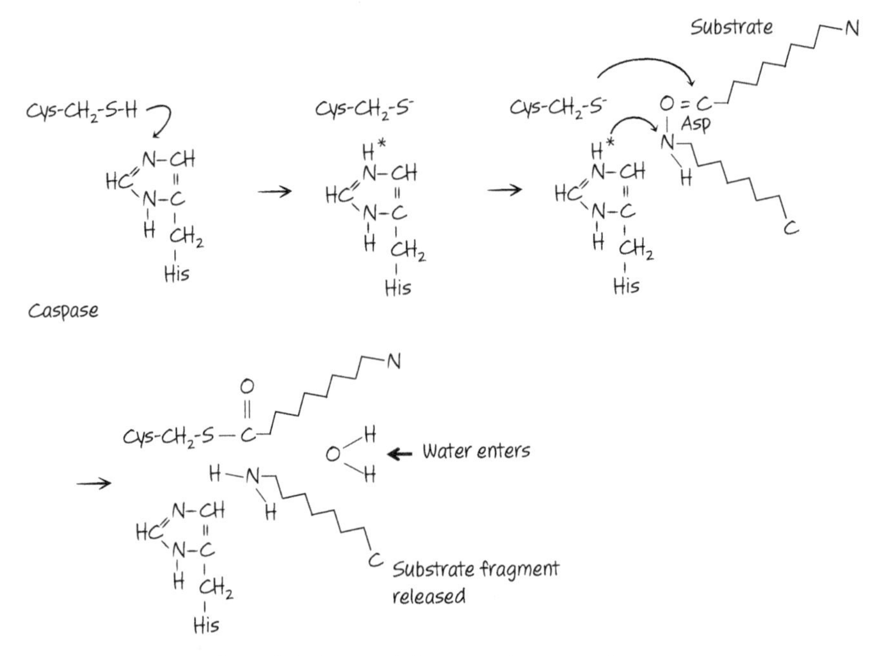

Figure 2.1. The acylation step of caspase cleavage.

action takes place. A crucial residue in the substrate pocket is an arginine, which holds the target aspartate in the substrate in position. This results in the situation shown in Figure 2.1, and the reaction follows. This step is called acylation; the next step is deacylation, shown in Figure 2.2. A water molecule is sacrificed, the peptide bond after aspartate is cut, and the caspase is now ready to cut another substrate protein.

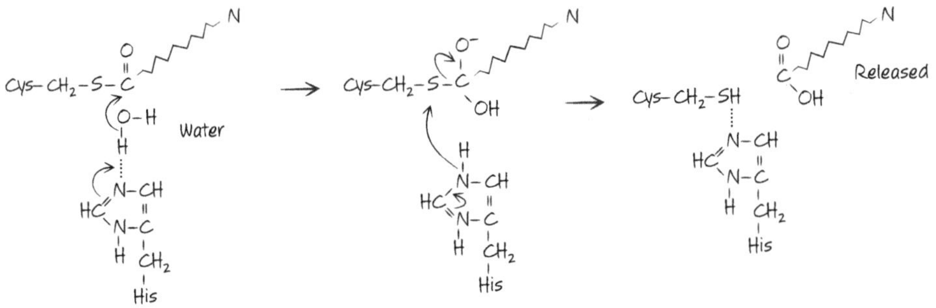

Figure 2.2. The deacylation step of caspase cleavage. The mechanism is based on the function of other cysteine proteases and has not been confirmed in caspases.

For a caspase to function, the catalytic dyad, cysteine–histidine, must be brought close to a target protein. This is the function of the substrate specificity pocket in the caspase. That pocket, with its arginine, only permits aspartates (and, to a much weaker extent, glutamate, in some cases) to gain proximity to the dyad. Inactive caspases ("procaspases"), however, do not allow access to the catalytic dyad. To form the pocket, changes to the procaspase must occur to form an active site. How this occurs depends on the type of caspase.

TYPES OF CASPASES

Caspases, like apoptosis, are found only in animals,[1] and, in most animals, there are several different caspases. In general, types of caspases can be distinguished by (1) their functions, (2) the structure of the procaspase, and (3) how they are activated. These are often interrelated, and the distinctions are not absolute.

In vertebrates and probably most other animals, at least two types of caspases are involved in apoptosis (the notable exception is nematodes, as discussed below): executioner caspases (such as in mammals, caspases-3, -6, and -7) and initiator caspases (in mammals, caspases-8 and -9). Initiator caspases are sometimes referred to as "apical" caspases. Another type of caspase, related to initiator caspases, includes inflammatory caspases (in humans, caspases-1, -4, and -5; in rodents, caspases-1 and -11). Additional caspases (in mammals, caspases-2, -10, -12, and -14, among others) are more difficult to place into one of these categories; their functions are less well understood.

All procaspases have three regions, termed the prodomain, large subunit, and small subunit. The latter two are always separated by one or more sites, where they are cleaved by the action of the caspase itself (self-cleavage), as it becomes activated, or by another protease (usually another caspase, but, as we will see in Chapter 3, at least one other protease can do this). Schematic examples of the domain structure of several human caspases are shown in Figure 2.3.

The prodomain and large subunit are often separated by a cleavage site targeted by the caspase itself. The functions of these cleavage sites vary in different caspases, as discussed in Chapter 3. For now, it is sufficient to be aware that there are different types of caspases with different biochemistries and functions. Those that concern us in this chapter are primarily executioner caspases, which, in vertebrates, are caspases-3, -6, and -7. These caspases become active during apoptosis and cleave substrates that bring about the characteristic features of apoptosis.

---

[1] Proteases are found throughout living things. Although caspases are only found in animals, proteases can have structural (but not sequence) similarity to caspases. Some of these have been called "metacaspases" and "paracaspases." Some studies have suggested that these proteases can have roles in cell death in plants and fungi, but, at present, the details of these processes remain obscure. Metacaspases are found in animals but are not known to have roles in apoptosis.

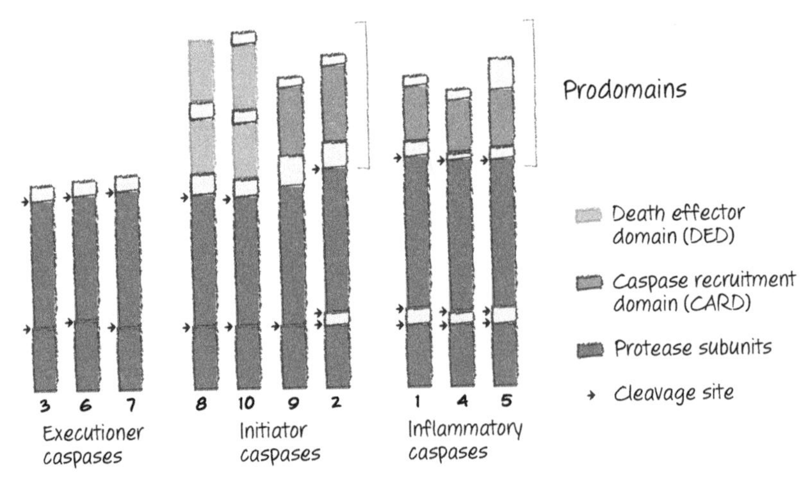

Figure 2.3. Schematics of several human caspases. Caspases-2 and -10 are grouped with the initiator caspases, although their classification is problematic. The representation aligns related sequences and should not be taken to indicate the actual structures of the proteins.

CASPASES IN OTHER ANIMALS

Nearly all animals for which we have genomic information appear to have caspases, and, in some invertebrate animals, we know that executioner caspases are important for apoptosis. As mentioned in the Introduction, the pathways of apoptosis in two invertebrates, namely, the nematode *Caenorhabditis elegans* and the fruitfly *Drosophila melanogaster*, have been characterized in detail.

In *C. elegans*, only one caspase, CED3, is important in apoptosis, and CED3 has features of an initiator caspase (e.g., a long prodomain), but it also acts as an executioner caspase, orchestrating apoptosis in the dying cells (Fig. 2.4).

In flies, several different types of caspases have been identified, and these are shown in Figure 2.5. Of these, DCP1 and Drice appear to be the most important executioner caspases. Two other caspases, Decay and Damm, might be executioner caspases, based on their sequences, but this is speculation.

In general, executioner caspases identified in other animals have essentially the same specificities as executioner caspases in vertebrates. In the discussion that follows, we focus on substrates of mammalian executioner caspases, which are the best characterized.

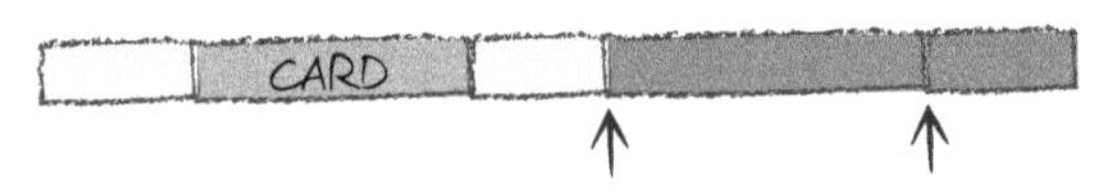

Figure 2.4. The linear organization of CED-3. CARD, caspase recruitment domain. The arrows are auto-cleavage sites.

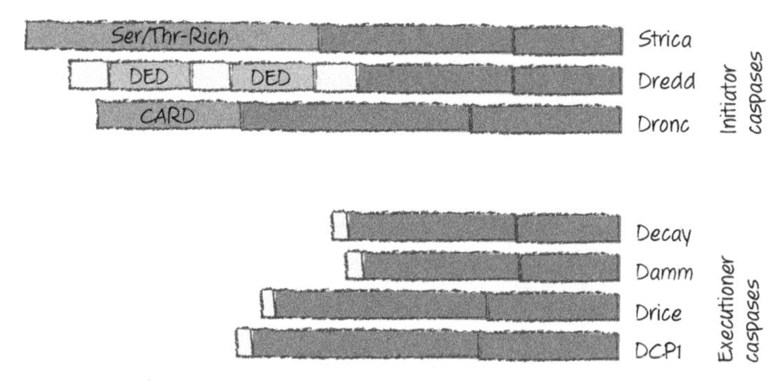

Figure 2.5. Caspases in *Drosophila*.

CASPASES SHOW PREFERENCES FOR THEIR SUBSTRATES

Caspase-mediated cleavage events can have catastrophic consequences that ensure not only that the cell dies, but that it does so quickly and "cleanly" (without inducing an inflammatory response). To understand this, we must discuss the sequences that executioner caspases recognize in proteins. And, for this, we need to become familiar with a bit of terminology.

We know that most caspases prefer to cut their substrate proteins just after aspartic acid residues, and this target amino acid after which the cut occurs is called the P1 residue of the substrate. The amino acid that follows the cut is called P1′. The residue just before P1 is P2, preceded by P3, P4, and so on (Fig. 2.6).

In nearly all caspase substrates, P1 is obviously aspartate, although very rarely it is glutamate. However, the presence of an aspartate in a substrate is not enough to predict that a caspase will cut at that site. First, the aspartate must be exposed so that the enzyme can access it, and, second, other amino acids contribute to recognition by the caspase. Based on the use of peptide libraries, caspases-3 and -7 share preferences for sequences that contain DXXD/G (or S or A), in which "X" is any amino acid, and "/" indicates the cleavage between P1 and P1′. Other caspases display different

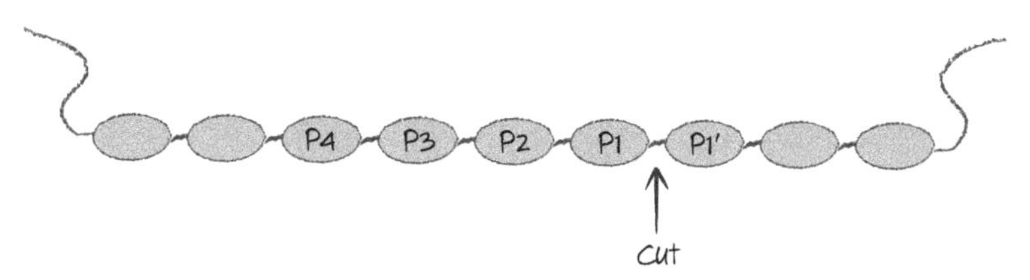

Figure 2.6. Numbering scheme for amino acids around a site in a substrate protein that can be cut by caspases.

IDENTIFYING CASPASE SPECIFICITY WITH COMBINATORIAL PEPTIDE LIBRARIES

One way to determine cleavage preference is to identify sites in substrates and compare them, as we will see. Another is through the use of combinatorial peptide libraries. This approach requires that we have a method to detect that cleavage of a substrate has occurred. This is most easily done by attaching a fluorescent molecule to the end of a peptide in such a way that it only becomes fluorescent when the bond between it and the peptide is cleaved, resulting in a measurable signal. This is a useful way to measure protease activity in general (Fig. 2.7).

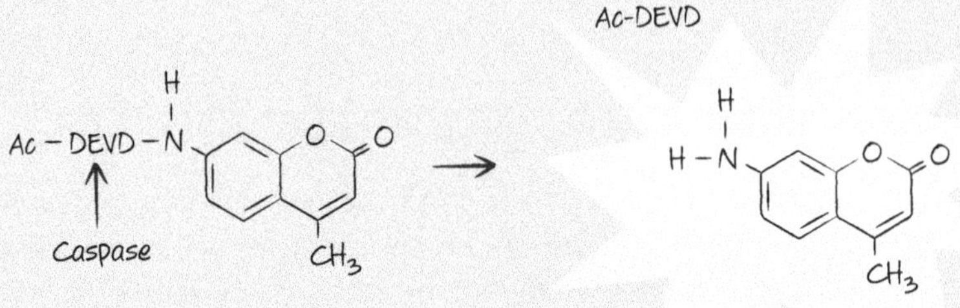

Ac – DEVD – N(H) → Ac-DEVD + H – N(H)

Caspase

CH_3

Figure 2.7. Caspase activity detected using a peptide substrate that generates a fluorescent signal after cleavage.

Because for most of the caspases P1 is D (aspartate), this position can be fixed, and then P1′, P1, P2, P3, and P4 (or more) can be varied with every possible amino acid to make a combinatorial library of the set. Each is then tested for its sensitivity to cleavage. An example of the preferences for caspase-3 found in this way is represented in Figure 2.8.

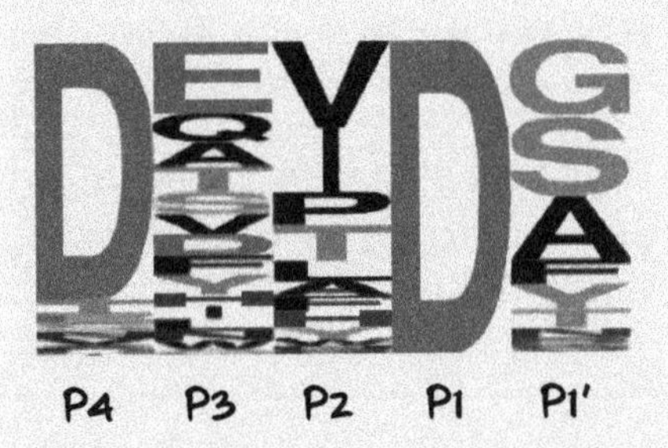

Figure 2.8. Representation of amino acid preferences for caspase-3. Amino acids are shown in single-letter code, and the size of each letter corresponds to its frequency in cleaved peptides. Cleavage occurs between P1 and P1′.

Caspase-1	WEHD
Caspase-2	VDQQD
Caspase-3	DEVD
Caspase-4	LEVD
Caspase-5	(W/L) EHD
Caspase-6	(T/V) QVD
Caspase-7	DEVD
Caspase-8	LETD
Caspase-9	LEHD

Figure 2.9. Caspase preferences, based on peptides. Sequences shown are optimally cleaved, although other peptide sequences can be cleaved nearly as well.

preferences, although D (aspartate) is generally preferred at P1. Some of these preferences[2] are shown in Figure 2.9.

How do these preferences apply to actual cellular substrates for caspases? We can identify protein substrates for the caspases by adding active caspases to cell extracts and looking for cleavage events. In this way, hundreds of substrates for caspase-3 have been identified, and investigators believe there could be a thousand or more. Some of these are listed in Table 2.1.

When all of the cleavage sites in caspase-3 substrates are compared, we get the pattern shown in Figure 2.10.

Clearly, the combinatorial libraries worked well, giving us a pretty good idea of the preferred sites for caspase-3 in native proteins. However, there could be another problem here: When we make a cell extract, we might allow caspase-3, for example, to access substrates it would not normally see in cells. For example, cells have many subcellular compartments from which caspases could be excluded, and this compartmentalization can be lost when lysing the cell to make an extract. Thus, it is important to know whether caspase-3 has a similar substrate preference during apoptosis in intact cells.

We can address this question by comparing the proteins that are cut during apoptosis. Although we cannot be sure that all of the cuts are due to caspase-3, an exhaustive analysis of substrate cleavage during apoptosis can be performed. The result of such a large-scale approach by mass spectroscopy is informative. This approach identified not only those substrates listed in Table 2.1, but more than 1000 substrates overall, and revealed a sequence preference (Fig. 2.11) that matched that predicted by the approaches mentioned above. Caspases with the preference of caspases-3 and -7 might therefore predominate during apoptosis to cut protein substrates. It is not unreasonable to suppose that these are, in fact, caspases-3 and -7.

[2] These preferences are not specificities. It is easy to fashion substrates that detect caspase cleavage. When an extract of apoptotic cells is added to such substrates, how much of each peptide substrate is cleaved can be determined, and this information is often extrapolated to define which caspases are active and to what extent. But this approach assumes specificity, a potentially fatal mistake: if a caspase, such as caspase-3, is active and in great excess, then even if it cleaves a peptide relatively poorly, this may be read as activity of a different caspase. Unfortunately, many conclusions have relied on such an approach to determine whether a particular caspase is active under a given condition. Such studies should be treated cautiously.

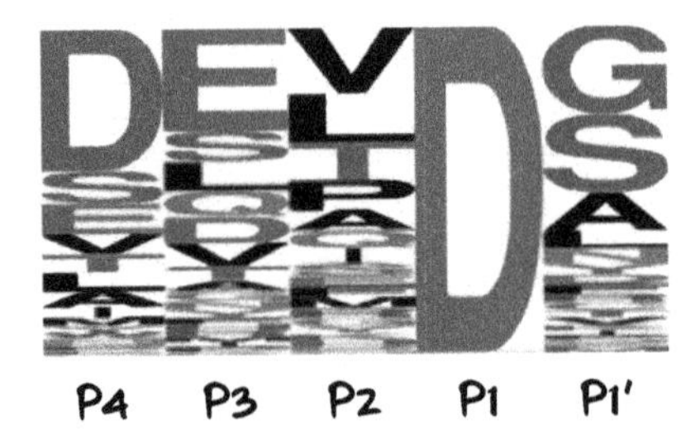

Figure 2.10. Representation of caspase-3 cleavage sites in known substrates.

KEY SUBSTRATES FOR EXECUTIONER CASPASES IN APOPTOSIS

We know what sort of sequences the executioner caspases like to cut during apoptosis, but how does this lead to the features of this form of cell death? Is the effect of executioner caspase activation really "death by a thousand cuts"? Or do only a few cuts in certain key substrates produce the phenomenon we call apoptosis?

Some insights come from considering how caspase cleavage of a substrate can affect a cell. We can envision four ways: Cleavage could (1) destroy an activity, provided that the cleavage is efficient; (2) trigger an activity by removing an inhibitor or an inhibitory domain in the substrate; (3) convert a protein into a dominant-negative version that inhibits the activity of the intact protein; or (4) have no relevant effect at all. The last is vexing and subtle. Many caspase cleavage events might simply "happen" in a cell that is doomed to die anyway. Therefore, even if an event seems likely to be important, on reflection, its importance could pale in the context of death. For example, the activation of an important transcription factor by an executioner caspase is unlikely to result in the production of a protein if, at the same time, the genome has been dismantled by DNA fragmentation.

The vast majority of proteins cleaved during apoptosis do not have established roles in the events that characterize this form of cell death. Therefore, to gain an understanding of how executioner caspases bring about apoptosis, a different but complementary approach is needed.

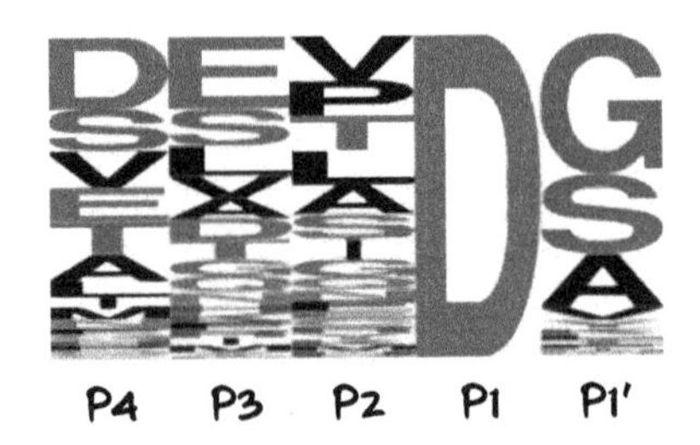

Figure 2.11. Representation of protein cleavage events in cells undergoing apoptosis.

Table 2.1. Examples of caspase substrates identified in apoptotic cells

Substrate	Putative or demonstrated functional consequence of cleavage	Cleavage site (D residue number)
Acinus	Involved in chromatin condensation	DELD (1093)
AKT/PKB	Loss of kinase activity. Putative—loss of survival signaling	ECVD (462), TVAD (108), EEMD (119)
ATP11C	Loss of phospholipid flippase activity	QEVD (439), SQTD (445), DAVD (481)
β-Catenin	Reduced α-catenin binding. Putative—loss of cell adhesion	YQDD (145), NDED (164), SYLD (32), ADID (83), TQFD (115), YPVD (751)
c-IAP1	Loss of signaling functions	ENAD (372)
E-Cadherin	Release of intracellular fragment	DTRD (750)
GATA-1	Loss of transcriptional activity, which leads to impaired erythropoiesis	EDLD (125)
Gelsolin	Loss of binding to monomeric actin and triggering of F-actin depolymerization, membrane blebbing	DQTD (403)
iCAD (DFF45)	Release of active CAD endonuclease	DETD (117), DAVD (224)
IL-33	Putative—activation of IL-33, which sensitizes toward a TH2 immune response	DGVD (178)
iPLA2	Increased phospholipid turnover; release of lysophosphatidylcholine (LPC), which attracts monocytic cells	DVTD (183)
Lamin A/C	Breakdown of nuclear envelope	VEID (230)
Lamin B1	Disassembly of nuclear lamina	VEID (231)
Stk4/Mst1	Kinase constitutively active; overexpression of amino-terminal fragment induces apoptotic morphology	DEMD (326)
NDUFS1	Disruption of electron transport (complex I) and transmembrane potential (ΔΨm), leading to production of reactive oxygen species (ROS), loss of ATP production, and mitochondrial damage	DVMD (255)
PAK-2	Kinase constitutively active and activates c-Jun amino-terminal kinase pathway; overexpression of cleaved fragment leads to apoptotic morphology (shrinkage and rounding up)	SHVD (212)
PARP-1	Loss of poly(ADP-ribose) polymerase activity	DEVD (214)
Procaspase-3	Activates protease activity	ESMD (28), IETD (175)
Procaspase-7	Activates protease activity	DSVD (23), IQAD (198)
RIPK-1	Inhibits activation of NF-κB	LQLD (324)
ROCK1	Kinase constitutively active —drives cell contraction and blebbing. Phosphorylates PTEN, which then inhibits Akt/PKB serine/threonine-protein kinase	DETD (1113)
TRAF-1	Inhibits activation of NF-κB	LEVD (163)
Vimentin	Disrupts intermediate filaments	DSVD (85)
Xkr8	Scrambles the lipids of the plasma membrane	DGVD (355)
XIAP	Amino-terminal fragments inhibit caspase-3 and caspase-7 activity. Carboxy-terminal fragment inhibits caspase-9 activity	SESD (242)

A useful way to identify caspase substrates that are important for apoptotic events is to dissect a process that depends on caspase activity and identify the key substrate responsible for that process. Below, we consider the substrates whose cleavage produces specific apoptotic events in different cellular compartments: the nucleus, plasma membrane, and mitochondria.

NUCLEAR EVENTS MEDIATED BY CASPASE CLEAVAGE OF SPECIFIC SUBSTRATES

A striking feature of apoptosis is the fragmentation of chromatin. The DNA is cut into pieces equivalent to one or more nucleosomes (multiples of 180 base-pairs), and this phenomenon depends on executioner caspases being active in the dying cell.

If we treat normal cytosol from living cells with caspase-3 and then add this to isolated nuclei, DNA fragmentation characteristic of apoptosis occurs. This approach allows identification of the enzyme responsible, as well as elucidation of how it is activated by caspase-3. The enzyme is a nuclease that preferentially cuts DNA at accessible sites between nucleosomes: "caspase-activated DNase" (CAD) or, alternatively, "DNA fragmentation factor of 40 kDa" (DFF40). It is present in healthy cells, but it is held in an inactive complex by an inhibitor, called iCAD (also called DFF45). The inhibitor is the caspase substrate, and, when cleaved by caspase-3, it releases the active nuclease to cut the DNA.

It would seem especially dangerous for cells to produce a nuclease capable of fragmenting the genome, but there is an additional safeguard. CAD is completely inactive unless properly folded by a chaperone, and this is a function of iCAD, which then holds the nuclease inactive unless iCAD is cleaved. The basic scheme is shown in Figure 2.12.

Cutting of the chromatin by CAD probably facilitates the degradation of the dead cell after it has been engulfed by a phagocytic cell (discussed in Chapter 9). In the absence of functional CAD (achieved by removing CAD or its chaperone, iCAD), DNA fragmentation does not occur in the cell before engulfment, but the cell dies nevertheless. Therefore, although it is unlikely that the cell can survive once caspases have cleaved iCAD and released active CAD, this is not required for cell death.[3]

So, are CAD and iCAD important? It is difficult to say. Both are present in most of the animals for which we have genomic sequences, with at least one exception. Among the animals, iCAD and CAD are found in hydra, sea anemones, insects, and of course vertebrates (to name a few), but they are not found in nematodes. Apparently, nematodes lost this substrate and its function in apoptosis along the way.

---

[3] Although widespread destruction of the DNA by active CAD is certainly lethal in most cells, there is evidence that cells can survive limited CAD activation (upon cleavage of iCAD by low levels of executioner caspases). In such cases, the DNA damage is repaired, and this can promote mutations in the surviving cell. We return to this concept in Chapter 11 in the context of apoptosis and cancer.

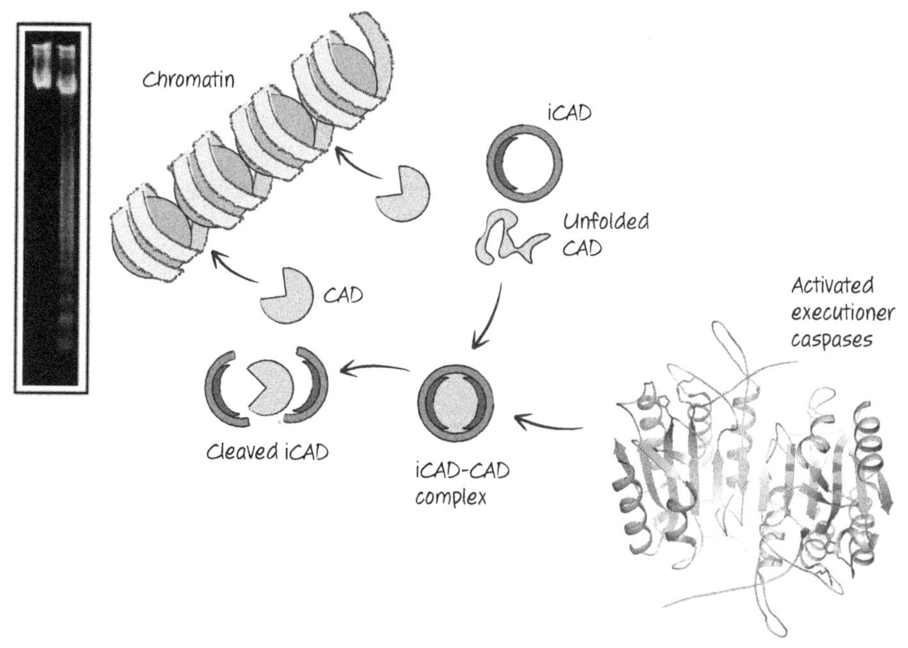

Figure 2.12. Cleavage of the iCAD–CAD complex by caspases is responsible for DNA fragmentation during apoptosis. Caspase-3 (*right*) cleaves iCAD (*center*), releasing the active CAD, which randomly cuts the chromatin at accessible sites between the nucleosomes (*top left*). When run on an agarose gel, the DNA forms a ladder (*far left*) composed of multiples of nucleosome-sized lengths.

Other nuclear events in apoptosis can be traced to the cleavage of additional caspase substrates. During apoptosis, the chromatin becomes condensed, and this has been linked to the cleavage of the protein acinus by caspase-3 (and probably caspase-7). Precisely how acinus causes chromatin condensation is unclear but might involve phosphorylation of histones or the participation of acinus in DNA fragmentation by CAD.

Disruption of the nuclear envelope also occurs in apoptosis, and this can be traced to the cleavage of lamins by caspase-6. Lamins are important structural elements that preserve nuclear integrity and are involved in the breakdown and regeneration of the nuclear envelope during mitosis. If mutant lamins lacking the cleavage site are introduced, the nuclear envelope remains intact during apoptosis; however, all other aspects of apoptosis seem to proceed normally.

EVENTS AT THE PLASMA MEMBRANE CAUSED BY CASPASES

During apoptosis, the cell undergoes extensive membrane blebbing, extending bulbous outgrowths that often break off as small, membrane-bound bodies. This is an effect, at least in part, of actin polymerization, and pharmacological inhibitors of actin polymerization block blebbing.

Three caspase substrates have been implicated in actin polymerization and blebbing during apoptosis. These are the actin regulator gelsolin and two kinases that function in signaling pathways that control actin organization in cells: p21-activated kinase (PAK) and ROCK1 kinase. In all three, caspase cleavage activates the protein by removing regulatory domains, which results in the dynamic changes in actin organization that cause blebbing. Pharmacological inhibitors of ROCK1 or silencing its expression prevent blebbing and can even stop it after it is under way (Fig. 2.13). As with the other related events that we have discussed, however, the cell still dies if blebbing is blocked.

One important consequence of apoptotic cell death is that the cell is eaten by other cells before the integrity of the plasma membrane is lost. The "eat me" signals that ensure that this will occur depend on caspase activation. The most important of these signals is the appearance on the cell surface of the lipid phosphatidylserine, which is recognized by the cells that do the eating (discussed in much more detail in Chapter 9). Phosphatidylserine is normally restricted to the inner leaflet of the plasma membrane, but, during apoptosis, the lipids scramble, and phosphatidylserine is exposed.

The molecule that is responsible for keeping phosphatidylserine (and other phospholipids) localized to the inner leaflet of the plasma membrane is adenosine triphosphatase type 11C (ATP11C), which uses ATP to "flip" phosphatidylserine from the outer to the inner leaflet. If ATP is depleted from a cell, phosphatidylserine can therefore accumulate on the surface. A second protein, CDC50A, is required to chaperone ATP11C to the plasma membrane, and cells without CDC50A constitutively expose phosphatidylserine. ATP11C is a caspase substrate; cleavage by executioner caspases destroys its flippase function, and phosphatidylserine is then exposed on the outer leaflet (Fig. 2.14). Mutation of the caspase cleavage sites in ATP11C results in cells that do not efficiently expose phosphatidylserine during apoptosis.

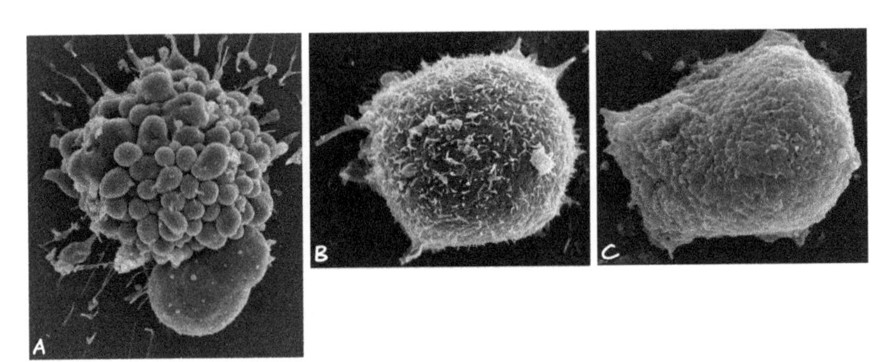

Figure 2.13. Silencing or inhibition of ROCK1 prevents blebbing during apoptosis. Blebbing seen during apoptosis (*A*), is eliminated when ROCK1 is silenced with small interfering RNA (siRNA) (*B*) or blocked with an inhibitor (*C*).

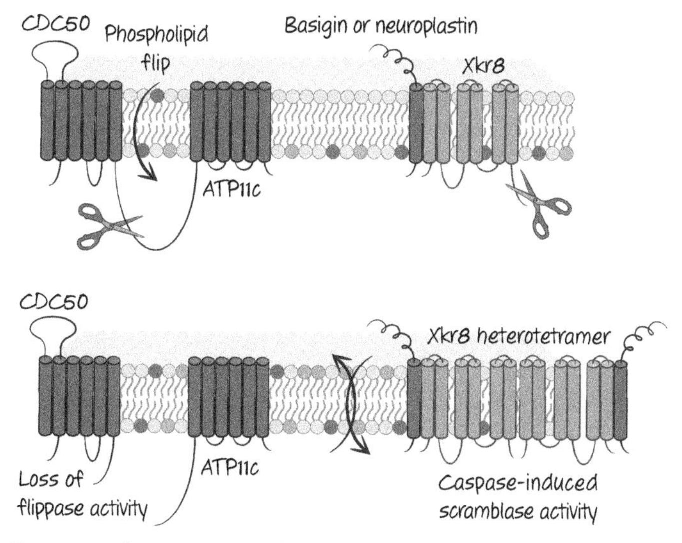

Figure 2.14. Caspases induce exposure of phosphatidylserine on the outer leaflet of the plasma membrane by two mechanisms. Caspase cleavage (scissors) disrupts the flippase activity of ATP11C (*left*) and induces the scramblase activity of Xkr8 (*right*). Both result in loss of phospholipid asymmetry, resulting in exposure of phosphatidylserine on the cell surface.

In addition to this loss of flippase function during apoptosis, the lipids of the plasma membrane also undergo an active scrambling to bring phosphatidylserine rapidly to the outer leaflet. Another caspase substrate is responsible for this event, a protein called Xkr8. Xkr8 associates with a partner protein, which can be either of the two related proteins basigin or neuroplastin. When Xkr8 is cleaved by an executioner caspase, it now forms heterotetramers of two Xkr8 and the two partner molecules. This higher-order complex is responsible for scrambling the lipids of the plasma membrane during apoptosis (Fig. 2.14).

The exposure of phosphatidylserine on the cell surface can occur without caspase activation, and therefore it is not specific to apoptosis. Because of the nature of the plasma membrane, any disruption in the membrane results in redistribution of phospholipids, including phosphatidylserine. An influx of calcium into the cell can activate another phospholipid scramblase, TMEM16F, resulting in transient exposure of phosphatidylserine. This is important, for example, in platelet activation. Cells without TMEM16F do not externalize phosphatidylserine in response to calcium influx, but they do in response to caspase activation during apoptosis.

MITOCHONDRIAL EFFECTS OF EXECUTIONER CASPASES

When the mitochondrial pathway of apoptosis is engaged (discussed in detail in Chapter 4), activated executioner caspases gain access to the inner membrane of mitochondria (because the outer mitochondrial membrane becomes permeable). Here, several caspase substrates on the inner membrane are found, including the oxidoreductase NDUFS1, which is an integral part of complex I of the electron-transport chain, the major source of energy in the cell.

When NDUFS1 is cut by caspases, several events rapidly follow. Electrons that would normally be used for oxidative phosphorylation are shuttled to oxygen to produce superoxides. These can then be converted to hydroxyl radicals that are damaging to many proteins and membranes. In addition, the proton gradient normally produced by electron transport dissipates, and ATP levels rapidly decrease. This loss of ATP has effects on the plasma membrane, because ion pumps require ATP to operate. Introduction of a noncleavable NDUFS1 mutant into cells prevents these rapid, caspase-dependent events, but it has no effect on other apoptotic changes, such as DNA fragmentation, blebbing, or cell death.

DEATH BY A THOUSAND CUTS?

As we noted, there are many hundreds of caspase substrates that are cleaved when executioner caspases become active. How many of these are actually responsible for cell death? Clearly, if a nuclease such as the CAD becomes active in a cell and fragments the DNA extensively, the cell will not survive. But some specialized cells, such as mammalian red blood cells, persist for more than 100 days without a nucleus (the elimination of the nucleus in red blood cells does not depend on CAD). And cells lacking iCAD or CAD still die by apoptosis. A glance at the substrates in Table 2.1 and those we discussed above give us an idea why. Indeed, it is remarkable that a few key substrates that have major roles in producing specific changes that we see during apoptosis can be identified. But it is unlikely that only a few key substrates (these or others) are responsible for cell death.[4]

CASPASES AS KILLERS

The short answer to the question of what kills the cell during apoptosis is, of course, "caspases." Introduction of an executioner caspase into any type of cell causes it to die, and, in animal cells, this occurs by apoptosis.

As we go into the pathways of apoptosis in this book, we will see several examples in which the activation of caspases is not essential for cell death, and other mechanisms come into play. For now, it is important to be aware of studies that reveal just how important caspases really are. For example, in *C. elegans*, one caspase,

[4] There are many reports that a favorite substrate is paramount in cell death. But in every case, introduction of high levels of such proteins (cleavable or mutated to avoid cleavage) can have other effects that alter the biology of the cell and its sensitivity to a given apoptotic challenge, and we should ask (as some investigators apparently have not) what happens to the other substrates and their functions when caspases are active.

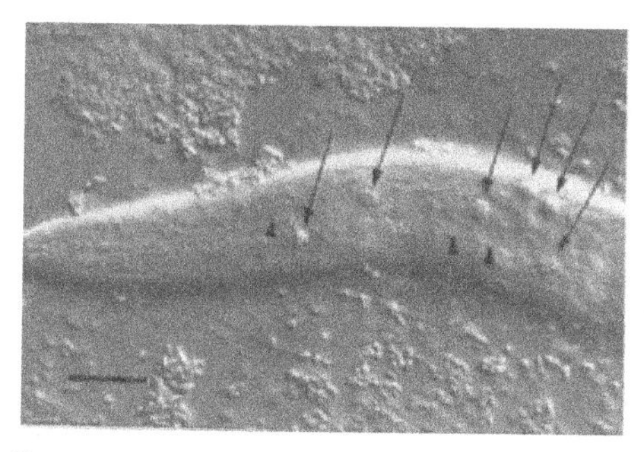
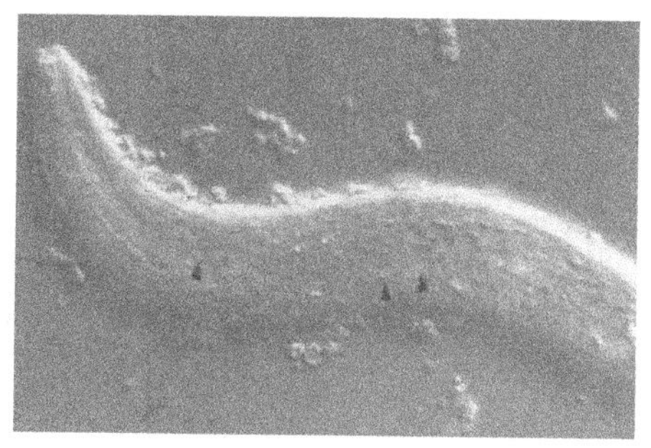

Figure 2.15. Extra cells in *CED3* mutant nematodes. Normal cell deaths in a wild-type larva (arrows, *left*) are not seen in a *CED3* mutant animal (*right*).

CED3, is required for most cell death during development and in germ cells of the adult worm that die in response to stress. Animals with mutations that prevent CED3 activation or function accumulate extra cells that differentiate and persist (Fig. 2.15). In this animal, cell death depends on completion of the apoptotic pathway and therefore depends on the caspase.

This is also true in *Drosophila*, although again not all cell deaths follow this rule. *Drosophila* lacking the initiator caspase Dronc accumulate extra cells, which is lethal for the developing fly (Fig. 2.16).

Cultured *Drosophila* cells that lack caspase activity continue to survive and proliferate when exposed to stresses that would normally cause apoptotic cell death, emphasizing the importance of these enzymes in cell demise. Mice lacking an initiator caspase (caspase-9), or an executioner caspase (caspase-3), also accumulate extra cells (Fig. 2.17). Elimination of caspase-7 as well as caspase-3 exacerbates this effect. However, as we will see in Chapter 10, this effect might not be due to the extra cells "not dying"—it is a more complex effect of the delayed death of cells in the developing neural tube permitting excessive proliferation of developing neurons.

Caspases must therefore be important for the death of some cells. However, a close look at the development of such mutant mice reveals that developmental cell

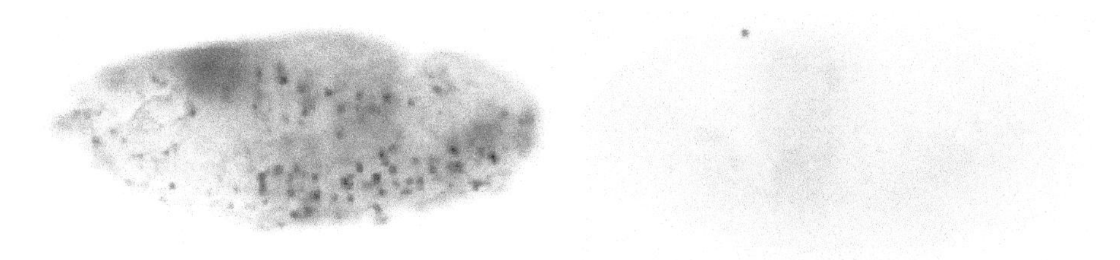

Figure 2.16. Extra cells in Dronc-deficient fly embryos. Cell deaths (stained blue) in wild-type embryos (*left*) are not seen in embryos lacking the caspase Dronc (*right*).

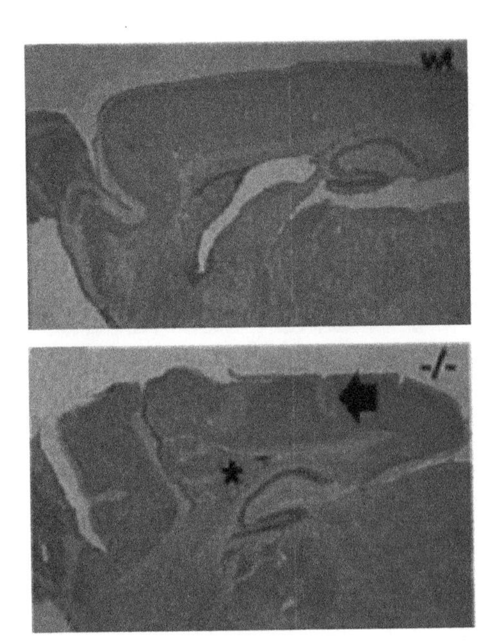

Figure 2.17. Extra cells (arrow and asterisk) in the brains of caspase-3-deficient mice (–/–; *lower*) compared with wild-type mice (wt; *upper*).

death still occurs. We return to this problem of caspase-independent cell death later (in Chapter 4) as we learn more about the pathways and other forms of cell death.

So far, we have discussed caspases and the way they contribute to apoptosis. But how are the caspases activated to cause this devastation? We consider this problem next.

C H A P T E R 3

Caspase Activation and Inhibition

ACTIVATION OF EXECUTIONER CASPASES

Executioner caspases, when activated, cleave hundreds or thousands of substrates in the cell to orchestrate apoptosis. In most animals, the executioner caspases have short prodomains lacking interaction sites for other proteins (the exception, as we have noted in Chapter 2, is in nematodes, which have an executioner caspase with a long, interactive prodomain). The inactive forms (or "proforms") of these caspases exist in the cell as dimers, with the potential to form two active sites. These are constrained from forming their active sites until they are cleaved between the large and small subunits. This cleavage permits the chain–chain interaction that snaps the two active sites into place, allowing the now-mature protease to be maximally functional. This is an important rule for understanding apoptosis that bears repeating:

> The executioner caspases that orchestrate apoptosis preexist in cells as inactive dimers that are activated by cleavage between the large and small subunits.

In the inactive proforms of the executioner caspases, the catalytic dyads are not in position to gain access to the target aspartate in the substrate protein. A look at one procaspase structure shows us why (Fig. 3.1, left). This is the structure of inactive procaspase-7, and we can see how it changes as it becomes activated (Fig. 3.1, right).

Note that both the inactive and active forms of the enzyme are dimers (remember the rule above), with two active sites. On activation, the active cysteine–histidine dyads do not move; what changes is the structure of the loop that ultimately forms the substrate-specificity pocket (the arginine that interacts with the target aspartate in the substrate is indicated). The reason that this shape change occurs on cleavage is shown in Figure 3.2.

In the inactive form, the center of the dimer is occupied by the region of each caspase monomer that serves as the linker between the large and small subunits.

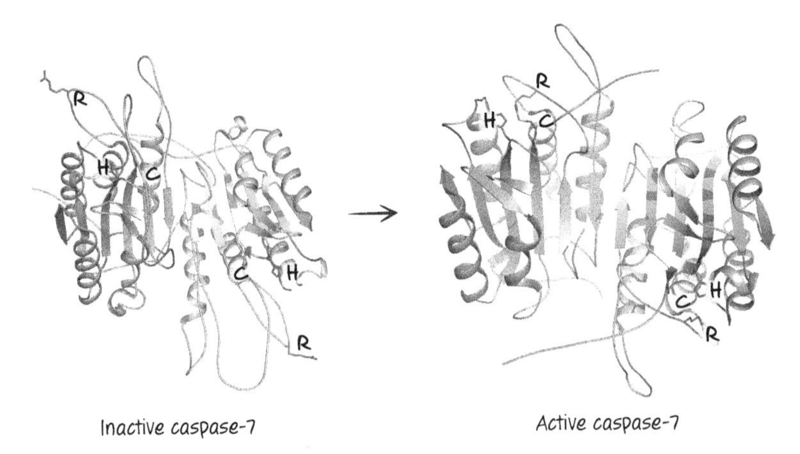

Figure 3.1. The structures of inactive and active caspase-7. The arginine (R) in the specificity loop is indicated, as is the cysteine–histidine (C–H) catalytic dyad.

When the linker is cleaved, it leaves the central region, and each end interacts with an end of the cleaved linker from the other chain, thus stabilizing the structure and holding the ends away from the center. Now, each loop comprising the specificity pockets can snap into position, drawn into place by a neighboring loop. The "elbow" of the specificity loop stretches into the center of the dimer, the region previously blocked by the linker. The result is the formation of two active sites in the mature caspase.

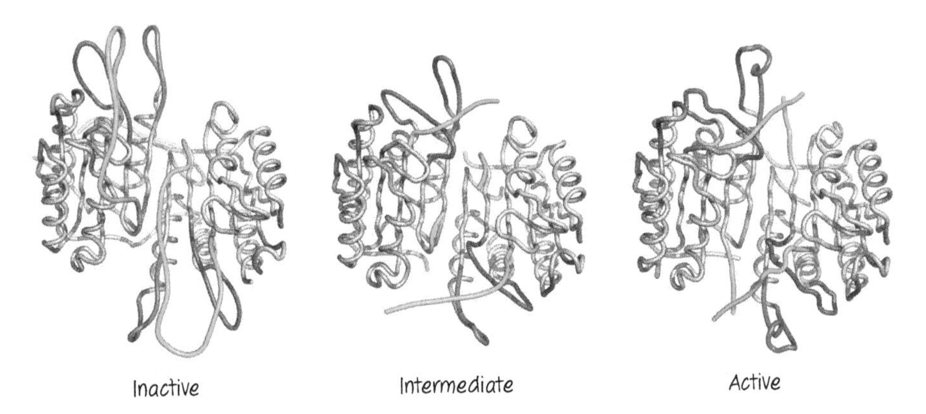

Figure 3.2. Close-up view of caspase-7 activation. Three loops mediate caspase-7 active site assembly: the specificity loop (red), a neighboring loop that draws the specificity loop into position (blue), and the linker between the large and small subunits (green). Structures are the inactive zymogen (*left*), the inhibitor-bound enzyme (*center*), and the active enzyme (*right*).

Once activated in this way, the caspases act in *trans* to remove the small prodomains as well. This changes the size of the mature protein, but it does not seem to be important for executioner caspase activation or function.

What mediates the cleavage event that activates the procaspases? In the case of caspase-6 in mammals, the cleavage responsible for activation is effected by caspases-3 and -7. Caspases-3 and -7 can also cleave and activate proforms of other caspases-3 and -7 dimers, although this is not an efficient process.

But this begs the question: What first cleaves and thereby activates the executioner caspases to cause apoptosis? The answer, in the vast majority of cases, is the initiator caspases—in particular, caspases-8 and -9 in vertebrate cells. However, before considering these caspases and how they work, we briefly consider another enzyme capable of cleaving and activating the executioner caspases. In doing so, we delineate our first, albeit simplified, apoptotic pathway.

INTERLUDE: GRANZYME B AND APOPTOSIS INDUCED BY CYTOTOXIC LYMPHOCYTES

Viruses are tricky little things. They quickly infect and subjugate cells for the production of more virus, posing special problems for multicellular organisms seeking to eliminate these pests. In vertebrates (and perhaps other animals), one approach to this problem is cytotoxic lymphocytes (cytotoxic T cells and natural killer cells) that identify virally infected cells and cause them to undergo apoptosis, destroying both the cells and the viruses that they harbor. Figure 3.3 shows the killing of a target cell by a cytotoxic lymphocyte. The cell death has all of the morphological and biochemical characteristics of apoptosis (the characteristic blebbing is apparent in Fig. 3.3).

Cytotoxic lymphocytes have several means at their disposal for dispatching their target cells, but here we concern ourselves with only one—the function of cytotoxic granules—and only one of the ways in which these granules work. During cell killing, the contents of the cytotoxic granules are released onto the membranes of the target

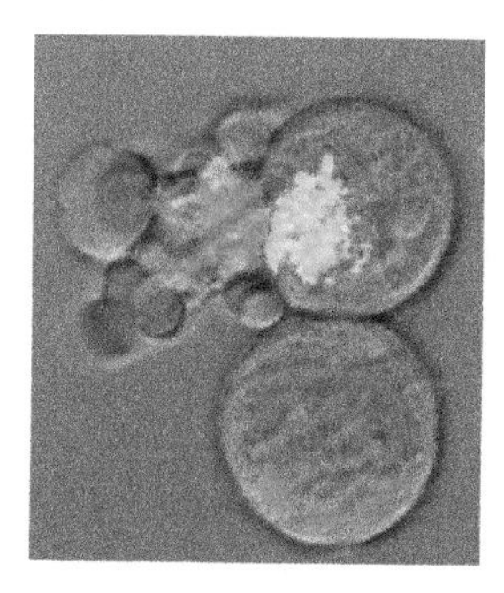

Figure 3.3. Cytotoxic lymphocyte killing a target (cell). The target cells are stained red, and the cell at *left* is undergoing apoptosis. Green: cytotoxic granules in the cytotoxic lymphocyte.

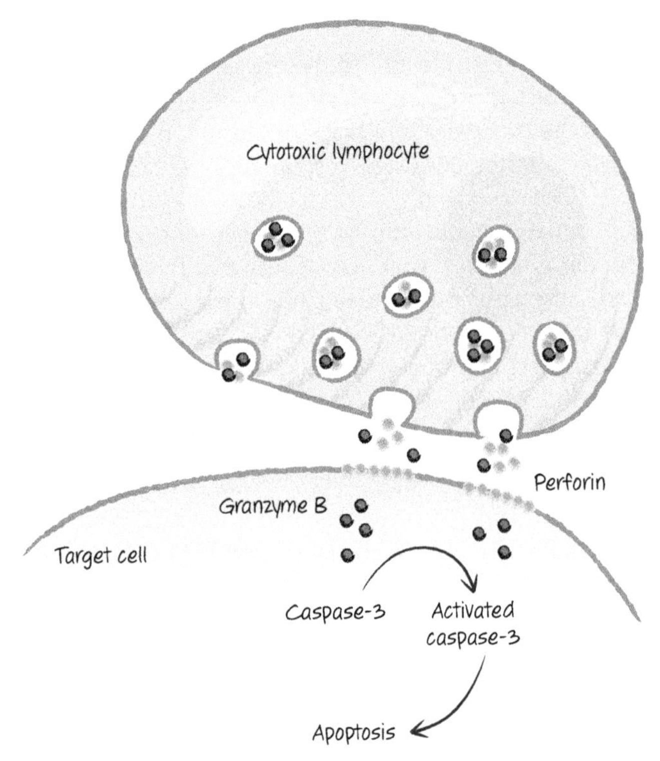

Figure 3.4. Simplified scheme of cytotoxic granule killing.

cells. The released molecules include a pore-forming protein, perforin, and a collection of proteases called granzymes. One of the latter is granzyme B. Granzyme B is a protease, but, unlike caspases, it is a serine protease (i.e., its active site contains a serine, rather than a cysteine). However, like the caspases, granzyme B is an endopeptidase that cleaves after aspartic acid residues, such as those involved in activating executioner caspases.

When the cytotoxic lymphocyte contacts a target cell and releases its granule contents, perforin permits the entry of the granzymes into the cell that is the victim of this murderous assault (Fig. 3.4).[1] If sufficient granzyme B is present, it cleaves caspases-3 and -7 at the aspartate between the large and small subunits of each, activating the caspases. The result is a feed-forward activation that kills the cell by apoptosis, effectively removing the compromised cell and the virus in the process.

This mechanism is not restricted to viral infections—it includes immune responses by cytotoxic lymphocytes to other intracellular infections as well as

[1] It is probably not as simple as this. Granzyme B is taken up by target cells, even without perforin, but perforin is necessary for the granzyme to gain access to the cytoplasm.

responses to tumor cells and foreign tissue grafts. This mechanism of killing is not the only way in which cytotoxic lymphocytes work, nor is it the only way granzyme B can function to trigger apoptosis. However, it illustrates how caspase activation can occur in this bona fide apoptotic pathway.

ACTIVATION OF INITIATOR CASPASES

In the vast majority of cases in which apoptosis occurs, the proteases that trigger executioner caspase activation are the initiator caspases. In mammals, these are predominantly caspases-8 and -9. When activated, they can cleave and thereby activate the executioner caspases, resulting in apoptosis.

Unlike executioner caspases, initiator caspases (as well as other caspases, such as inflammatory caspases and caspase-2) exist in cells as inactive monomers. And also unlike executioner caspases, these monomers are not activated by cleavage but, instead, by dimerization. This is a second key rule of caspase activation that bears repeating:

> Initiator caspases preexist in cells as inactive monomers. They can only be activated by dimerization.

Once the initiator caspases dimerize, they become active and can cleave themselves between the large and small subunits, which can act to stabilize the dimer. Therefore, although cleavage does not activate the caspase, it does have a function.

The idea that the activation of initiator caspases involves bringing monomers together to form dimers is called the "induced proximity" model. The principle can be shown experimentally using recombinant caspases that lack the prodomains. For example, when such a truncated caspase-2, -8, or -9 is produced, each is enzymatically inactive. But if salt conditions are altered to promote aggregate formation, caspase activity quickly appears (Fig. 3.5, upper middle). This happens regardless of whether the caspases can undergo cleavage—that is, even if the cleavage sites between the large and small subunits have been mutated (Fig. 3.5, lower middle).

When the salt is removed, the cleaved active dimers remain active (Fig. 3.5, upper right). But if the caspase was uncleavable (because of mutation), its activity is rapidly lost (Fig. 3.5, lower right). This is why we say that cleavage of initiator and related caspases is not required for activity but serves to stabilize the dimers. The stabilizing effect of cleavage has been shown for caspase-2 and caspase-8 and might apply to other caspases as well. An exception is caspase-9, in which cleavage does not stabilize the dimer (as we will see, the effect of caspase-9 cleavage is more complex).

This stability, however, is important. Although a noncleavable mutant of caspase-8, for example, can demonstrably be activated by dimerization, it does not

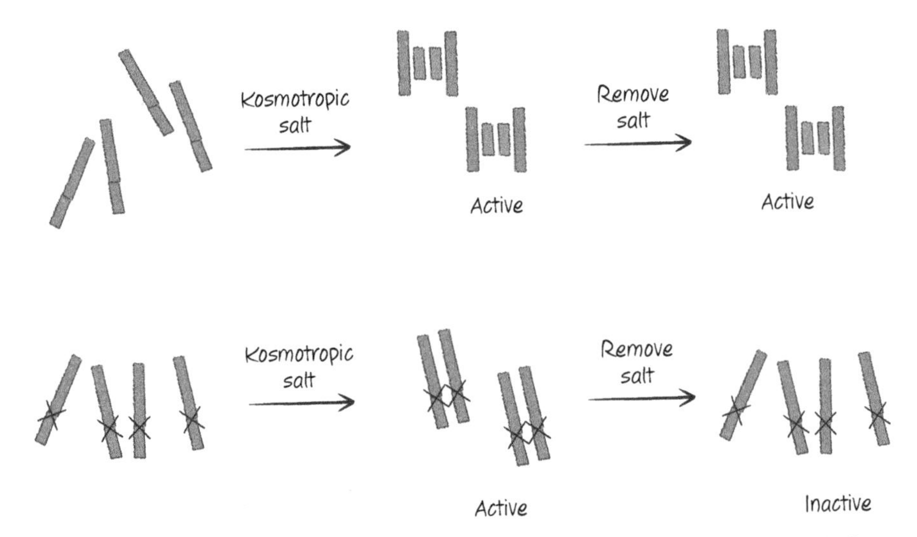

Figure 3.5. Initiator caspases can be activated by induced proximity. Kosmotropic salts, which cause protein aggregation, activate initiator caspases that remain active after the salt is removed. If the cleavage sites between the protease subunits are made uncleavable by mutation, the salts can still activate the enzyme, but the enzyme becomes inactive when the salts are removed.

efficiently promote apoptosis. We return to this when we consider the biological functions of caspase-8 in apoptosis and other phenomena.

But now, there arises a major question: What dimerizes the initiator caspases (and the other caspases that behave in this way)? The answer is adapter proteins.

ADAPTER PROTEINS INTERACT WITH CASPASES VIA DEATH FOLDS

All caspases that are not executioner caspases have long prodomains containing regions that interact with other proteins—the adapter proteins. To activate caspases, however, these adapter proteins must not only bind the caspases but also dimerize them. In general, this occurs through processes that dimerize or oligomerize the adapters. How this happens is through distinct interactions, depending on the adapter. These interactions, the adapters involved, and the initiator caspase engaged all define the different apoptotic (or related) pathways that are the subjects of the following chapters. All of these apoptotic pathways, however, have the basic form shown in Figure 3.6.

The protein–protein interaction regions in the prodomains of initiator, inflammatory, or other caspases are of different types, depending on the caspase. Caspases-8 and -10 have two death effector domains (DEDs). Caspases-1, -4, and -5 (and in

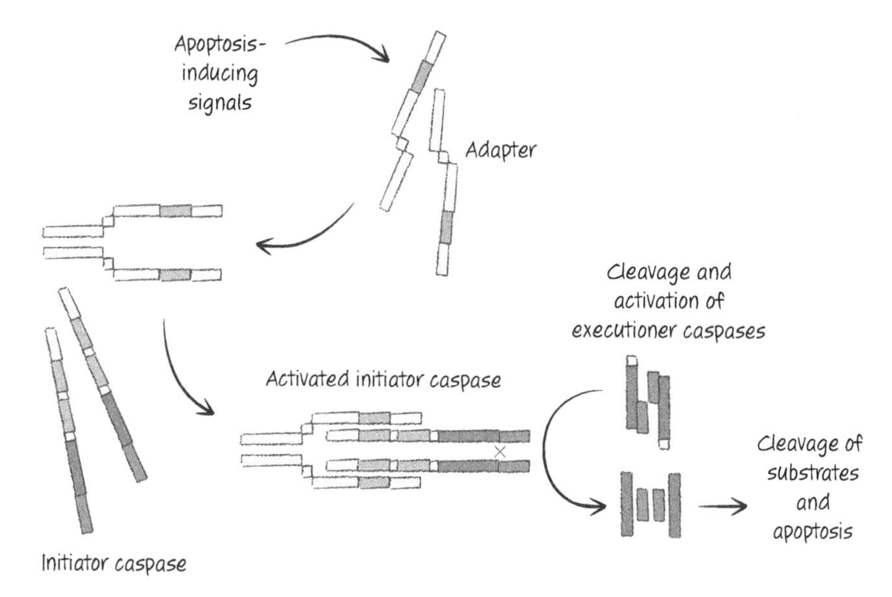

Figure 3.6. General apoptotic pathways.

rodents, caspase-11), caspase-2, and caspase-9 (the initiator caspase) all have caspase-recruitment domains (CARDs). DEDs and CARDs are also involved in other protein–protein interactions not involving caspases, and so the presence of such a domain is not itself a demonstration that the protein is involved in caspase activation or apoptosis.

Although DEDs and CARDs are not related by sequence homology, they are structurally similar (Fig. 3.7). This type of structure has been termed a "death fold." Other sequences that form death folds are also shown in Figure 3.7, including a death domain (DD) and a pyrin domain (PyD). Although these are not generally found in mammalian caspases, they are found in some of the molecules leading to caspase activation. As an intriguing aside, one of the caspases identified in zebrafish has a PyD (not seen in the prodomains of mammalian caspases) in its prodomain.

In general, these protein–protein interaction regions work by like–like recognition. That is, a protein that binds to a CARD in a caspase will do so through its own CARD. Although not every CARD (DED, DD, or PyD) will interact with every similar domain (in fact, the motif generally interacts with only one other protein, although there are exceptions), the binding will be through like–like domain recognition.

This theme frequently recurs in the chapters that follow, with sentences that take on the form "the *X* domain in protein A binds to the X domain of protein B." Why things have turned out this way is not at all obvious. It does, however, seem as though there is a deep, but elusive, evolutionary principle at work in these interactions.

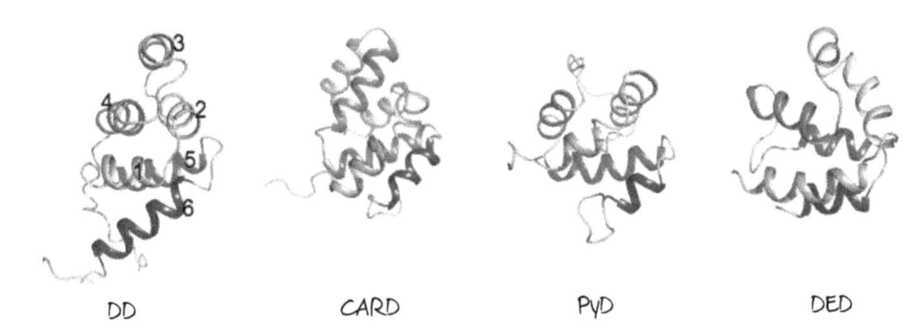

Figure 3.7. Representative death folds. Although different death folds do not have sequence similarity, they are structurally related. Each has six globular helical bundles, shown here colored gray to dark purple from the amino to carboxyl termini.[2]

ACTIVATION OF INITIATOR CASPASES BY DIMERIZATION

How does dimerization activate initiator caspases, and why is cleavage not required for proteolytic function but only for stabilization of the dimer? The proforms of the initiator (and other) caspases, as with the executioner caspases, are inactive because the specificity-determining pocket in the enzyme is disorganized and out of position for bringing the substrate to the catalytic cysteine–histidine dyad. However, unlike the case for executioner caspases, for initiator caspases this has nothing to do with the linker region between the large and small subunits, which is not cleaved until after the caspase is activated. Instead, another loop from the opposite chain in the dimer is needed to effectively pull the loop containing the substrate-specificity pocket into position. In caspase-9, this event can only happen at one protease site (of the pair) at a time, owing to constraints in the dimer. Figure 3.8 shows this interaction in the caspase-9 dimer that produces an active site.

Clearly, this formation of an active site in the protein depends on the formation of the dimer. When the inactive procaspase monomers are brought together by the appropriate adapter protein, the interaction results in the formation of an active site.

This has practical consequences for our understanding of apoptotic pathways because, although initiator caspases undergo cleavage on activation, cleavage does not itself activate them. Unfortunately, many studies have relied on detection of cleaved initiator caspases as an indication that the caspase has been activated en route to apoptosis. Such studies must be treated with caution. Similarly, a misunderstanding of the role of cleavage in initiator caspase activation has led some investigators to suggest that proteases that cleave initiator caspases thereby activate them, but

[2] The molecules from which these particular structures were obtained are FADD (DD and DED), APAF1 (CARD), and NLRP1 (PyD). These molecules are considered in later chapters, but for now the main point is that the different death folds are structurally related.

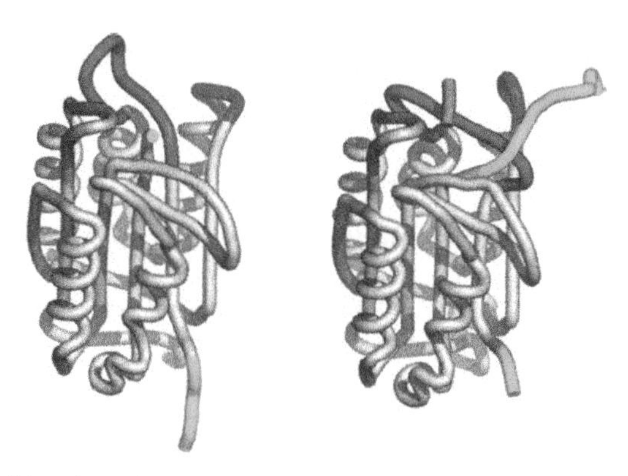

Figure 3.8. The transition from inactive to active caspase-9. When caspase-9 is dimerized, the specificity-determining loop (red) is brought into position by a loop (blue) from the other monomer. Cleavage of the region between the protease subunits (green) is not necessary for activation. Only one site in the dimer can form at a time: One is inactive (*left*) and the other is active (*right*).

we can now see that cleavage of the inactive monomer cannot do so. As with many areas of science, the fact that a conclusion is published does not make it true!

INITIATOR CASPASES CAN BE ACTIVATED BY DIMERIZATION WITH OTHER MOLECULES

The way in which initiator caspases are activated makes it possible for other proteins to induce the conformational changes necessary to form an active site in the caspase. Two such proteins are FLIP[3] and MALT, both of which can be brought by adapter proteins to form heterodimers with the initiator caspase caspase-8.

FLIP is closely related to caspase-8 at the sequence level but lacks the cysteine necessary for proteolytic activity. When dimerized with caspase-8, FLIP interacts with the caspase so that it forms an active site, but, because FLIP is not a protease, caspase-8 is not cleaved. When FLIP is present, it prevents caspase-8-mediated apoptosis. This dimer, however, is important for a signaling event that is needed for cell survival in certain circumstances (see Chapter 9).

MALT is not related to caspases at the sequence level, but it has predicted structural similarities that classify it as a paracaspase.[4] Like FLIP, MALT can be brought by adapter proteins into forming heterodimers with caspase-8. Again, the activated caspase-8 is not cleaved, and the function of this heterodimer remains obscure.

[3] FLIP is a protein encoded by the gene *CFLAR* (for "CASP8 and FADD-like apoptosis regulator") and exists in two isoforms. The protein we are discussing here is the long isoform, c-FLIP$_L$. To keep things simpler, we refer to it as FLIP.

[4] See Footnote 1, Chapter 2.

Nevertheless, it illustrates the principle that caspases need not necessarily be *homodimers* to be active.

OTHER INITIATORS AND EXECUTIONERS IN THE CASPASE COLLECTION

So far, we have focused on mammalian caspases in our consideration of how they work. But, as we know, apoptosis is not restricted to vertebrates. A brief tour of the caspases in other organisms is warranted.

Two initiator caspases have been identified in *Drosophila*. Dredd, which contains DED regions in its prodomain, appears to have a nonapoptotic function that remains elusive. The other, Dronc, which has a CARD-containing prodomain, clearly functions as an initiator caspase, activating the executioner caspases Drice and DCP1 by cleaving them. Like other initiator caspases, Dronc is activated by dimerization. Another caspase, Strica, seems to have an unusual prodomain and might be an initiator caspase, although this is not proven.

In *Caenorhabditis elegans*, only one caspase is known to function in apoptosis. This caspase, CED3, has a long prodomain containing a CARD (Fig. 2.4) and appears to be activated by dimerization. Once dimerized, CED3 then functions as an executioner caspase, causing the cell to die.

Is the process in *C. elegans* representative of the primordial apoptotic pathway, where activating one caspase causes death? Perhaps not. Cnidaria (including the hydroids, jellyfish, and corals), which are probably older in evolutionary terms than the nematodes, have both initiator- and executioner-type caspases, based on sequence. This might suggest that, somewhere along the way, nematodes lost caspases from the genome. This is not unlikely—the purple sea urchin (*Strongylocentrotus purpuratus*), an echinoderm, has 31 different caspases; compare that with the relatively small number we find in vertebrate species. In time, we hope to learn how these novel caspases function in cell death or other phenomena in such organisms.

INFLAMMATORY CASPASES FUNCTION IN SECRETION AND CELL DEATH

In mammals (and other vertebrates), one caspase has a well-characterized role that is distinct from induction of cell death (although, as we will see, it can cause cell death). This is caspase-1. Like initiator caspases, it exists in cells as an inactive monomer and is activated by dimerization. However, at low levels of the active caspase, caspase-1 functions not as a killer but, instead, in inflammation. This is because caspase-1 processes (by cleavage) the precursor forms of two related cytokines: interleukin-1 and interleukin-18. These cytokines have important roles in inflammatory responses. The pathways involved in caspase-1 activation are discussed in Chapter 7, but, for now, it is worth noting that the same principles we have learned that apply to apoptosis apply to other caspases as well, even if the result is not apoptosis.

Caspase-1 occupies a position in the mammalian genome in close proximity to other related caspases, including caspases-4, -5, and -12 in humans and caspases-11 and -12 in rodents. For this reason, these are often also referred to as inflammatory caspases. Some of these—caspases-4 and -5 (human) and caspase-11 (rodent)—can participate in caspase-1 activation, as we will see later, but others such as caspase-12 do not.

Caspase-12 is particularly intriguing. It is found throughout mammals and primates, until we examine humans. Among sub-Saharan Africans, caspase-12 is expressed in about 20% of individuals. In contrast, most humans have a stop codon in the third exon that results in an unstable truncated protein. Evidence suggests that this null allele was positively selected during the African migration about 67,000 years ago. To get an idea of why this might have occurred, we have to consider what caspase-12 does. In general, the caspase itself appears to have very little activity. When coexpressed with caspase-1, it can dampen the activity of the latter. It is possible that, under some conditions in which strong caspase-1 function is a problem, this dampening function is favored, whereas, under other conditions, increased caspase-1 activity provides protection (e.g., from some infections). In any case, it might be correct to think about caspase-12 more as a regulator of protease activity than as a protease per se.

Caspase-1, in addition to processing interleukin-18 and interleukin-1, appears to have another role in inflammation that is not well understood, functioning in an unusual mode of secretion in the cells that express it. This secretion does not involve signal peptides or the endoplasmic reticulum (as conventional secretion does), but, beyond that, we do not yet know how such secretion occurs. Proteins that are secreted by this mechanism include interleukin-1β, interleukin-18, and proteins that are not processed by caspase-1, such as interleukin-1α and peptides with antibacterial activities.

Under conditions of intense activation, caspase-1 can trigger cell death. Like initiator caspases, caspase-1 can cleave and activate at least one executioner caspase (caspase-7). But, in addition, it can itself cleave another substrate in the cell, resulting in a form of cell death that resembles necrosis. This caspase-1-triggered cell death is often referred to as "pyroptosis" (Fig. 3.9). The inflammatory caspases—caspases-4 and -5 in humans (and caspase-11 in rodents)—also cleave this substrate and induce pyroptosis. Pyroptosis and the secretion of inflammatory cytokines is discussed in much more detail in Chapter 7. It remains a possibility that it is this nonapoptotic cell lysis that is actually responsible for the mysterious "secretion" associated with caspase-1 activation, although this is a matter of debate.

In *Drosophila*, the initiator-like caspase Dredd does not seem to function in cell death but is involved in the secretion of peptides that have antibacterial activity. For this reason, it is therefore tempting to think of Dredd as a sort of inflammatory caspase.

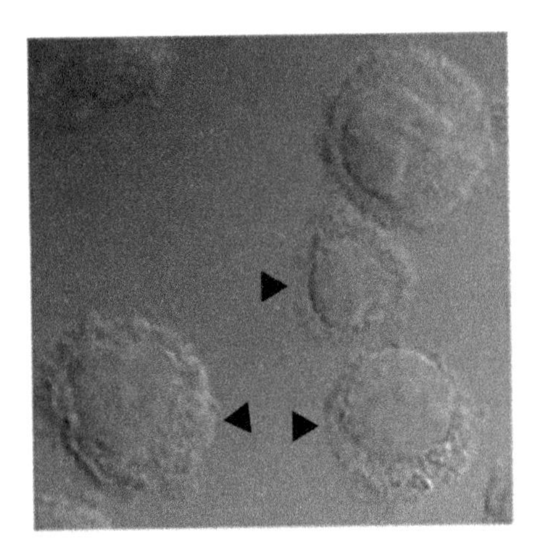

Figure 3.9. Pyroptosis. Cells die in a caspase-1-dependent manner, indicated by arrows.

Caspase-2: The "Orphan" Caspase

Caspase-2 was the second mammalian caspase discovered, and, by sequence, it is the most highly conserved among animals. Like the initiator and inflammatory caspases, it is activated by induced proximity of the inactive monomers, and cleavage serves to stabilize the mature enzyme. However, unlike the initiator caspases and caspase-1, caspase-2 does not seem to be very good at cleaving and activating the executioner caspases. It can cause apoptosis indirectly by engaging the mitochondrial pathway (discussed in Chapter 7), but this does not seem to be a primary function for this enzyme. We just do not know its primary function.

Caspase-2 has been implicated in the response to a number of stressors, but in no case is it absolutely required for apoptosis. What caspase-2 does and what it is "for" are unknown at present. Is caspase-2 important? In a mouse model of lymphoma, the absence of caspase-2 accelerates the appearance of tumors. No other caspase has such a profound effect in this model. As we will see in Chapter 7, caspase-2 may have an important role in monitoring the cell cycle.

ON BEYOND 12

Vertebrates have a number of other caspases that have been characterized, and some have multiple copies of the ones that we have discussed. In most cases, their functions are unknown, as are the adapter molecules that activate them.[5] Figure 3.10 shows a list of these vertebrate caspases.[6] The figure is provided more as a point of interest and for the reader to get a feeling of the evolutionary plasticity in the system than as

[5] There is evidence that caspase-14 has a role in the differentiation of skin.

[6] Caspase-13 is not included in the figure because this caspase turned out to be the bovine homolog of caspase-4.

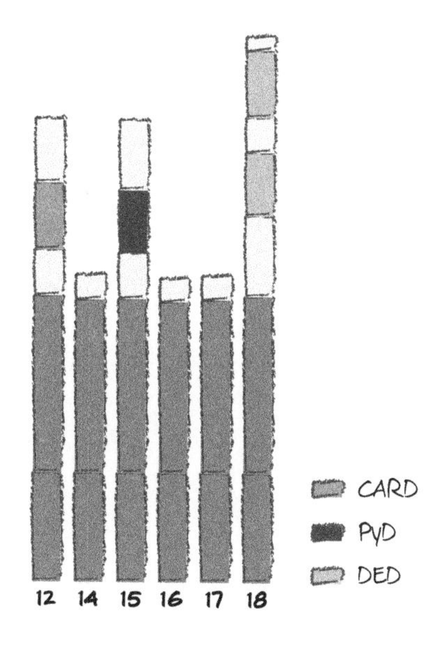

Figure 3.10. Caspase-12 and beyond in mammals. Lengths shown are arbitrary. Caspases-15, -17, and -18 are not found in humans, and caspase-12 is frequently not found in humans. Caspases-11 (mouse) and -13 (cow) are homologs of caspase-3 and caspases-4, -5, respectively, and are not shown.

important information for the discussions that follow. That said, we might want to ask ourselves why a set of molecules that are so well conserved among the animals show so much variability, duplication, deletion, and modification, if the only role is to orchestrate apoptotic cell death.

INHIBITION OF CASPASES IN CELLS

The importance of caspases in cell death and apoptosis leads us to the question of how they are regulated in cells, because it is almost axiomatic that anything so devastating must be inhibited at several levels. As we will see, however, endogenous caspase inhibitors are essential in some cases, but not all.

VIRAL AND ENDOGENOUS CASPASE INHIBITORS: INHIBITOR OF APOPTOSIS PROTEINS

When a virus infects a cell, it uses its host to make more virus, whereas, if the cell can undergo apoptosis before more virus is made, the infection can be stopped. But viruses have evolved strategies for preventing apoptosis. One strategy is to inhibit caspases.

The insect virus baculovirus actually makes two different caspase inhibitors that work by different mechanisms. One is termed an "inhibitor of apoptosis protein" (IAP). When this was first identified, it quickly became apparent that insects have their own "IAP proteins" (the expanded term "inhibitor of apoptosis protein proteins" is, of course, redundant, but this phrase persists, and we will not buck the trend here). The

SYNTHETIC CASPASE INHIBITORS

Peptides that are recognized by caspases can be fashioned into caspase inhibitors if a reactive "warhead" is appended to the P1 aspartate that binds (reversibly or irreversibly) to the active cysteine in the caspase. Again, as with the substrates, such inhibitors are not specific, although they are often effective as general caspase inhibitors (that can also inhibit other proteases as well). Therefore, although these can be useful for research, any conclusions that come from using them have to be regarded with care. Examples of synthetic inhibitors based on peptides are shown in Figure 3.11. The structure of caspase-3 bound to a peptide inhibitor is shown in Figure 3.12.

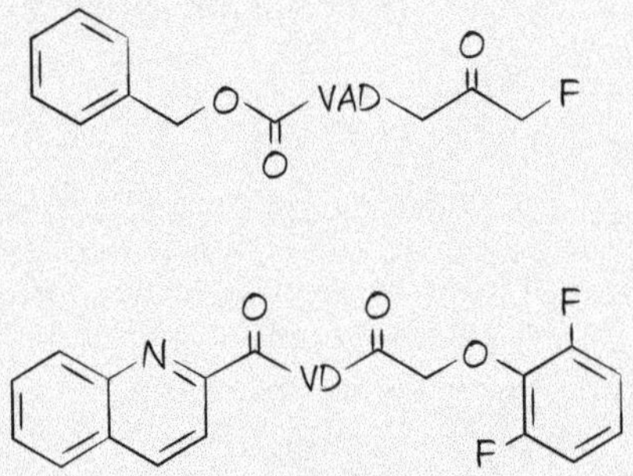

zVAD-fmk

Q-VD-oph

Figure 3.11. Caspase inhibitors. The peptide portion is shown with a single-letter code.

Other inhibitors of caspases that are not peptide based have been developed. These work by either binding to the active site or inhibiting the conformational changes needed for caspase activation.

If we induce apoptosis, for example, by stressing cells, and add agents that inhibit the executioner caspases, all of the characteristic morphological features of apoptosis are blocked (Fig. 3.13).

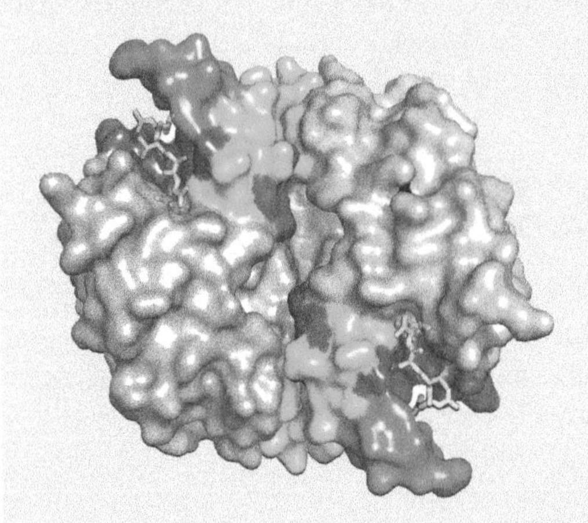

Figure 3.12. Caspase-3 bound to the inhibitor zVAD-fmk.

(Continued)

(Continued from previous page)

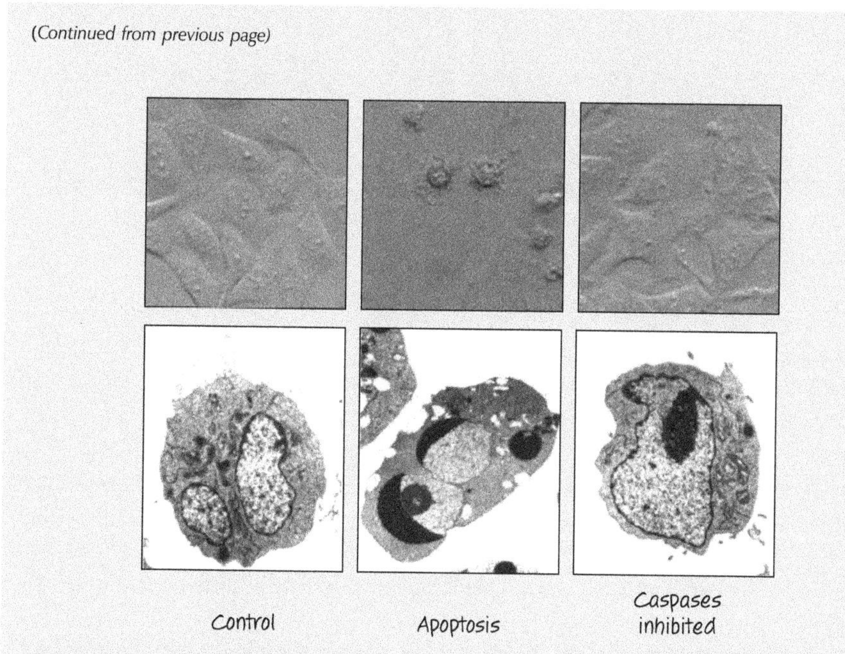

Figure 3.13. Caspase inhibitors block the morphological features of apoptosis. Appearance of different cell types treated to induce apoptosis in the presence or absence of the caspase inhibitor zVAD-fmk. (*Upper*) Bright-field images of fibroblasts. (*Lower*) Electron micrographs of tumor cells.

most important of these is called DIAP1 (*Drosophila* IAP1, so named before a moratorium was called on *Drosophila* proteins starting with a "D"). DIAP1 binds to and inhibits the initiator Dronc and the executioner caspases. Experimental removal of DIAP1 is sufficient to cause apoptosis in fly cells and embryos (Fig. 3.14).

It was also quickly realized that IAP proteins are found in many animals, including humans, based on sequence similarities. All IAP proteins share one or more copies of a motif called the baculovirus IAP repeat (BIR). *But*—and this is very important—despite their somewhat unfortunate name, most IAP proteins do not inhibit caspases or block apoptosis. Some IAP proteins are listed in Figure 3.15. Those few capable of directly inhibiting caspases are noted.

It should be immediately apparent that, of the mammalian IAP proteins, only XIAP is a direct inhibitor of caspases.[7] In fact, different regions of the protein inhibit

[7] Some studies suggest that cIAP1, cIAP2, NAIP, and survivin also inhibit caspases. All can associate with some caspases (directly or indirectly) and can ubiquitinate them, which might influence activity and/or turnover. However, none of these is a direct inhibitor of caspases, and any inhibition by these other mechanisms appears to be minor.

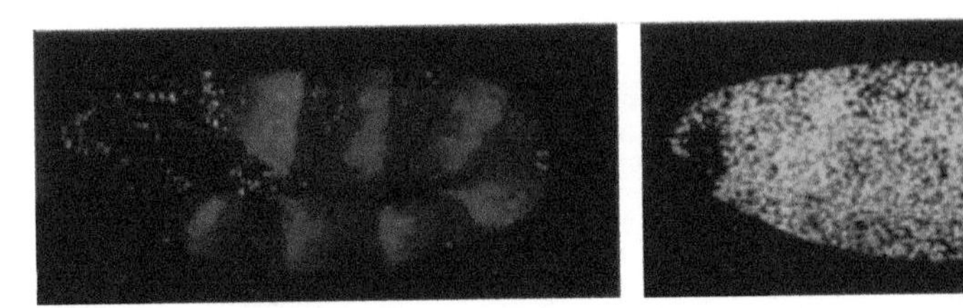

Figure 3.14. Loss of DIAP1 causes apoptosis. Extensive cell death, staining green, is not seen in a wild-type embryo (*left*), but is observed in a *DIAP1* mutant embryo (*right*).

different caspases: BIR3 (and the region adjacent to it, the RING; see Fig. 3.15) inhibits the initiator caspase-9, whereas BIR2 (and the linker between it and BIR1) inhibits the executioner caspases, caspases-3 and -7. Figure 3.16 shows the binding of XIAP to caspase-3.

In addition to BIR domains, many IAP proteins have another domain, called RING, that functions as an E3-ubiquitin ligase, responsible for putting ubiquitin chains on target proteins and targeting them for degradation by the proteasome or changing their signaling properties. XIAP and DIAP1 cause degradation of caspases by this mechanism.

Is XIAP essential? That is, if it is absent, do caspases spontaneously activate? Mice lacking XIAP have no developmental defects, and so the answer appears to be "no." However, we will see later (in Chapter 6) an example of XIAP function in apoptosis.

So, we have a mystery: DIAP1 is essential in the control of *Drosophila* apoptosis, but IAP proteins as inhibitors of caspases do not appear to be crucial in other

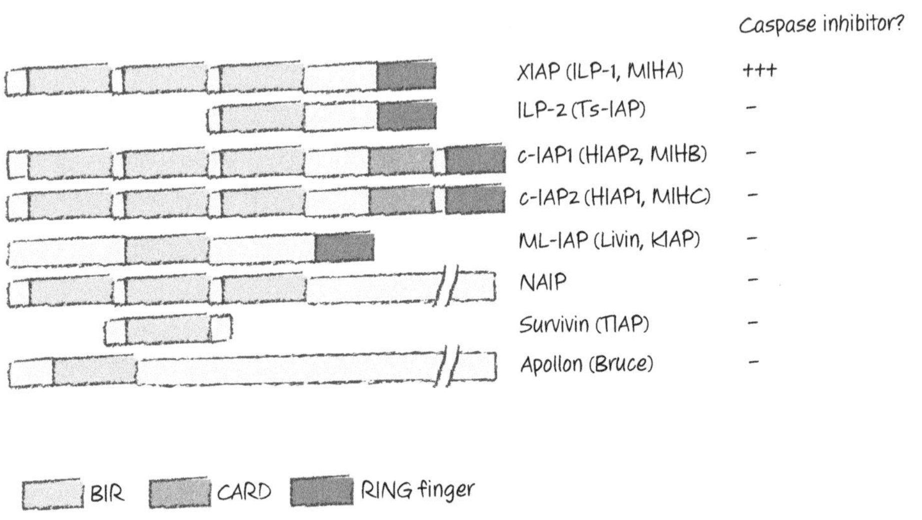

Figure 3.15. Some IAPs.

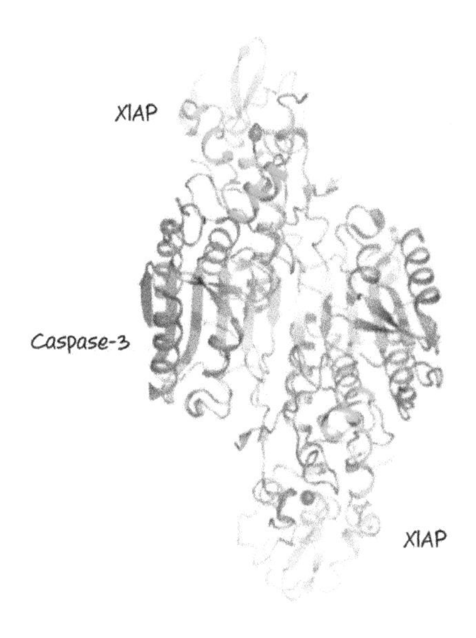

Figure 3.16. XIAP binds to the active sites in caspase-3. Only the BIR2 domain in XIAP (green) is shown. (Purple circles indicate zinc atoms that coordinate the folding of XIAP.)

organisms. Nematodes lack caspase-inhibitory IAP proteins altogether, and XIAP appears, at first pass, to be dispensable, at least in mice. It could be that the control of apoptosis upstream of initiator caspase activation is so tight that control by IAP proteins in most animals is unnecessary. As we delve further into the different apoptotic pathways, we will see why this could be so. Alternatively, there might be caspase inhibitors other than XIAP that we have not discovered or sufficiently appreciated.

At this point, we can begin to piece together what we have discussed so far in terms of caspase activation, inhibition, and apoptosis in animals (see Fig. 3.17).

OTHER VIRAL INHIBITORS OF CASPASES

Viruses that infect mammalian cells sometimes carry caspase inhibitors, but these are distinct from the IAP proteins. Poxviruses, for example, express inhibitors belonging to the serpin family. Most serpins act as inhibitors of serine proteases (remember that caspases are a different type—cysteine proteases), but some viral serpins instead act as caspase inhibitors. For example, one serpin, CrmA, expressed by cowpox virus, does not inhibit caspase-9 or the executioner caspases in cells, but it is effective in blocking inflammatory caspase-1 and initiator caspase-8, and it also inhibits the serine protease granzyme B (recall that granzyme B is made by cytotoxic lymphocytes and triggers apoptosis in target cells). As we will see, caspase-8 has roles in some immune effector mechanisms, and therefore this inhibitor might function in immune evasion.

In discussing baculovirus, we mentioned that there are actually two different caspase inhibitors that it expresses. In addition to its IAP, baculovirus also makes p35, an

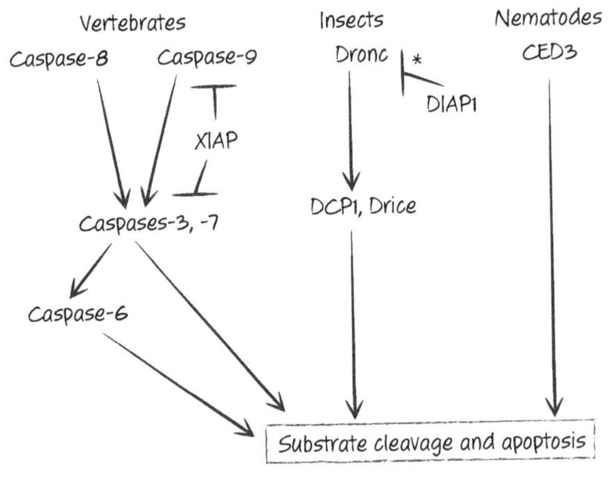

Figure 3.17. Caspases and IAPs in apoptosis.

inhibitor not found in any form in animals. This protein is remarkably specific for caspases and acts as a caspase substrate. When it is cut, it becomes irreversibly bound to the caspase. Although unique to insect viruses, p35 can effectively inhibit mammalian (and other) caspases, making it useful for experimental purposes.

So, we have reached the point in our discussion where we know how caspases work, what they cut, and how they are inhibited. And to a first approximation, we know that apoptosis is caused by key substrates being cleaved by executioner caspases that are themselves activated when they are cut by initiator caspases. The latter are activated when adapter proteins bring them together so that they activate themselves by "induced proximity." But what are these adapters and how do they come to engage the initiator caspases? The answer is apoptotic pathways. We will begin in the following chapter with the pathway responsible for most apoptosis occurring in humans (and vertebrates in general)—the mitochondrial pathway.

CHAPTER 4

The Mitochondrial Pathway of Apoptosis, Part I

MOMP and Beyond

MITOCHONDRIA AND CELL DEATH

The major mode of apoptosis in vertebrates is the mitochondrial pathway. This pathway of cell death is engaged by a vast array of cell stresses, including deprivation of growth factors, disruption of the cytoskeleton, DNA damage, accumulation of unfolded proteins, hypoxia, and many others. It is also activated by developmental signals, such as hormones, that instruct cells to die. In this chapter, we discuss how mitochondria are involved in the activation of caspases in this type of cell death.

A JUST-SO STORY

Before examining the remarkable way in which mitochondria effect apoptosis, it might be worth considering how this organelle might have become involved in cell death—after all, in nearly all eukaryotes, mitochondria are essential for life, providing not only energy via the tricarboxylic acid cycle and oxidative phosphorylation, but also many other essential services to the cell. These include lipid metabolism and the ability to live in the toxic world of oxygen.

Rudyard Kipling, in his *Just So Stories*, provided fanciful explanations of biological phenomena, such as how the elephant got its trunk, that were scientifically untestable. In evolutionary biology, too, we have similarly untestable just-so stories, such as the one that follows. But it may have value in helping to frame what is to come, and we will return to this fantasy at later stages in this and other chapters.

Approximately 2 billion years ago, an α-proteobacterium invaded an archeon cell. The subsequent endosymbiotic relationship produced what we now know as a eukaryotic cell, and the bacterium became the first mitochondrion.[1] Although this idea of a grand symbiosis is strongly supported by data, the initial relationship might well have started off far from cooperatively. After all, the bacterium was infecting the archeon.

As we have noted, a very good strategy for host defense against intracellular pathogens is for the infected cell to die. This even seems to apply to single-cell organisms that die altruistically to avoid spreading infection to their identical clone mates.[2] Of course, a pathogen that can prevent such proactive suicide gains the upper hand. The initial infection that led to the formation of the first mitochondrion might therefore have triggered a suicide response in the invaded cell (probably not apoptosis, but a precursor to this form of cell death). In turn, this was checked by the infecting bacterium (Fig. 4.1).

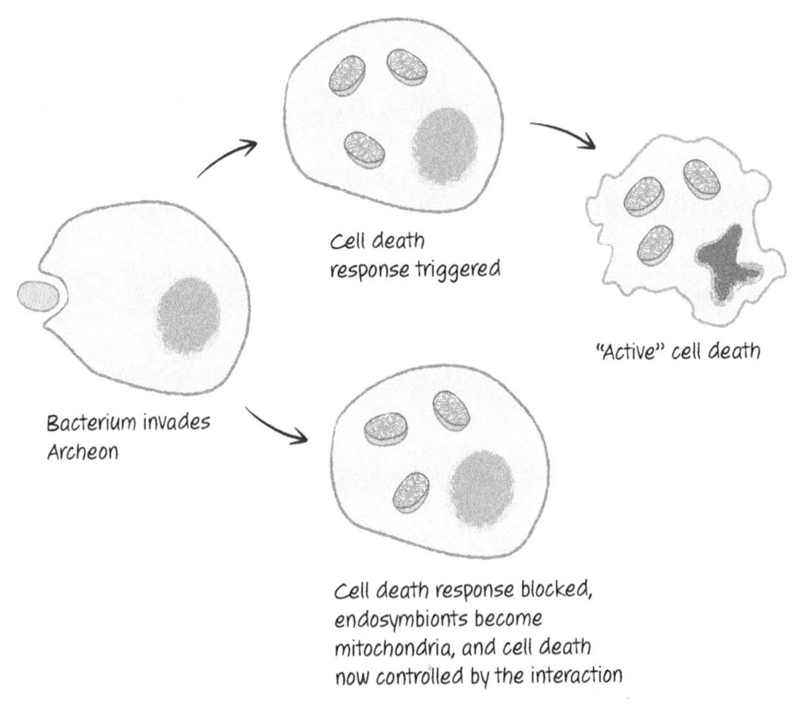

Figure 4.1. A just-so story about mitochondria and cell death. According to this tale, an early archaeal cell defends itself by committing suicide when infected (*upper pathway*), but, when faced by a variant bacterium possessing by chance the wherewithal to interfere with the archaeal cell's defenses (*lower pathway*), the archaeal cell is subjugated by the invader. Over time, the invader repays the cell by developing into the cell's mitochondrial powerhouses that still retain the ability to act as lynchpins of cell death regulation.

[1] This is the endosymbiont hypothesis of mitochondrial evolution.

[2] There is a literature on active cell death (although not apoptosis) in bacteria and evidence that this is an effective response to infection by phage.

As the bacteria became mitochondria, decisions regarding the life and death of the cell might have resided under the control of this evolving organelle. In time, the control moved from mitochondria to the eukaryotic cell as a whole. The mitochondria, however, remained the focus for the effects.

HOW MITOCHONDRIA ACTIVATE CASPASES

The mitochondrial pathway of apoptosis has been described primarily in vertebrates, although there is compelling evidence for its existence in echinoderms (such as sea urchins and sand dollars) and in the platyhelminthes (flatworms). As we know, apoptotic pathways involve mechanisms that activate an initiator caspase, which in turn activates the executioner caspases to orchestrate cell death. In the mitochondrial pathway, the initiator caspase is caspase-9.

In Chapter 3, we discussed how activation of initiator caspases by dimerization of inactive monomers requires an adapter protein to bind to their prodomains. In the case of caspase-9, the adapter that activates the caspase is APAF1 (apoptotic protease activating factor-1). APAF1 binds to caspase-9 by means of the APAF1 caspase-recruitment domain (CARD), which binds to the CARD domain in caspase-9 (Fig. 4.2).

APAF1, like caspase-9, preexists in the cell as a cytosolic, inactive monomer that cannot bind to or cause dimerization of the caspase. Several events must occur to

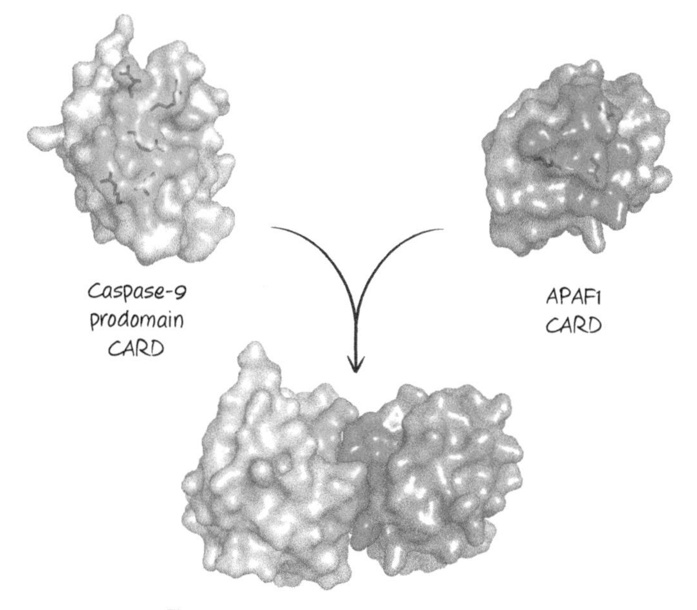

Figure 4.2. A CARD–CARD interaction.

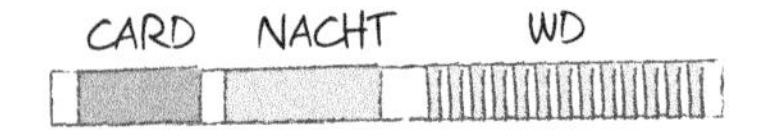

Figure 4.3. Domains of APAF1. A CARD and NACHT domain are followed by a series of WD repeats at the carboxy-terminal end.

change this during the initiation of apoptosis. To understand these, it is useful to look at the different domains of APAF1 (Fig. 4.3).

The CARD region of APAF1, which binds to caspase-9 (via its own CARD), and the oligomerization domain are both buried in the inactive APAF1 monomer. The NACHT[3] domain contains a nucleotide-binding site that is inaccessible, because access to it is blocked by the WD domain.[4] If the WD domain is removed, the nucleotide deoxy-ATP (dATP) can bind to the nucleotide-binding site in the NACHT domain. This triggers a conformational change, exposing the oligomerization and CARD domains. Binding of cytochrome *c* (see below) to the WD domain has the same effect, changing the conformation of APAF1, allowing the binding of dATP. APAF1 molecules then assemble into the complex shown in Figure 4.4. The center of the APAF1 oligomeric complex contains the CARD domains that recruit caspase-9 molecules to activate the caspase. This APAF1–caspase-9 complex is called the "apoptosome."

Removal of the WD region results in formation of the apoptosome, but this is not what occurs in cells undergoing apoptosis. Instead, another protein binds to the WD region, producing a conformational change that exposes the nucleotide-binding site (presumably by moving the WD region away from it). The protein responsible is cytochrome *c*, a protein present in mitochondria, where it has a central role in electron transport and energy production.

Cytochrome *c* is a nuclear-encoded protein that is synthesized in the cytosol as apocytochrome *c* and then transported into mitochondria to the space between the inner and outer mitochondrial membranes (the intermembrane space). Here, the enzyme heme lyase attaches a heme group to create the mature protein (holocytochrome *c*). Apocytochrome *c* cannot activate APAF1, but holocytochrome *c* (which can do so) is normally sequestered away from the cytosol (and APAF1) by the mitochondrial outer membrane. For apoptosis to occur by the mitochondrial pathway, this barrier must be disrupted.

The ability of cytochrome *c* to engage APAF1 to induce apoptosome formation and caspase activation is independent of its function in electron transport. We can

[3] NACHT is an acronym based on several proteins that contain this domain: NAIP, CIITA, HET-E, and TP1. Although these proteins do not concern us here, APAF1 contains this domain. It is also referred to as a nucleotide-binding domain (NBD).

[4] WD domains (also called WD40) contain multiple motifs of approximately 40 amino acids that often end with a tryptophan–aspartic acid ("WD" in single-letter amino acid code). Many proteins have WD domains that have functions in signaling and cell cycle control (and in the case of APAF1, apoptosis).

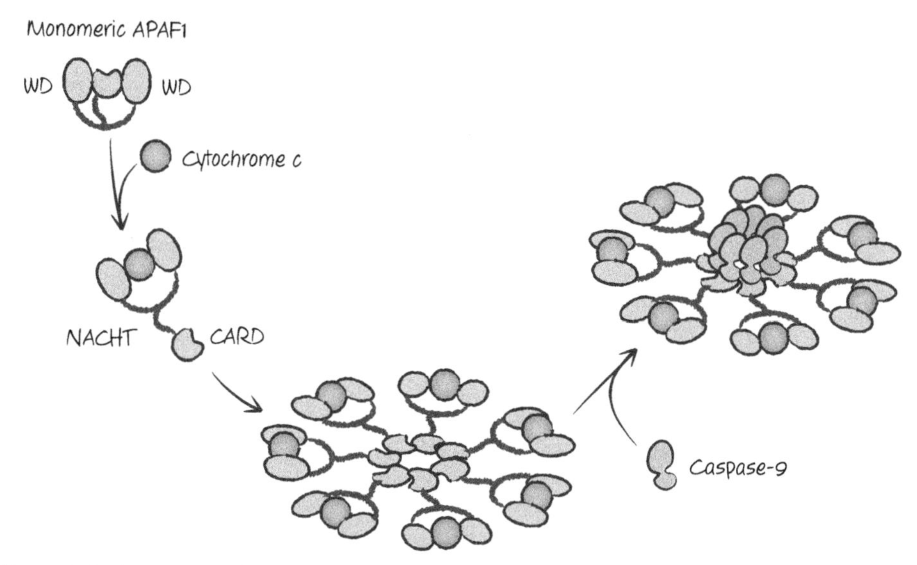

Figure 4.4. Apoptosome formation in vertebrates. Interaction of cytochrome *c* with the WD region of APAF1 allows dATP binding to the NACHT, resulting in a conformational change. This drives oligomerization and exposure of the CARD that interacts with the CARD of caspase-9, resulting in the activation of the caspase.

introduce mutations in a particular amino acid (lysine 72) that do not affect electron transport but impair apoptosome formation. Mammalian cells engineered to lack cytochrome *c* do not activate caspases when the mitochondrial pathway is engaged, nor do cells in which cytochrome *c* is mutated at this key residue.

Mice engineered to lack APAF1 or caspase-9 often die during development or just after birth, owing to extensive developmental abnormalities. These include large outgrowths of the brain occurring because of the presence of excess neurons. Cells from these mice do not activate caspases in response to signals that normally would engage the mitochondrial pathway of apoptosis. These defects in development are also seen in mice lacking caspase-3. In addition, a mouse has been generated in which cytochrome *c* was mutated at lysine 72. This mutation produced an animal with the same defects as in those lacking APAF1, caspase-9, or caspase-3 (Fig. 4.5). This serves as a formal demonstration that the mitochondrial pathway of apoptosis requires the function of cytochrome *c*.

DISRUPTING THE MITOCHONDRIAL OUTER MEMBRANE

If cytochrome *c* is to trigger apoptosis by engaging APAF1, it must move from the mitochondrial intermembrane space to the cytosol, where both APAF1 and caspase-9 reside. It is able to do this during apoptosis because upstream signals that induce

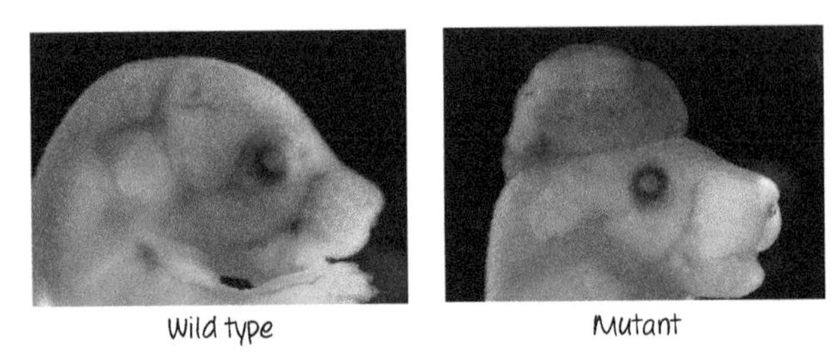

Figure 4.5. Mouse fetuses engineered to lack expression of APAF1, caspase-9, or caspase-3 or with a cytochrome K72A (lysine-to-alanine) mutation experience overgrowth of the forebrain (*right*) compared with a wild-type control (*left*).

cell death cause the outer membranes of all (or nearly all) mitochondria in the cell to become permeable by a process called "mitochondrial outer membrane permeabilization" (MOMP). This results in the release, by diffusion, of any soluble molecules residing in the intermembrane space, including cytochrome *c*. An example is shown in Figure 4.6.

Through time-lapse imaging of cells expressing fluorescent fusion proteins, we know that, during apoptosis, MOMP is usually sudden, rapid, and irreversible. That is, when a cell is induced to undergo apoptosis, an indeterminate time passes, and then suddenly nearly all of the mitochondria undergo MOMP within a very

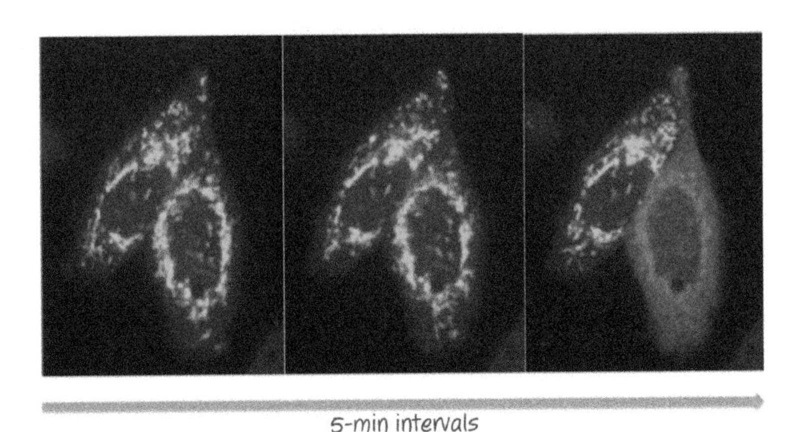

Figure 4.6. Mitochondrial outer membrane permeabilization (MOMP). Cells expressing a fusion between cytochrome *c* and green fluorescent protein (GFP) undergoing MOMP in response to an apoptosis-inducing stress (*left to right*). As MOMP occurs (cell *on the right* in each pair), the distribution of fluorescence changes from localization in the mitochondria to being diffuse throughout the cytoplasm. (The cell *to the left* underwent MOMP at a later time.) The time between images is 5 min, with the first image taken several hours after the initial stress.

short time period (~5–10 min). Shortly after this, caspases become active, leading to apoptosis.

Because MOMP does not involve a loss of integrity of the inner mitochondrial membrane, mitochondrial function is not destroyed by this process, although electron transport is greatly reduced because cytochrome *c* becomes diluted by diffusion into the cell cytosol. However, as we saw in Chapter 2, as executioner caspases become active, they gain access to proteins that are exposed on the inner membrane. Consequently, complex I of the electron-transport chain is destroyed and mitochondrial physiology is altered dramatically.

OTHER PROTEINS ARE RELEASED TOGETHER WITH CYTOCHROME *c*

When MOMP occurs during apoptosis upstream of caspase activation by the apoptosome, all soluble proteins of the intermembrane space are free to diffuse out of the mitochondria and into the cytosol. Of these hundreds of different proteins, some have roles in apoptosis and possibly other forms of cell death.

Among these is a protein called "second mitochondrial activator of caspases" (Smac, also called Diablo). This protein performs a function in caspase activation that is distinct from that of cytochrome *c*. Like the latter, it is produced by a nuclear gene, and the protein is imported into the mitochondrial intermembrane space. In the process, the amino-terminal region is proteolytically removed, revealing a short sequence at the new amino terminus with an important function.

Remember that vertebrate cells express XIAP, an endogenous inhibitor of caspase-9 and the executioner caspases. This can block caspase activation and apoptosis even if cytochrome *c* is released to trigger the formation of the APAF1 apoptosome. The function of Smac is to prevent XIAP from exerting this inhibitory effect. The amino terminus of mature Smac binds to the same region of XIAP that binds to caspase-9, preventing the inhibition of the caspase by XIAP and allowing caspase activation to proceed. This is illustrated in Figure 4.7.

Smac is a large protein dimer. Is the only function of this protein to present the small amino-terminal peptide to XIAP following MOMP? It seems unlikely, but no other mitochondrial functions of this protein have been identified. Peptides and drugs that mimic the amino terminus of Smac bind to and inhibit two other IAPs (in addition to XIAP)—cIAP1 and cIAP2—that have roles in other types of (nonapoptotic) signaling. However, at this point, we simply do not know whether Smac has a function beyond that in apoptosis.

Smac is not the only protein with this XIAP-neutralizing activity that is released following MOMP. Another is the serine protease Omi (also called HtrA2), which like Smac has an amino-terminal sequence that inhibits XIAP. Omi has other functions in the mitochondria that appear to be conserved not only in animals but in other

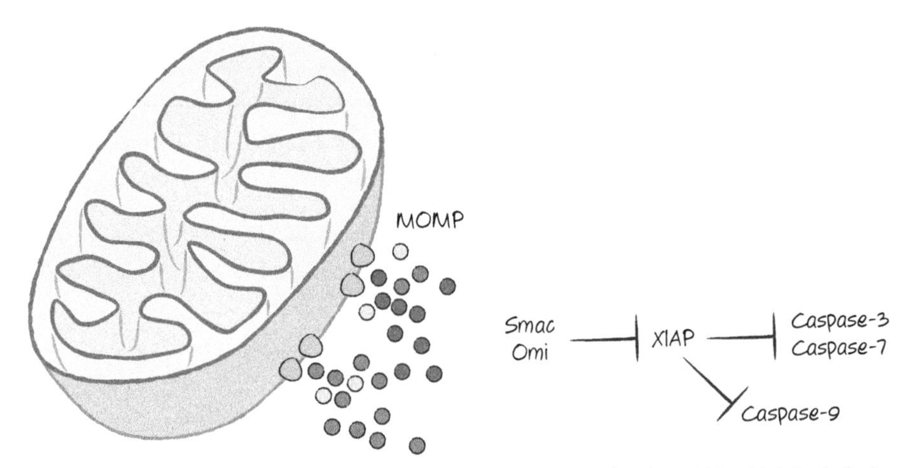

Figure 4.7. MOMP releases proteins that inhibit XIAP. Proteins that bind to XIAP, which include Smac and Omi, reside in the mitochondrial intermembrane space and are released upon MOMP. These block the ability of XIAP to inhibit caspase-3, caspase-7, and caspase-9.

eukaryotes as well.[5] In some animals, Omi does not carry the XIAP-neutralizing sequence at all (one example is the Omi found in bovids). We return to Omi and Smac in Chapter 6 when we consider another pathway of vertebrate apoptosis.

MOMP CAN CAUSE CASPASE-INDEPENDENT CELL DEATH

Cells in which the mitochondrial pathway of apoptosis is engaged, but in which caspase activation is disrupted or blocked, can still die as a consequence of MOMP. This mode of cell death is often referred to as "caspase-independent cell death" (CICD). The name is a bit problematic because any cell death that is not apoptotic is generally "caspase independent." But CICD has come to imply that MOMP is involved. In conditions under which MOMP is inhibited, CICD does not occur (we discuss the regulation of MOMP in detail in the next chapter).

Mice lacking APAF1 have a number of developmental abnormalities, as we have seen (Fig. 4.5). However, in many tissues in which developmental apoptosis[6] would normally occur, these mice also display cell death, but it has a different appearance. In particular, the cells die without the characteristic appearance of apoptosis. An example is cell death that is seen in the interdigital webs of the developing mice, an event required for the formation of digits (Fig. 4.8).

[5] This is based on studies in organisms lacking the gene. In knockout mice, for example, the animals display cell death in the brain and immune systems ~1 month after birth, but the reasons for this are not fully elucidated. It is clear, however, that this is not due to the role of Omi discussed here.

[6] Discussed in more detail in Chapter 10.

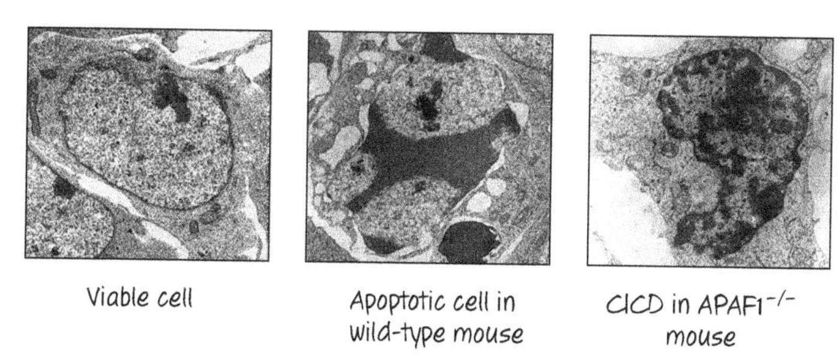

Figure 4.8. Apoptosis and caspase-independent cell death (CICD) in interdigital webs of developing mice. During development, cells in the region between what will become digits die with the characteristics of apoptosis (note the chromatin condensation in the *center* panel). In mice engineered to lack APAF1, these cells still die, but without apoptotic morphology (*right* panel).

In fact, when we look more closely at developmental cell death in wild-type animals, some cell deaths more closely resemble CICD than apoptosis. It might be that, in some cells, APAF1 or the caspases are not efficiently expressed or engaged following MOMP, but, because the cell dies anyway, it does not matter.

Not all cells undergo CICD in response to MOMP if caspases are not active. In neurons, for example, the cells can survive MOMP if caspases are not activated, and the cells can eventually recover. In culture, most cells that undergo MOMP are generally doomed, but, under certain conditions, cells can recover if caspase activation is blocked or disrupted. CICD therefore appears to be less efficient than apoptosis, which might help to explain the developmental abnormalities in knockout mice that cannot engage caspase activation through the mitochondrial pathway.

MECHANISMS OF CICD

One likely explanation for CICD is "mitochondrial catastrophe." Once MOMP occurs, the outer membranes of the mitochondria are compromised, and therefore all of the soluble proteins from the intermembrane space become severely diluted, affecting mitochondrial function. ATP production, lipid biogenesis, and other important functions of mitochondria no longer occur efficiently, and the cell reaches a point of no return and expires.

An alternative explanation for CICD is that some of the proteins released following MOMP can kill the cell regardless of whether caspases are activated. Two of these potential killers are endonuclease G and apoptosis-inducing factor (AIF).

Endonuclease G is a mitochondrial enzyme that can cleave DNA between nucleosomes, similarly to CAD/DFF40 (discussed in Chapter 2). Cells lacking CAD or iCAD, or cells in which caspases are inhibited, generally fail to fragment their DNA during cell death. It is not clear, therefore, that endonuclease G cleaves nuclear chromatin during either apoptosis or CICD, and the role of endonuclease G in CICD is not well established. Nevertheless, it remains possible that limited cutting of DNA by this enzyme contributes to CICD, but, at this point, we cannot say with certainty that it is involved.

The case for AIF is more intriguing. AIF is an essential protein that appears to be important for the proper transport and function of the mitochondrial electron-transport chain, and homologs of AIF are found throughout the eukaryotes. It resides in the intermembrane space and is tethered to the inner membrane. Following MOMP, proteases gain access to the intermembrane space as we described and then free AIF from its tether. Caspases can do this, as well as other proteases, such as calpain. It has been suggested that AIF then locates to the nucleus to effect CICD, possibly by causing DNA fragmentation (Fig. 4.9).

Lack of AIF appears to be incompatible with tissue development, and most cells lacking AIF are severely compromised, which makes it difficult to rigorously test the importance of AIF in CICD. Although there are many studies concluding that the release of AIF from mitochondria is a cause of cell death in various systems, this

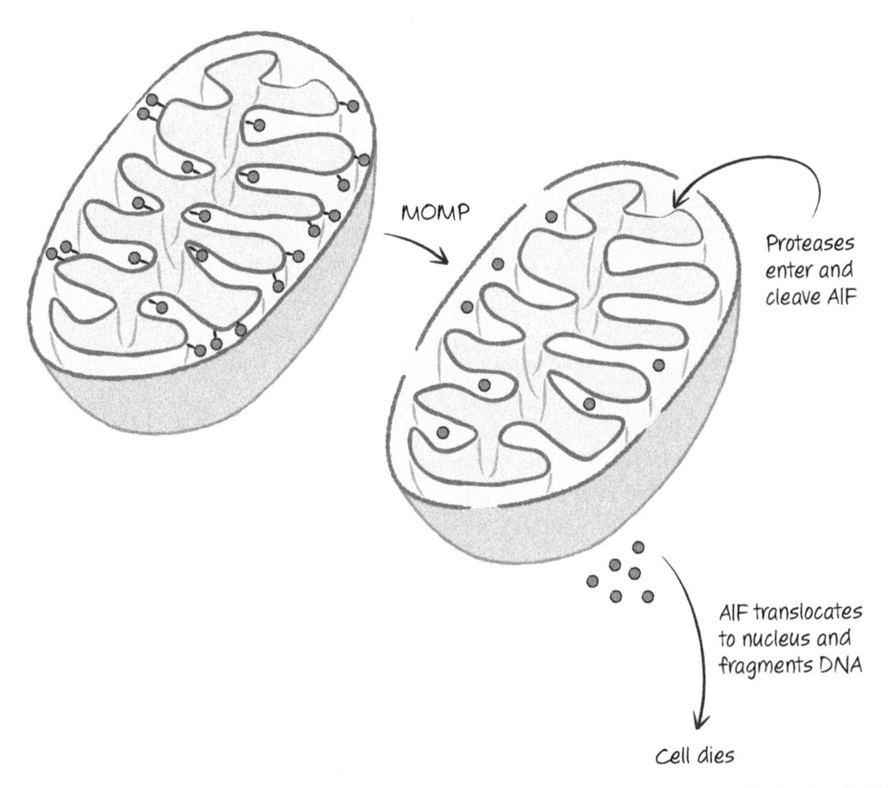

Figure 4.9. Apoptosis-inducing factor (AIF) model of caspase-independent cell death (CICD). MOMP, mitochondrial outer membrane permeabilization.

remains controversial as other studies report no role for AIF in cell death of cells that do not require it for survival.

MITOCHONDRIAL PERMEABILITY TRANSITION

How does MOMP occur? If isolated mitochondria are exposed to high concentrations of calcium, this causes a channel—the permeability transition pore (PTP)—to open in the mitochondrial inner membrane. As a result, the transmembrane voltage potential across the inner membrane immediately dissipates, solutes enter the central matrix, and water swells the mitochondria until the inner membrane ruptures the outer membrane. This change in the inner membrane is referred to as the mitochondrial permeability transition (MPT).

Many signals in addition to calcium fluxes can cause the MPT, including reactive oxygen species (ROS), changes in cellular pH, and certain drugs. Because of this, it was widely believed that the MPT was the event that initiates apoptosis by breaking the outer membrane and releasing cytochrome *c* (and other proteins). This view persists in some quarters, and could well be correct in some settings. However, although the MPT is likely to have important roles in some forms of cell death, the available evidence is that it has no (or little) role in most forms of apoptosis, including the mitochondrial pathway of apoptosis, as we will see.

Unfortunately, we know very little about what comprises the PTP. Most schemes representing the PTP show several characterized molecules involved in forming a channel through both the inner and outer membranes, such as that depicted in Figure 4.10.

In this view, the PTP is mostly composed of the adenosine nucleotide transporter (ANT, which shuttles ADP and ATP across the inner membrane), with roles for the voltage-dependent anion channel (VDAC) in the outer membrane. The PTP forms when ANT opens a channel in the inner membrane. However, mitochondria from cells lacking different ANTs display normal permeability transitions (as well as apoptosis), as do cells lacking different VDACs. So currently, we do not know which membrane molecules comprise the PTP. Some evidence suggests that proteins comprising the ATP synthase in mitochondria are themselves responsible for the PTP. However, as this complex is generally essential for cell survival, it is difficult to prove this idea, or to test the importance of the PTP in different forms of cell death.

One component of the PTP for which there is some consensus is a matrix protein called cyclophilin D. Cyclophilin D is a peptidylprolyl isomerase, an enzyme that reconfigures proline residues in proteins.[7] Mitochondria from mice lacking cyclophilin D display a defective MPT response to calcium (although, at much higher

[7] Cyclophilin D interacts with ANT and other proteins, and it has been suggested that its reconfiguration of prolines in the target proteins is what opens the PTP.

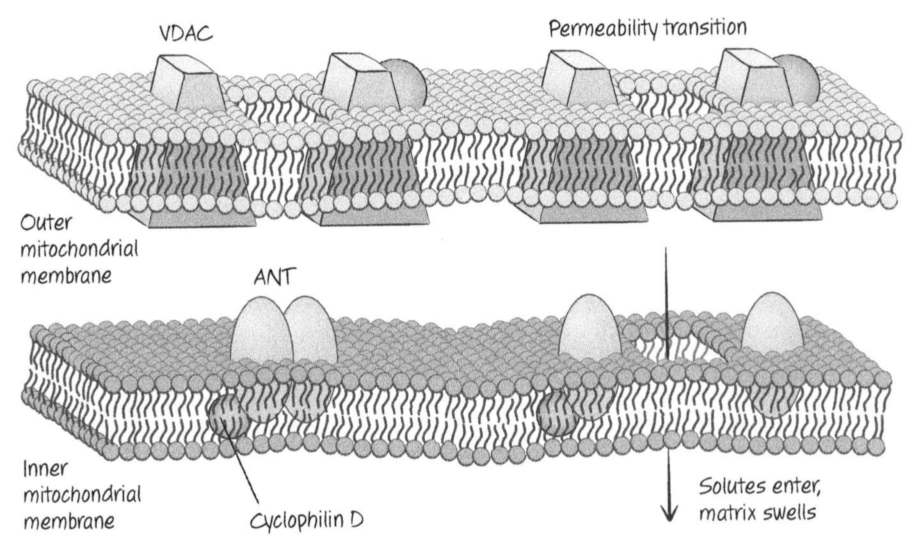

Figure 4.10. Not the PTP. This is often the scheme used to explain the phenomenology of the PTP, but experimental evidence strongly suggests that VDAC and ANT are unlikely to represent key elements of the mechanism. How the PTP actually occurs remains controversial.

concentrations of calcium, an MPT can be detected). Significantly, however, no defects in apoptosis, either developmental or induced, can be detected in these mice, and they are developmentally normal. Interestingly, the mice are somewhat resistant to a nonapoptotic form of cell death arising from ischemic injury (discussed in more detail in Chapter 8).

Given these observations, the MPT is unlikely to be a major mechanism of MOMP in apoptosis. The calcium levels needed to trigger MPT are not achieved in routine cell signaling but might arise under pathological conditions. A more likely set of molecules that control MOMP in the mitochondrial pathway of apoptosis is the subject of the next chapter.

APOPTOSOMES OF FLIES, WORMS, AND OTHER BEASTS

The adapter proteins responsible for the activation of initiator caspases in the fly (Dronc) and nematode (CED3) are homologs of APAF1 (Fig. 4.11). In *Drosophila*, this homolog is called APAF1-related killer (ARK), and, in *Caenorhabditis elegans*, it is CED4.

In each case, a CARD in the adapter protein binds to the CARD in the prodomain of the caspase to activate the latter. However, unlike APAF1, neither of these appears to be activated by cytochrome *c*.

For CED4, the reason for this is pretty clear. CED4 does not possess a WD region (see above), and, as a consequence, it can spontaneously oligomerize to form an

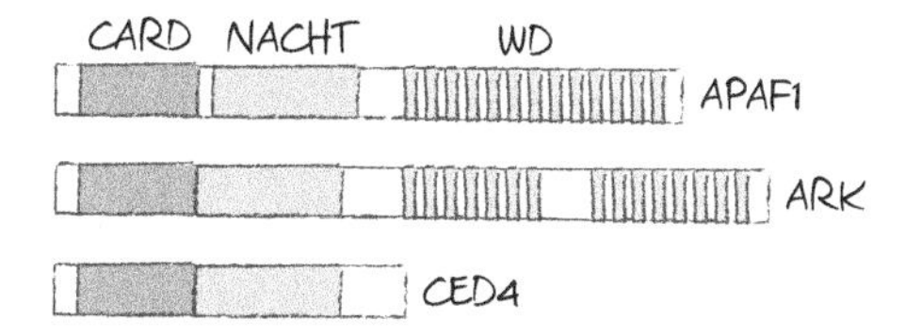

Figure 4.11. Domain structures of apoptotic protease activating factors (APAFs) in humans (*top*), flies (*center*), and worms (*lower*).

octameric apoptosome and activate CED3. In healthy cells, this is prevented by another protein, CED9, that holds CED4 as an inactive monomer. Curiously, this occurs on the surfaces of mitochondria for reasons that are not known (but we speculate on this in the next chapter). During apoptosis, CED4 is released from CED9. CED4 then oligomerizes, recruits CED3, and activates the caspase to promote apoptosis (Fig. 4.12).

In flies, the APAF1 homolog, ARK, appears to be constitutively active. That is, the protein might spontaneously oligomerize to bind and activate the initiator caspase Dronc. We say "might" because ARK, like APAF1, has a WD region, and therefore another molecule (a protein?) might have a role in the activation of ARK, but currently this is not known.[8] However, if ARK is constitutively active, what holds apoptosis in check? The answer is the *Drosophila* inhibitor of apoptosis protein, DIAP1, which prevents Dronc activation. It is the disruption of the DIAP1–Dronc interaction that triggers apoptosis. For now, we can create the scheme shown in Figure 4.13 for apoptosis in the fly.

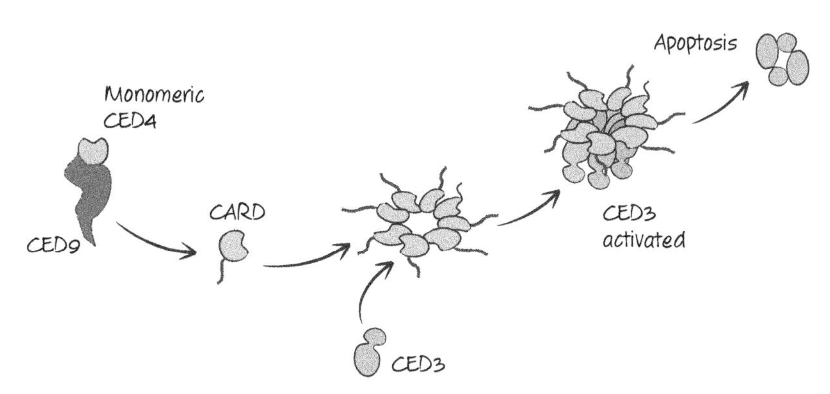

Figure 4.12. Caspase activation in nematodes. Remember that this illustration is misleading: CED4 oligomerizes near its nucleotide-binding region into an octameric apoptosome; the CED4 CARD binds to the CARD of CED3, which is then activated by proximity.

[8] There is some controversy about this. Genetic evidence supports a role for cytochrome *c* in the activation of caspases in spermatogenesis in the fly and perhaps apoptosis in some cells. However, experiments with *Drosophila* cells and cell extracts do not support a role for cytochrome *c* or its release from mitochondria in apoptosis.

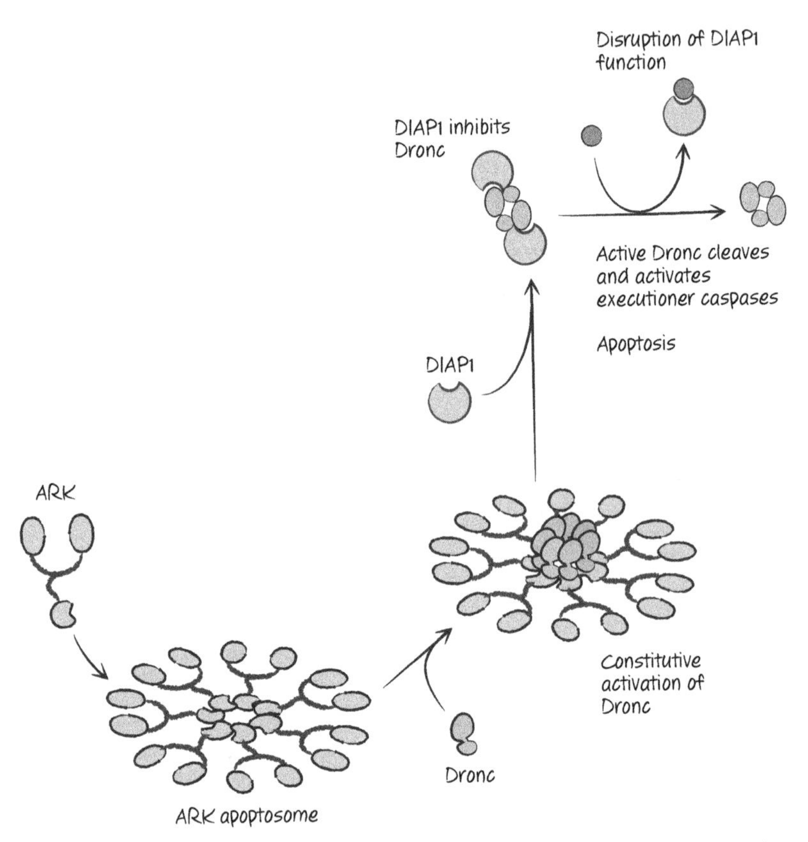

Figure 4.13. Caspase activation in *Drosophila*. ARK is constitutively active, and forms the apoptosome, which activates the initiator caspase Dronc. Dronc is held inactive by the inhibitor of apoptosis protein DIAP1. Disruption of DIAP1 permits the active Dronc caspase to cleave executioner caspases, and apoptosis proceeds.

Furthermore, what do the apoptotic pathways in these animals say about our just-so story of an ancient role for mitochondria in controlling cell death? Is the nematode pathway ancestral, giving rise to the arthropod pathway in insects, and then the mitochondrial pathway in vertebrates? If so, then our story is simply a fantasy without value.

Flies did not evolve from nematodes, however, and humans did not evolve from flies. Thus, it is possible that the mitochondrial pathway, as it appears in vertebrates, is the ancestral mechanism. For some reason, nematodes and insects might have subsequently lost it.

Currently, to guide us we have only sequence data from different animal phyla, with limited functional information. APAF1 homologs are found throughout the animal kingdom, and, in those examined so far, only in nematodes do any of these lack the WD region that interacts with cytochrome *c* in the vertebrate pathway. At this point, we do not know whether most of these actually interact with cytochrome *c*. As noted at the beginning of this chapter, there is compelling evidence that MOMP occurs and cytochrome *c* activates APAF1 in echinoderms and platyhelminthes, and therefore it remains possible that the mitochondrial pathway of apoptosis, as we have described it, is widespread in the animals. However, what we can say is that the pathway in nematodes is unlikely to be ancestral.

IAP INHIBITION IN *DROSOPHILA* APOPTOSIS

As we have discussed, the APAF1 homolog in flies appears to be constitutively active, and apoptosis is controlled by the action of DIAP1 to block the initiator caspase Dronc. In flies, MOMP does not occur upstream of caspase activation, and therefore no IAP antagonist is released from mitochondria to promote cell death. Instead, proteins that use a strategy similar to that of Smac and Omi are transcriptionally expressed and neutralize DIAP1 to cause apoptosis. These proteins are encoded within a complex of functionally related genes that all have suitably morbid names, including Grim, Reaper, Sickle, and the less evocative Hid (because it was originally identified in a different setting). These proteins, like Smac (and other IAP antagonists), share the small amino-terminal region that allows them to bind to the IAP and take it out of action. The resulting *Drosophila* pathway for apoptosis is shown in Figure 4.14.

Smac, Omi, and these *Drosophila* proteins all interact with IAP proteins in similar ways, through the binding of their amino-terminal sequences; Figure 4.15 shows these sequences for comparison. The same sequence is found in caspase-9, and the binding of this sequence to XIAP is necessary for inhibition of the caspase. The ability of proapoptotic proteins to neutralize IAP proteins therefore appears to be simple competition with the caspases for binding. If so, this small peptide region should be sufficient to produce the effect. This appears to be true in experimental systems.

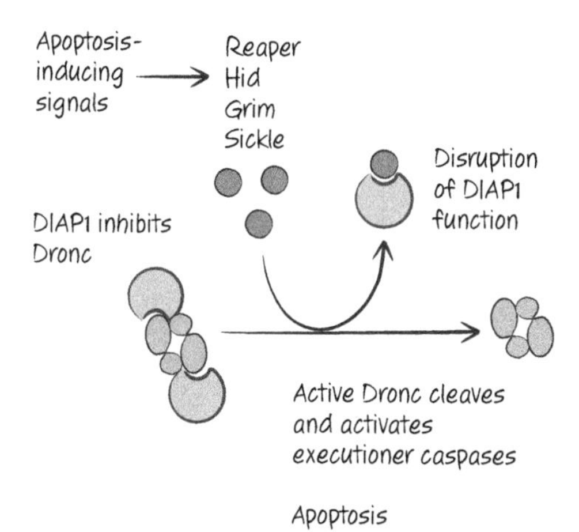

Figure 4.14. Apoptotic signals in flies induce the expression of inhibitors of DIAP1, including Reaper, Hid, Grim, and Sickle.

Smac	AVPI
Omi	AVPS
Reaper	AVPI
Hid	AVPF
Grim	AIAY
Sickle	AIPF
Caspase-9	ATPF

Figure 4.15. IAP-binding sequences (single-letter amino acid code) in humans and *Drosophila*. The amino-terminal sequences of Smac and Omi are produced following cleavage of the mitochondrial-localizing sequence in mitochondria. That of caspase-9 is produced by caspase cleavage. The similarities among these sequences is evidence for an evolutionarily conserved mechanism of IAP binding.

So far, we have considered only the part of the mitochondrial pathway of apoptosis that is downstream from MOMP, and MOMP is clearly an important event that determines life or death, at least in vertebrate cells. What causes MOMP and how it is regulated are considered next.

CHAPTER 5

The Mitochondrial Pathway of Apoptosis, Part II

The BCL-2 Protein Family

CONTROLLING MITOCHONDRIAL OUTER MEMBRANE PERMEABILIZATION

We are now near the top of the mitochondrial pathway of apoptosis, at the point before the mitochondrial outer membrane permeabilizes to release proteins such as cytochrome *c* into the cytosol. This is where the major decisions are made that determine whether a cell will die by engaging this pathway. And these decisions depend on interactions among members of the BCL-2 family of proteins.

The BCL-2 proteins, named for the first family member to be described (B cell lymphoma-2), are a collection of related molecules found throughout the animal kingdom. They share only limited sequence similarity, except in short regions called BCL-2 homology (BH) domains, and can be grouped according to which of these domains they carry and, as we will see, by their functions. A list of several BCL-2 proteins in mammals and their BH domains is shown in Figure 5.1.

There are three "flavors" of BCL-2 proteins. The pro-apoptotic BCL-2 effectors promote apoptosis by causing mitochondrial outer membrane permeabilization (MOMP). These are the proteins that essentially make the holes in the outer mitochondrial membrane. The anti-apoptotic BCL-2 proteins, which include BCL-2 itself, prevent apoptosis by preventing MOMP. The third group is a subfamily of proteins that promote apoptosis by regulating the other two types of BCL-2 molecules. These are the BH3-only proteins, which share only the BH3 domain, hence their somewhat unfortunate designation.

The BCL-2 proteins are an alphabet soup of names, some sounding nearly identical and others bordering on the unpronounceable (BMF is "bimf," if that helps). The acronyms have ceased to have any real meaning, and we will treat them here as

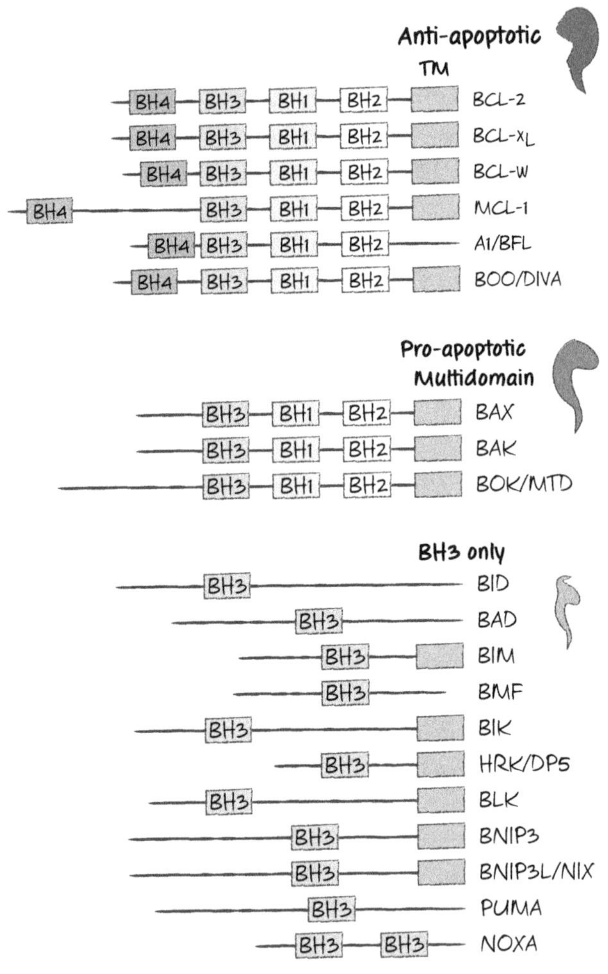

Figure 5.1. BH domain organization of selected members of the BCL-2 family. TM, transmembrane domain (although in some cases, this is sequestered by the BH groove).

simple (if confusing) names.[1] Nevertheless, these molecules are essential to the mitochondrial pathway of apoptosis.

BAX AND BAK ARE THE EFFECTORS OF MOMP

At first glance, BAX and BAK seem to be quite different proteins, sharing little sequence similarity outside the short BH1, BH2, and BH3 domains. BAX is generally

[1] To make matters a bit worse, many of the BH3-only proteins have distinct gene names (the gene encoding PUMA is *BBC3*).

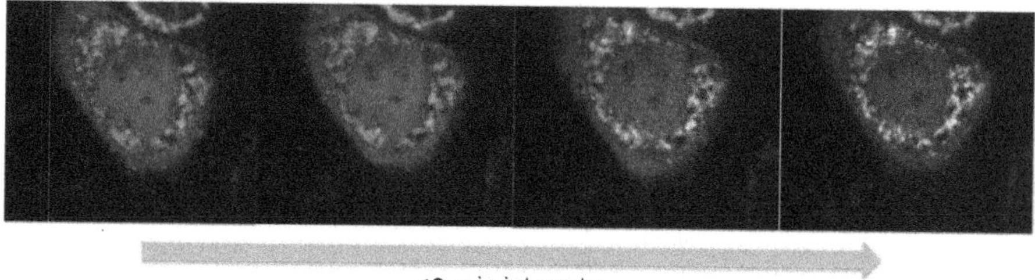

Figure 5.2. BAX moves from the cytosol to the mitochondrial outer membrane during apoptosis. A cell engineered to express both BAX linked to green fluorescent protein (BAX–GFP) and a red fluorescent protein (RFP) that localizes to mitochondria was induced to undergo apoptosis. BAX (green) moves onto the mitochondria (red), revealed as the development of a yellow signal (i.e., merging of red and green emissions) over time. The first image was taken several hours after the initial stress.

soluble in the cytosol but moves to mitochondria when apoptosis is induced (Fig. 5.2). In contrast, BAK is tethered to mitochondria in the cell by its carboxy-terminal region.

When BAX and BAK are activated, they bury themselves in the mitochondrial outer membrane, where they form oligomers of multiple sizes. Studies using artificial membranes have shown that these oligomers can form holes in membranes (technically, these are not pores or channels, both of which are much more organized). These openings are capable of allowing large molecules to pass through the membrane and effect the process of MOMP.

Mice engineered to lack either BAX or BAK develop normally, although BAX-deficient mice have reproductive problems, and cells from these animals undergo MOMP and apoptosis normally. However, animals lacking both BAX and BAK are another story—these have severe developmental defects that are usually lethal to the embryo. Cells from such double-deficient animals do not undergo MOMP or engage the mitochondrial pathway of apoptosis. Therefore, this seems to be a striking case of molecular redundancy—either BAX or BAK can effect MOMP, but at least one must be present.

Another BCL-2 family protein, BOK, is also a pro-apoptotic effector (together with BAX and BAK). Like BAX and BAK, BOK can effect MOMP to cause apoptosis. However, it is regulated in a distinct manner from that of the other effector proteins, and it is not involved in the responses to most of the stimuli that engage BAX and BAK.

The pro-apoptotic BCL-2 effector proteins directly cause MOMP by oligomerizing and inserting into the mitochondrial outer membrane. There are two models for how they do this. One model suggests that the activated proteins form a pore that is lined with the effector molecules arranged in their oligomeric complexes. An alternative model suggests instead that hydrophobic amino acids that are exposed in the activated effectors disrupt the lipids in the membrane, such that they produce a "lipidic pore" (Fig. 5.3). Although there is evidence for both models, most favors the lipidic pore—although this remains controversial.

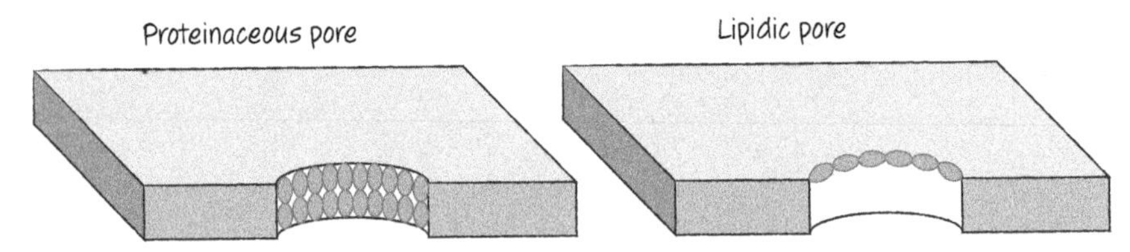

Figure 5.3. Proteinaceous versus lipidic pores. In a proteinaceous pore, the pore-forming protein "coats" the pore, penetrating the membrane. In a lipidic pore, the protein causes changes in the membrane lipids to produce the opening. Although some models propose a proteinaceous pore comprising the effector proteins (BAX, BAK, or BOK) for MOMP, most evidence supports the idea that these proteins induce the formation of lipidic pores to effect MOMP.

As we will see in Chapter 11, apoptosis is activated by oncogenic transformation and suppresses cancer unless the apoptotic pathway is disrupted. From this perspective, it is not surprising that tumors sometimes mutate BAX or BAK and may delete BOK. This is not, however, the only way a cancer can avoid cell death.

ANTI-APOPTOTIC BCL-2 PROTEINS PREVENT MOMP

The anti-apoptotic proteins BCL-2, BCL-xL, MCL-1, and A1 (among others) all act to prevent MOMP and block the mitochondrial pathway of apoptosis. These could work simply by preventing the oligomerization of BAX and BAK, which indeed they do. But *how* they do this brings us to the edge of waters that were made murky by a controversy that is now resolving into clarity.

At first, everything seems pretty simple. The anti-apoptotic BCL-2 proteins bind to active BAX and BAK. More specifically, they bind to the BH3 regions of BAX and BAK. The structures of BCL-xL bound to a peptide corresponding to the BH3 of BAX or BAK have been solved and are fairly informative (Fig. 5.4).

The BH3 domains of BAX and BAK bind to a groove in BCL-xL formed from α-helices containing its BH1, BH2, and BH3 domains (the so-called BH groove[2]). This binding pocket is present in all of the anti-apoptotic BCL-2 proteins and is clearly crucial for their anti-apoptotic function because mutations affecting it can destroy the ability to block MOMP and apoptosis.

Here is the first bit of murk. Given that the BH3 regions of BAX and BAK are not exposed in the native proteins (Fig. 5.5), how could anti-apoptotic BCL-2 proteins bind to them? Indeed, if anti-apoptotic BCL-2 proteins are mixed with BAX or BAK, they do not bind them. However, if detergents are added, they now bind readily. This is a useful hint to what is going on: Anti-apoptotic BCL-2 proteins and

[2] This is also called a "BC groove," with the "B" referring to the binding of BH3 regions of target proteins and the "C" referring to the binding of the protein's own carboxyl terminus in some of the structured BCL-2 proteins. Here, we use "BH groove" because it is formed from the BH1, BH2, and BH3 domains.

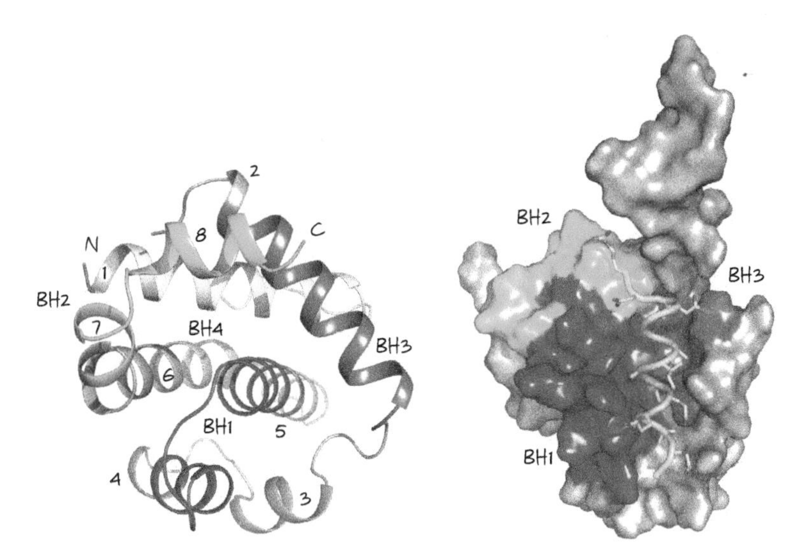

Figure 5.4. Structure of BCL-xL, a globular protein with eight α-helices (numbered 1–8, *left*). The four BCL-2 homology (BH) regions, BH1–BH4, are also shown. The structure on the *right* shows the groove formed by BH1, BH2, and BH3, bound to a BH3 peptide from BAK (pale blue).

pro-apoptotic effectors can only interact when they are embedded in a hydrophobic environment, such as the outer mitochondrial membrane.

So far, so good. Biochemical evidence indicates that, as BAX or BAK are activated, this exposes the BH3 domain of the protein, which opens a groove into which the BH3 domain of another BAX or BAK molecule can bind, forming a dimer. BCL-xL, then, can prevent this step in oligomerization by binding to this exposed BH3 domain and thereby preventing the BAX–BAX or BAK–BAK interactions required for MOMP (Fig. 5.6).

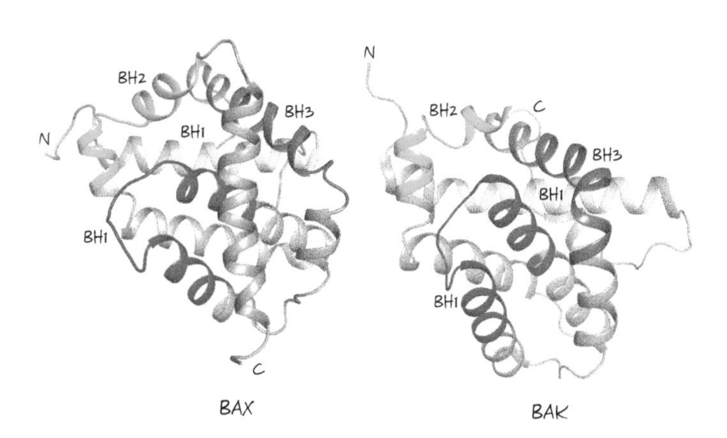

Figure 5.5. Effector BCL-2 family members BAX and BAK. The BCL-2 homology (BH) regions BH1, BH2, and BH3 are shown, as is the carboxy-terminal tail of BAX (gold).

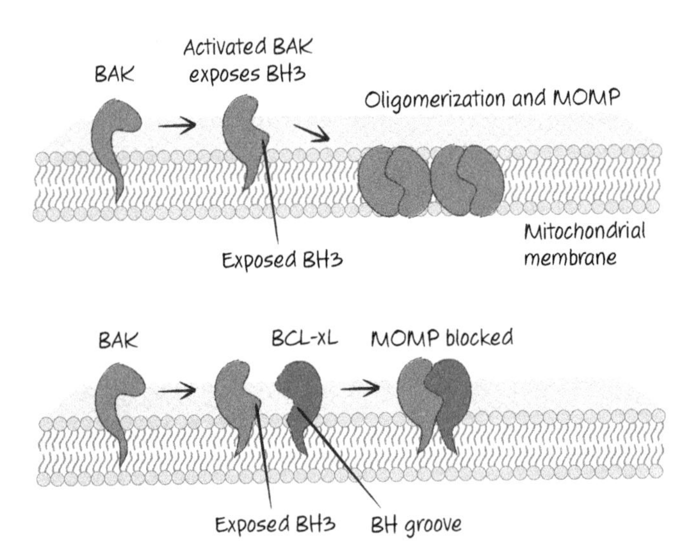

Figure 5.6. BAK activation and inhibition. When BAK is activated, it exposes its BH3 region and appears to create a BH groove. This allows BAK–BAK oligomerization (*upper*) or binding of anti-apoptotic BCL-2 proteins that block oligomerization (*lower*).

The binding of anti-apoptotic BCL-2 proteins to BAX and BAK is not universal—there is some specificity to the interactions. The BH3 domain of BAK binds to the BH pockets of BCL-xL and MCL-1 very well, but much less effectively to BCL-2. Conversely, the BH3 domain of BAX binds very well to BCL-xL and BCL-2, but poorly to MCL-1 (Fig. 5.7).

But here the already murky waters deepen. MCL-1, despite binding poorly to BAX, can block apoptosis very well when apoptosis is mediated by BAX (e.g., when there is no BAK). Similarly, BCL-2 binds poorly to BAK but prevents apoptosis that is mediated by BAK (e.g., when there is no BAX). We can conclude from this that apoptosis is not simply controlled by the balance of anti-apoptotic BCL-2 proteins and the pro-apoptotic effectors BAX and BAK. The latter idea, once called the "rheostat model" (Fig. 5.8), is not quite right—something is missing.

BH3-ONLY PROTEINS PROMOTE MOMP AND APOPTOSIS

The "somethings" missing are the BH3-only proteins. Although the BH3 domains of BAX and BAK bind to the anti-apoptotic BCL-2 proteins, the BH3-only proteins also

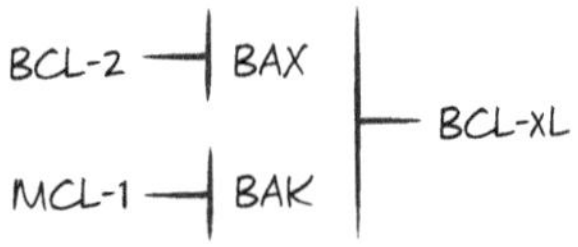

BCL-2 ⊣ BAX
MCL-1 ⊣ BAK
BCL-xL

Figure 5.7. BAX/BAK specificity of anti-apoptotic BCL-2 proteins. The binding differences are relative, not absolute.

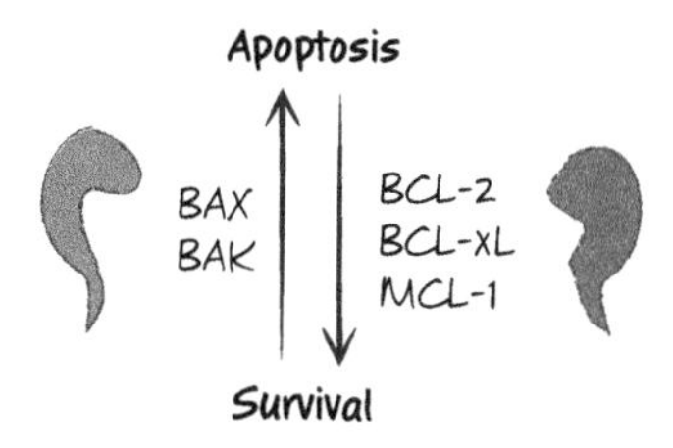

Figure 5.8. The simple rheostat model. An increase in BAX or BAK causes apoptosis, whereas an increase in BCL2, BCL-xL, or MCL-1 results in survival. Problems with this simple model include the specificities of anti-apoptotic proteins for binding to the pro-apoptotic effectors.

bind to the BH grooves of the anti-apoptotic molecules. The BH3 domain is not well conserved, however, and does not have sufficient sequence characteristics to permit its simple identification by straightforward bioinformatic approaches. Consequently, BH3-only proteins are generally identified by their functions (i.e., binding to anti-apoptotic BCL-2 proteins via a BH3-like region). Therefore, we do not know whether there are more BH3-only proteins waiting to be elucidated or whether other unrelated sequences can have similar functions (we return to this idea later). Some examples of bona fide BH3 sequences in BH3-only proteins are shown in Figure 5.9.

Unlike the pro-apoptotic effectors BAX and BAK, most of the BH3-only proteins are intrinsically unstructured, and their BH3 regions might be readily available for binding once the BH3 proteins are synthesized, rather than requiring activation by another protein. (An exception to this is the BH3-only protein BID, discussed in more detail below.) These BH3-only proteins can therefore interfere with the ability

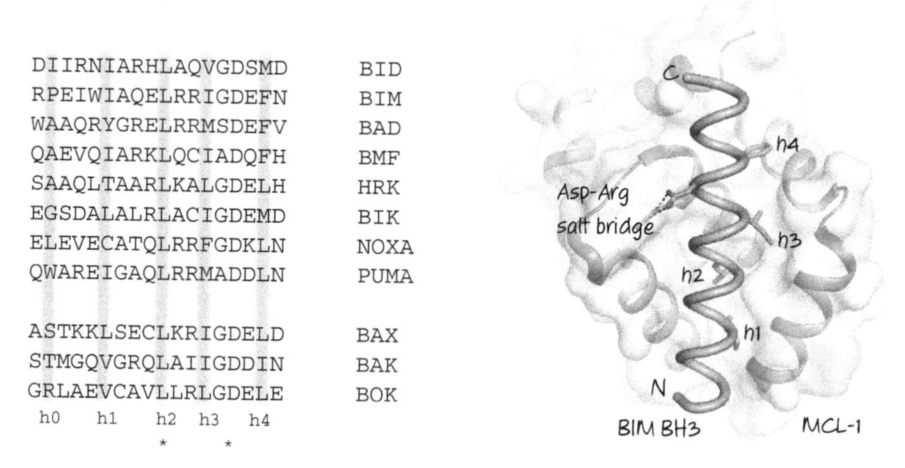

Figure 5.9. (*Left*) BH3 regions of several human BH3-only proteins. BH3 regions of BAX, BAK, and BOK are shown for comparison. Note the conserved leucine (L) and aspartate (D) residues (*). Hydrophobic residues (h0–h4) are often (but not always) present where indicated. (*Right*) The binding of the BH3 region of BIM to MCL-1. The hydrophobic residues and the conserved aspartate interact with the BH groove of MCL-1.

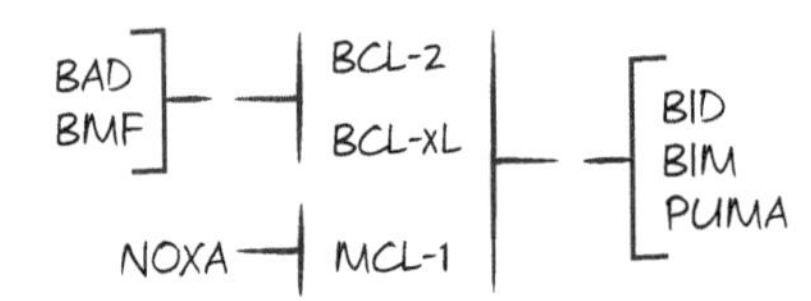

Figure 5.10. Specificities of some BH3-only proteins for binding to anti-apoptotic BCL-2 proteins Bcl-2, BCL-xL, and MCL-1.

of an anti-apoptotic protein to bind to the BH3 region of other proteins and thus neutralize them. This depends, of course, on how well the BH3-only protein binds to a particular anti-apoptotic BCL-2 protein (and how much of the BH3-only protein is available). Figure 5.10 shows the relative specificities of several BH3 peptides from BH3-only proteins for different anti-apoptotic BCL-2 proteins.

In a cell expressing both BCL-2 and MCL-1, neutralization of only BCL-2 (e.g., by BAD) might not be sufficient to promote apoptosis unless MCL-1 is also blocked (e.g., by NOXA). This tells us that there could be cooperation between the BH3-only proteins in controlling apoptosis by the mitochondrial pathway. As we will see, however, inhibiting the anti-apoptotic BCL-2 proteins might not be enough to induce MOMP and apoptosis.

SOME BH3-ONLY PROTEINS ACTIVATE BAX AND BAK

As we saw above, the pro-apoptotic BCL-2 effectors BAX and BAK undergo conformational changes, insert into membranes, and oligomerize when they are activated. This results in MOMP unless anti-apoptotic BCL-2 proteins block them. So, if BH3-only proteins are available to block the anti-apoptotic BCL-2 proteins, what activates the effectors? It turns out that some BH3-only proteins have this function too.

BH3 peptides from the BH3-only proteins BID and BIM can trigger BAX or BAK oligomerization and permeabilization of synthetic membranes or isolated mitochondria. In addition, active forms of BID and BIM proteins can activate BAX and BAK to cause MOMP. For this reason, they are referred to as "direct activators."

The BH3-only protein PUMA is another direct activator of BAX and BAK. This might be true of other BH3-only proteins as well, but we have yet to show this definitively. However, mice lacking PUMA, BIM, and BID do not have the developmental defects seen in BAX–BAK double-knockout mice. Even more remarkably, cells lacking all known BH3-only proteins can undergo MOMP and engage the mitochondrial pathway of apoptosis. Therefore, it is clear that there are other ways to activate BAX and BAK in addition to the action of the BH3-only proteins. There might also be other proteins or other types of molecular interactions that activate BAX and BAK. One of these non-BCL-2-family direct activators is considered in Chapter 11, but there is clearly more to the activation of BAX and BAK than we know.

ACTIVATION OF BAX AND BAK IS BY A "HIT-AND-RUN" MECHANISM

When BAX or BAK is activated by BID or BIM, the BH3-only direct activators do not remain associated with the pro-apoptotic effector proteins. A physical and transient interaction between BAX and active BID protein has been shown using Förster resonance energy transfer (FRET), a method that can detect interactions between proteins by monitoring the energy transfer between fluorescent tags attached to them. These studies showed that BID and BAX interact only when they are in membranes and that this is rapidly followed by the interaction of BAX with other BAX molecules (i.e., oligomerization) releasing the BH3-only direct activator. This is then followed by permeabilization of the membranes, presumably by BAX oligomers.

Insights into how the interaction of a direct activator with BAX or BAK results in their activation have come from structural studies. A BH3 peptide from the direct activator BID binds to BAX or BAK in solution, causing conformational changes in the effector molecules. The direct activator BID BH3 peptide binds to the BH groove of BAX or BAK, similar in structure to the BH grooves in anti-apoptotic BCL-2 proteins, but, unlike the latter, this moves the "latch" region of the protein ($\alpha6$–$\alpha8$) away from the core region defined by $\alpha1$–$\alpha5$ in BAX or BAK and exposes the BH3 of the effector (in $\alpha2$). The bent core then straightens and interacts with the membrane (as do exposed hydrophobic amino acids in $\alpha2$–$\alpha5$) (Fig. 5.11). Meanwhile, the exposed BH3 of the effector binds to the groove of an adjacent effector molecule, forming a dimer (Fig. 5.12). It is likely that this binding displaces the BH3 of the direct activator BH3-only protein, accounting for the hit-and-run kinetics discussed above. Additional interactions (probably $\alpha6$–$\alpha6$, based on biochemical studies) allow these dimers to form higher-order oligomers in the mitochondrial outer membranes.

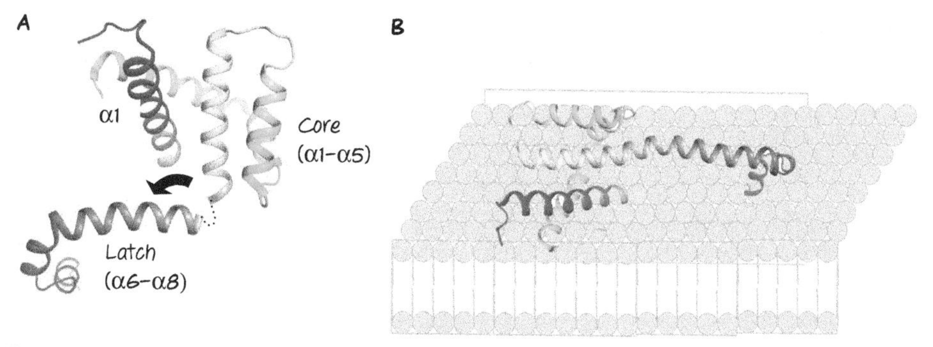

Figure 5.11. (*A*) Structure of activated BAX. Following activation, the "latch" of the protein ($\alpha6$–$\alpha8$) moves away from the core ($\alpha2$–$\alpha5$), ultimately straightening $\alpha5$–$\alpha6$, and (*B*) exposed hydrophobic residues in the protein presumably interact with the lipids of the outer mitochondrial membrane. Similar events occur during activation of BAK.

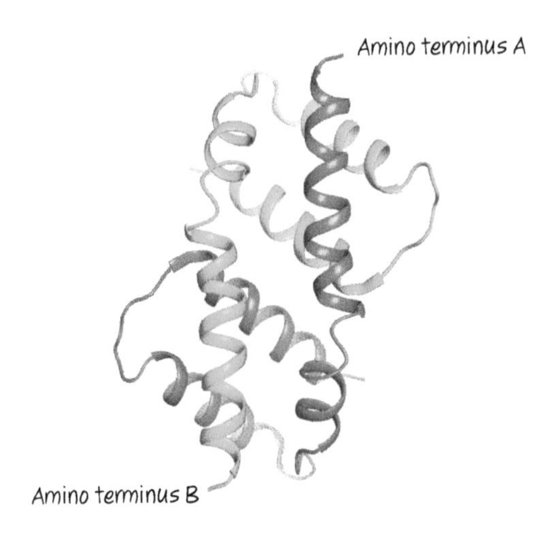

Figure 5.12. Activated BAX forms dimers by mutual BH3–groove interactions. The structure of the dimer formed by two BAX α2–α5 proteins (blue and green, respectively) showing the symmetrical binding of the BH3 from one in the groove of the other.

But why does this only happen on the mitochondria? There is evidence that specific lipids on the mitochondrial outer membrane participate in the activation process. Cardiolipin, present in the cell only on the mitochondrial inner membrane, can facilitate this activation of effector molecules, but its presence on the mitochondrial outer membrane is controversial. Conversely, a lipid formed on the mitochondrial outer membrane, sphingosine-1-phosphate, promotes the BH3-induced activation of BAK, whereas another product of the same pathway (the fatty aldehyde 2-*trans*-hexadecanal) promotes the activation of BAX. Interestingly, this lipid pathway is initiated by enzymes present in the endoplasmic reticulum (ER), suggesting that contact sites between the mitochondria and the ER are important for MOMP, and indeed there is evidence that these contact sites are where MOMP begins.

Unlike BAK, BAX is soluble in the cytosol. In this state, the BH groove of BAX is filled with the carboxy-terminal α9 helix of the protein. Another interaction exposes this α9 helix, which then interacts with the mitochondrial outer membrane, tethering the protein and preparing it for the activation events that follow. Other structural studies suggest a mechanism; the BIM BH3 domain does not bind to BAX in the region corresponding to the BH groove of an anti-apoptotic protein. Instead, it binds to the "back" of BAX (Fig. 5.13). This causes conformational changes in the BAX protein that might serve to expose the carboxy-terminal α9 helix of BAX, allowing it to tether to the mitochondria, while making the BH groove accessible to the BH3 region of direct activator proteins (Fig. 5.14). Therefore, it appears that there are two different interactions of the activating BH3-only proteins with BAX—one at the "back" to expose the BH groove and tether BAX to the membrane, then binding to the BH groove to induce the opening of the protein and its dimerization and subsequent oligomerization.

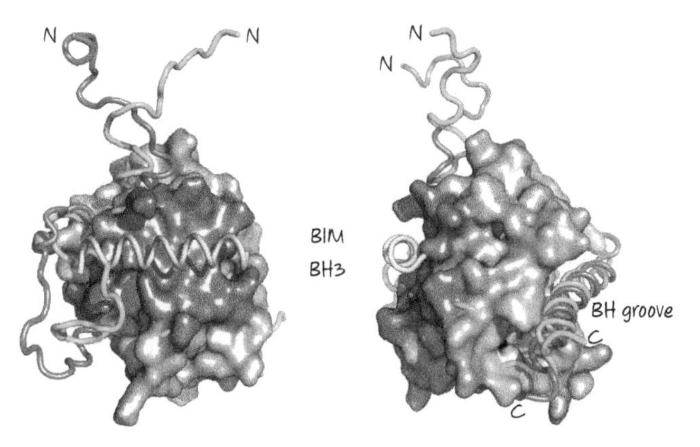

Figure 5.13. The BIM BH3 region binds to BAX and induces conformational changes. Two views are shown. A BIM BH3 peptide (yellow) binds to the "back" face of BAX (brown), opposite the BH groove, where it displaces a loop to cause several subtle rearrangements in BAX. Illustrated are regions of the free (green) and BIM-bound (blue) BAX that undergo conformational changes.

TWO MODELS OF BH3-ONLY PROTEIN FUNCTION IN APOPTOSIS

Currently, there are at least two different models for how BH3-only proteins cause MOMP and apoptosis. As we will see, these models are fairly similar on reflection, but they might have profoundly different consequences regarding our ability to manipulate apoptosis for therapeutic benefit.

In the first model, apoptosis is triggered when BH3-only proteins disrupt the interactions between anti-apoptotic BCL-2 proteins and the pro-apoptotic effectors BAX and BAK. This model is based on the idea that activation of BAX and/or BAK is neither rate limiting nor a decision point in cell death or survival. When active BAX or BAK are

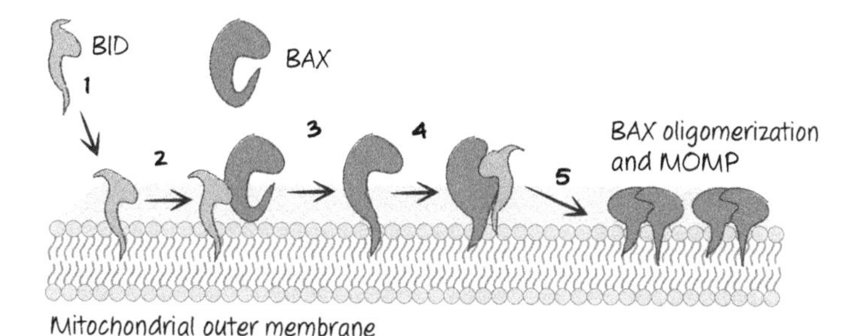

Figure 5.14. Order of events for BAX activation and MOMP. (1) Active BID binds very rapidly to the mitochondrial outer membrane, and then (2) BID binds to the back of BAX, whereupon (3) BAX inserts its α-helix 9 (α9) into the membrane. Then (4) BID binds to the BH groove of BAX, "unlatching" BAX and exposing its BH3, and finally (5) BAX oligomerizes by binding to additional, active BAX molecules, inducing MOMP.

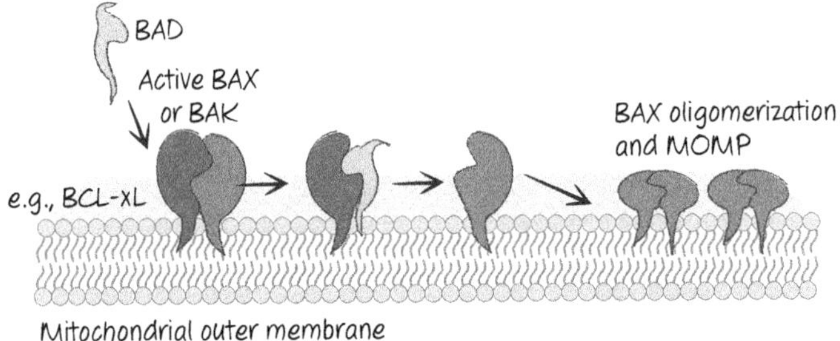

Figure 5.15. Neutralization model for BH3-only function. BH3-only proteins such as BAD bind to anti-apoptotic BCL-2 proteins (e.g., BCL-xL) to prevent or disrupt the binding of the latter to active BAX or BAK, and MOMP ensues.

displaced from the anti-apoptotic proteins by the appropriate BH3-only proteins, BAX and/or BAK self-associate and promote MOMP and apoptosis. We refer to this as the "neutralization model" (Fig. 5.15).

The neutralization model is essentially an update of the rheostat model (Fig. 5.8) discussed earlier, because the relative levels of functional pro-apoptotic and anti-apoptotic BCL-2 proteins determine whether and when apoptosis occurs. In it, the BH3-only proteins drive apoptosis by reducing anti-apoptotic activity. Because BH3-only proteins bind to anti-apoptotic proteins with differing efficiency, the relationships can be complex, but ultimately life and death in this model are determined by the net anti-apoptotic activity in the cell.

In the second model ("direct activator/derepressor model"), activation of BAX and BAK is a crucial component of the process that helps to make the decision to undergo MOMP. Here, the anti-apoptotic BCL-2 proteins sequester direct activators of BAX and BAK if such activators are induced. Other BH3-only proteins drive apoptosis by displacing the direct activators, freeing the direct activators to trigger BAX and BAK. The latter BH3-only proteins, which lack direct activator function, act as sensitizers or derepressors in this model. This scheme is shown in Figure 5.16.

According to this second model, inhibition of anti-apoptotic functionality will not necessarily cause MOMP and apoptosis unless molecules with direct activator function are present. The major problem with this model is that, other than BID and BIM (and a few others), we do not know what else activates BAX and BAK.

Although the distinctions between these models are important, the two really are rather similar. BH3-only proteins either displace direct activators of BAX and BAK from anti-apoptotic BCL-2 proteins (model 2) or they displace active BAX or BAK from such proteins (model 1). Unified models of the action of the BH3-only proteins and the inhibition of apoptosis by the anti-apoptotic proteins combine the neutralization and direct activator/derepressor models by suggesting that both happen. Interestingly, the two modes of action of the anti-apoptotic proteins have different consequences for the cell. When anti-apoptotic BCL-2 proteins act predominantly

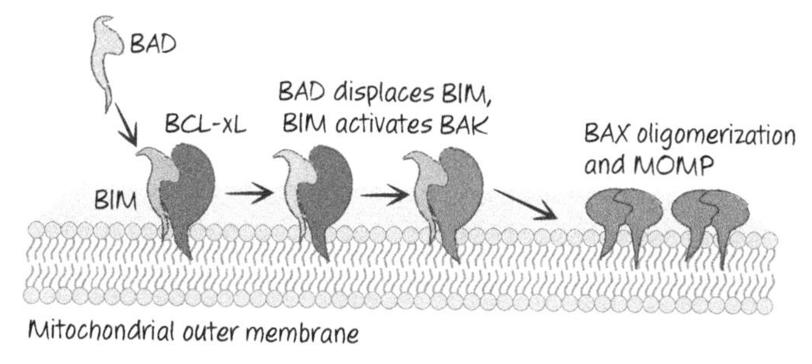

Figure 5.16. Direct activator/derepressor model of BH3-only function. Derepressor BH3-only proteins, such as BAD, bind to anti-apoptotic BCL-2 proteins, preventing or disrupting the binding of the latter to direct activators of BAX and BAK (such as BIM). BAX and BAK are then activated, and MOMP ensues.

to sequester the direct activator BH3-only proteins (MODE 1), the cells are then poised to rapidly undergo MOMP and apoptosis when this interaction is disrupted (e.g., by other BH3-only proteins). However, when the anti-apoptotic proteins act to sequester the active forms of the effectors, BAX and BAK (MODE 2), the cells appear to be more resistant to derepression (Fig. 5.17). It is possible that the complexes that form in MODE 2 are more stable than those in MODE 1, perhaps owing to additional interactions between the anti-apoptotic proteins and the effectors.

BOK IS REGULATED INDEPENDENTLY OF OTHER BCL-2 PROTEINS

Unlike BAX and BAK, BOK appears to be constitutively active, and its activity is not affected by the presence or absence of BH3-only proteins. Also, unlike BAX or

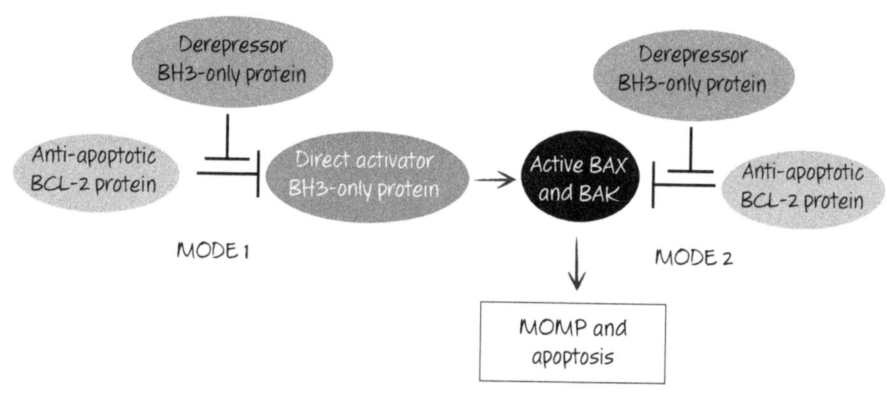

Figure 5.17. Two modes of action of anti-apoptotic BCL-2 proteins. Anti-apoptotic BCL-2 proteins can prevent apoptosis either by binding to BH3-only direct activator proteins (MODE 1) or by binding to active BAX or BAK (MODE 2). Additional BH3-only protein interactions can neutralize the function of the anti-apoptotic BCL-2 protein, allowing apoptosis to proceed. MODE 1 is more readily derepressed than MODE 2 (i.e., MODE 2 is functionally more stable).

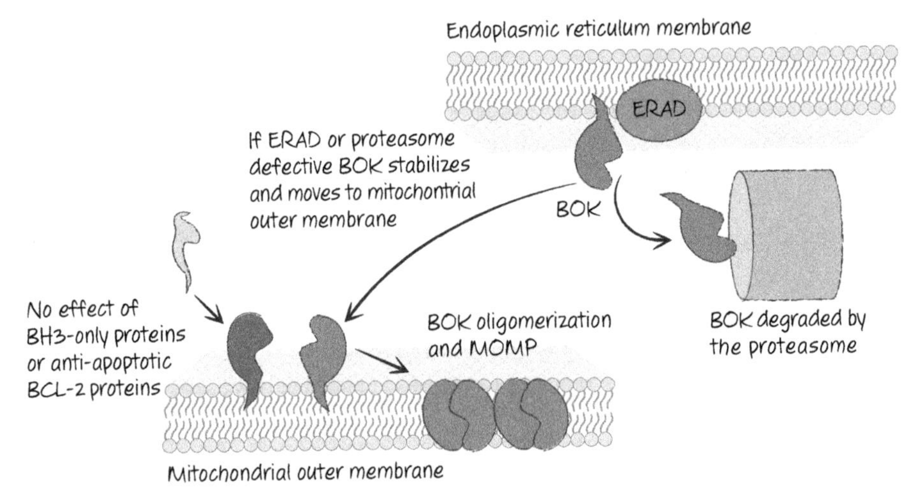

Figure 5.18. Regulation of BOK. BOK is not regulated by other BCL-2 proteins, but is controlled by endoplasmic reticulum–associated degradation (ERAD), which induces degradation of BOK. Disruption of ERAD or the degradation machinery allows BOK to accumulate and oligomerize on the outer mitochondrial membrane to cause mitochondrial outer membrane permeabilization (MOMP).

BAK, it is not inhibited by BCL-2 or BCL-xL and binds to MCL-1 only very weakly. Instead, BOK seems to be controlled at the level of protein stability. When it is expressed, BOK mostly localizes to the ER, where components of a disposal system, called the endoplasmic reticulum–associated degradation (ERAD) apparatus targets BOK for degradation by the proteasome. If ERAD or the proteasome machinery becomes compromised, BOK accumulates and moves to the mitochondria, where it effects MOMP and apoptosis (Fig. 5.18).

This helps to explain an apparent paradox. Cells that express high levels of anti-apoptotic proteins, such as some cancers, can nevertheless undergo apoptosis upon treatment with inhibitors of the proteasome. Because BOK functions independently of the anti-apoptotic proteins, it can promote MOMP and apoptosis under conditions in which it is stabilized.

We do not know the physiological conditions under which BOK promotes apoptosis,[3] but, as we will see in Chapter 10, BOK functions in development to remove unwanted cells.

BH3-ONLY PROTEINS ACT AS "STRESS SENSORS"

We have seen that BH3-only proteins can promote apoptosis in two ways: by directly activating the pro-apoptotic effectors BAX and BAK and by inhibiting the anti-

[3] It is likely that the regulation of BOK is more complex than indicated here. Some cells in the body (as well as some tumor cells) express stable BOK protein, yet do not spontaneously die. How this can occur is currently unknown.

apoptotic BCL-2 proteins. But when do they do this? The different BH3-only proteins have varied tissue distributions, are expressed under different conditions, and are regulated in different ways. They are targets of signal transduction pathways, and therefore we can think of them as "sensors" that connect the environment to the mitochondrial pathway of apoptosis.

The following are a few examples of BH3-only proteins functioning as sensors. In the chapters that follow, we return to specific BH3-only proteins in the context of different physiological or pathological situations.

BID IS A PROTEASE SENSOR

Unlike the other BH3-only proteins that have been examined, BID is structured, and it looks similar to anti-apoptotic BCL-2 proteins and pro-apoptotic effectors (Fig. 5.19). The BH3 domain of BID in its native state is unavailable for interaction with other BCL-2 family proteins.

BID has a large flexible loop that can be cleaved by a variety of proteases, including lysosomal proteases (cathepsins), the calcium-activated protease calpain, granzyme B, and caspases (Fig. 5.20). If BID is cut in this linker, the protein can now insert into mitochondrial membranes, and the BH3 domain presumably becomes exposed. BID can now interact with anti-apoptotic BCL-2 proteins that sequester it, or it can function to activate BAX and BAK, as we have seen. Therefore, if activating proteases appear in the cytoplasm of a cell, BID cleavage can engage the mitochondrial pathway of apoptosis.

BIM AND BAD ARE SENSORS FOR GROWTH FACTOR SIGNALING

When cells are deprived of growth factors, they often undergo apoptosis. In many cases, growth factor receptor signaling activates the serine/threonine-protein kinase

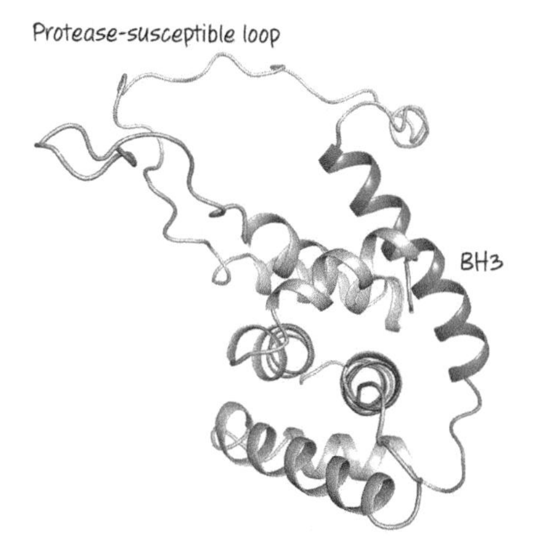

Figure 5.19. Structure of BID, a structured BH3-only protein with one of the most divergent BCL-2 cores. The BH3 region (red) is next to a protease-susceptible loop, the digestion of which is required for activation.

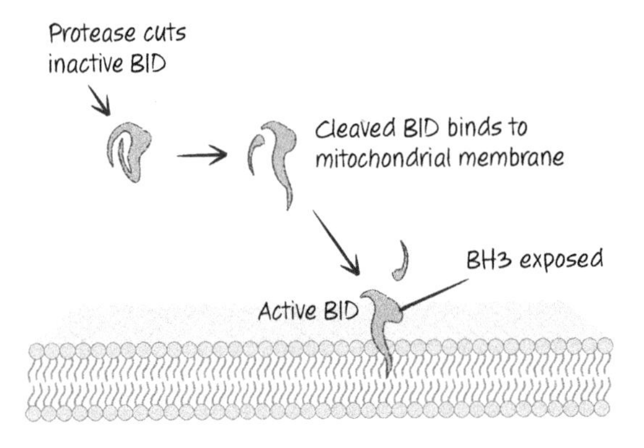

Figure 5.20. The pathway for cleavage and activation of BID.

AKT that phosphorylates (among other things) the transcription factor FOXO3a. The phosphorylated FOXO3a is sequestered in the cytosol by one of the 14-3-3 proteins that regulate the availability of signaling proteins in the cell, preventing FOXO3a from going to the cell nucleus. When growth factor signaling is disrupted, such as when growth factors become limiting, FOXO3a is released, transits into the nucleus, and triggers the transcription of the gene encoding BIM (Fig. 5.21). Lymphocytes from mice that lack BIM resist apoptosis caused by growth factor deprivation.

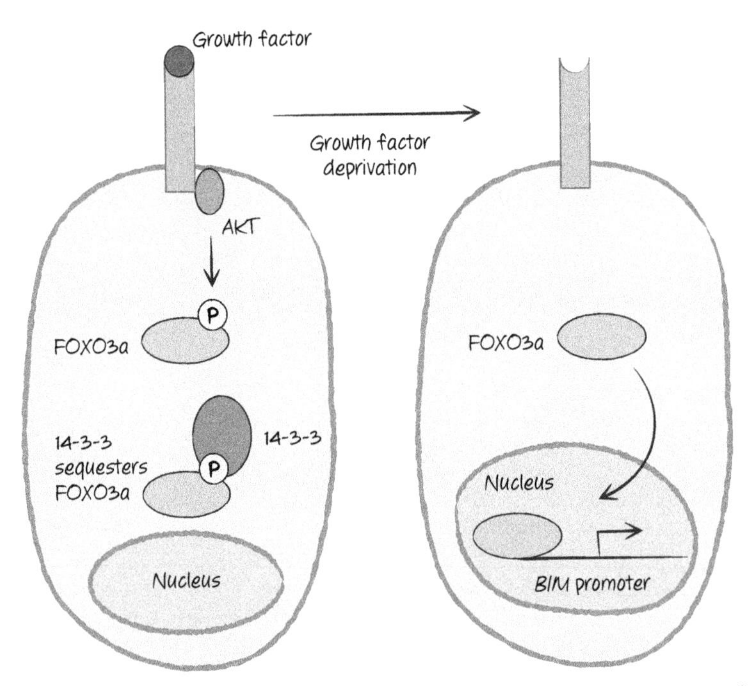

Figure 5.21. Growth factor signaling controls BIM expression through the sequestration of FOX3a, a transcription factor required for transcriptional activation of the promoter for the gene encoding BIM.

This is not the only way in which BIM is regulated: The mitogen-activated protein kinase (MAP kinase) ERK is activated by a variety of signaling mechanisms (including growth factor receptor signaling), and this protein phosphorylates BIM, which targets it for degradation, thereby promoting cell survival in some settings. Another MAP kinase, called c-Jun amino-terminal kinase (JNK), has the opposite effect—when JNK phosphorylates BIM, this prevents BIM degradation, and therefore JNK can promote apoptosis by BIM. The interplay of the MAP kinases, therefore, can dictate how extracellular signals cause apoptosis through regulating BIM.

The BH3-only protein BAD is directly phosphorylated by AKT, and the phospho-BAD is then sequestered by 14-3-3 proteins (as we saw for the transcription factor FOXO3a). Following growth factor withdrawal, BAD is released and can now neutralize BCL-2 and BCL-xL, releasing active BIM (Fig. 5.22).

BIM and BAD are not the only sensors of growth factor deprivation or related forms of cell stress. As we will see, there are other mechanisms whereby anti-apoptotic proteins, MCL-1 in particular, are regulated under growth factor deprivation, and these too contribute to the decision to die.

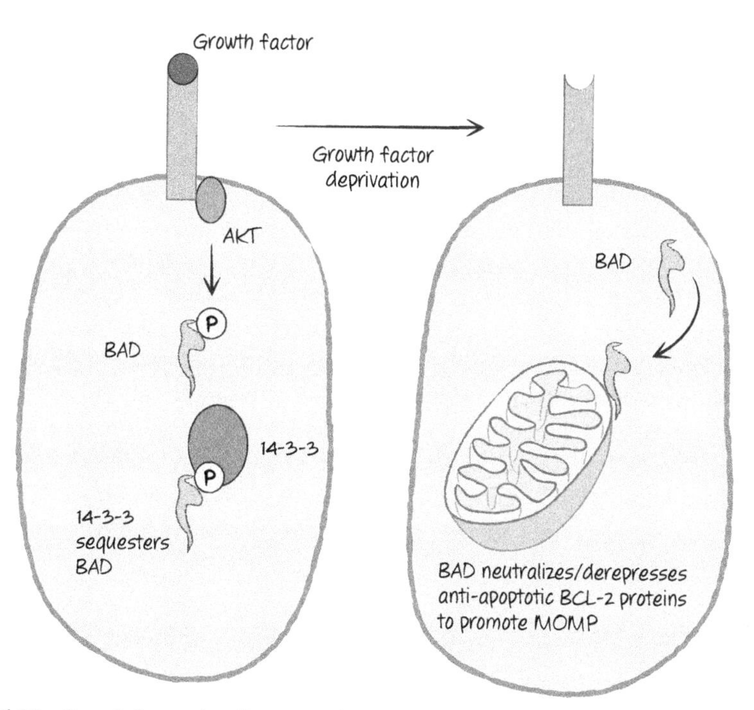

Figure 5.22. Growth factor signaling controls BAD function through the sequestration of BAD.

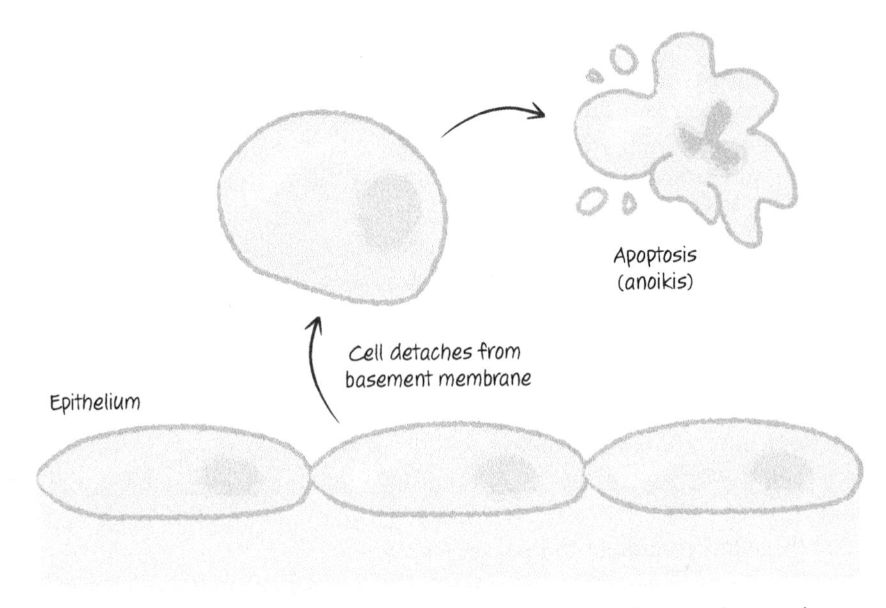

Figure 5.23. The pathway of anoikis allowing detached cells to be eliminated.

BIM AND BMF MEDIATE ANOIKIS

Severe disruption of the cytoskeleton induces apoptosis in many cells, and BH3-only proteins are involved in sensing such disturbance. This can happen when adherent cells lose their attachment to a surface or to basement membranes. The resulting apoptosis is called anoikis ("homelessness") (Fig. 5.23).

Dynein motor complexes move along the cytoskeleton in cells. Two BH3-only proteins, BIM and BMF, associate with specific motor complexes, binding to different dynein light chains. BIM appears to associate with cytoskeletal microtubules through this interaction, whereas BMF associates with another component of the cytoskeleton, the actin microfilaments. During anoikis, or in response to pharmacologic agents that disrupt the cytoskeleton, BIM and BMF are released to promote MOMP and apoptosis.

The control of BIM and BMF by the cytoskeleton is not universal, and in some cells, these BH3-only proteins are not associated in this way. Furthermore, it is not established that the cytoskeletal control of BIM and BMF is the only way in which anoikis occurs.[4] Nevertheless, the interactions with dynein light chains illustrate one way in which BH3-only proteins act as potential sensors.

[4] There is at least one other possibility. When adherent cells lose contact with their substrate, they often stick together (this also applies to daughters of a dividing adherent cell lacking substrate attachment). As a consequence, one cell often engulfs the other, a process called "entosis." The engulfed cell dies. This could be why many adherent cells in vitro do not grow in semisolid medium; whenever they divide, one cell is killed by entosis. We consider entosis in more detail in Chapter 9.

The above are only a few ways in which BH3-only proteins sense apoptotic signals and transduce them to the mitochondrial pathway. Additional examples are discussed in Chapters 10 and 11.

ANTI-APOPTOTIC BCL-2 PROTEINS ARE ALSO CONTROLLED BY SIGNALING

The anti-apoptotic protein MCL-1 is rapidly turned over in cells, and its levels appear to be actively regulated by phosphorylation, ubiquitylation, and degradation. One kinase that phosphorylates MCL-1 to target it for degradation is glycogen synthase kinase 3 (GSK3), which itself is inhibited by AKT. Therefore, when AKT activity is reduced on growth factor withdrawal, MCL-1 levels decline as a consequence of GSK3 function. This is illustrated in Figure 5.24. There are other kinases that similarly impact the stability of MCL-1 as well.

Once MCL-1 is phosphorylated, it interacts with FBW7, an E3-ubiquitin ligase that places the small protein ubiquitin onto proteins, directing them to be degraded by the proteasome. Many cancers lose FBW7, and as a result, these cancers resist

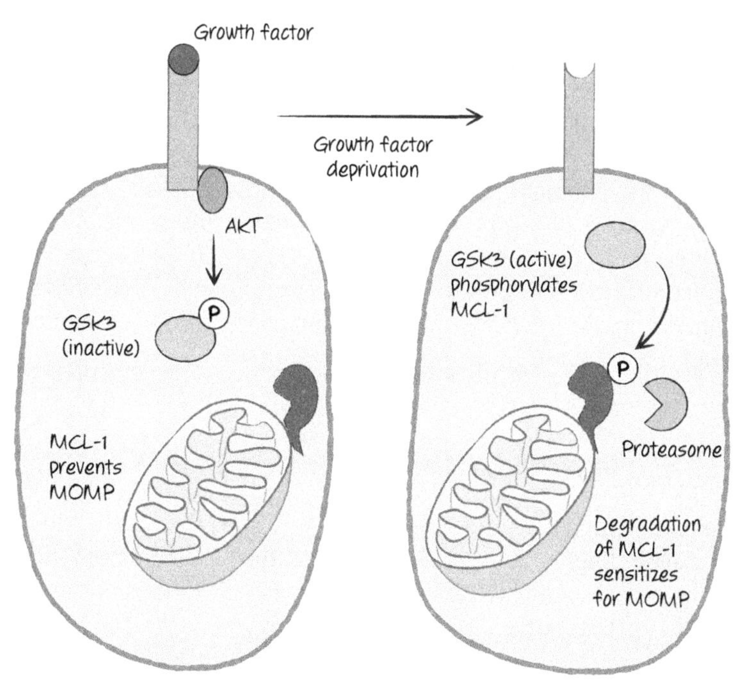

Figure 5.24. Growth factor receptor signaling regulates the stability of MCL-1, either preventing or enhancing mitochondrial outer membrane permeabilization (MOMP), according to the presence (*left*) or absence (*right*) of growth factors.

apoptosis by increasing MCL-1 levels. Conversely, inhibitors of mTOR, a kinase that acts in complexes to promote cell growth, can sensitize cells for apoptosis by derepressing the activity of GSK3, leading to increased MCL-1 degradation.

Other proteins are involved in control of the stability of MCL-1. MULE (also called HUWE1) is another of the E3-ubiquitin ligases that ubiquitylate MCL-1, targeting it for degradation. Intriguingly, MULE has a BH3 domain that binds to MCL-1, and this might interfere with its enzymatic activity. When it is displaced from MCL-1 by a BH3-only protein such as NOXA, MULE can then cause the degradation of MCL-1 by ubiquitylating it.

BCL-2 and BCL-xL are also phosphorylated, but in this case the role for phosphorylation is less clear. Phosphorylation of BCL-2 on one particular serine (serine 70) has been reported to increase its activity, but phosphorylation at other sites effectively inhibits it. These phosphorylation events occur in the unstructured region of the BCL-2 proteins between BH4 and BH3 (Figs 5.1 and 5.3). This affects the interaction of the disordered region with the core of the protein, influencing the affinity of the BH groove for the BH3 regions of pro-apoptotic BCL-2 proteins.

Another interesting modification of BCL-xL has been described, involving conversion of two asparagine residues into aspartate or isoaspartate in the loop between the BH4 and BH1 regions. This reduces the affinity of the protein for BH3-only proteins, and might represent another mode of regulation. This process, called deamidation, is nonenzymatic, and is affected by intracellular pH, dependent on the sodium-proton antiporter NHE-1. Some cancers have defects in BCL-xL deamidation, making them resistant to DNA-damage-induced apoptosis. Again, this process occurs in the unstructured region of BCL-xL, between BH4 and BH3, affecting its interaction with the core of the protein and the function of the BH groove.

The anti-apoptotic BCL-2 proteins are regulated at the transcriptional level as well, and this might be the major way in which BCL-2 and BCL-xL are controlled. Several transcription factors induce expression of these proteins in different cell types, including nuclear factor-κB (NF-κB), which can produce anti-apoptotic effects in some settings.

BCL-2 PROTEINS IN OTHER ANIMALS

BCL-2 proteins are found throughout the animal kingdom. The situation in nematodes is particularly interesting. In *Caenorhabditis elegans*, the APAF1 homolog CED4 is held inactive on the mitochondrial outer membrane by CED9. It turns out that CED9 is a homolog of BCL-2 (Fig. 5.25).

Despite this similarity, mammalian BCL-2 proteins do not sequester APAF1 or affect its activity. Therefore, although BCL-2 and CED9 have similar overt functions (inhibition of apoptosis), they perform these functions through different biochemical mechanisms.

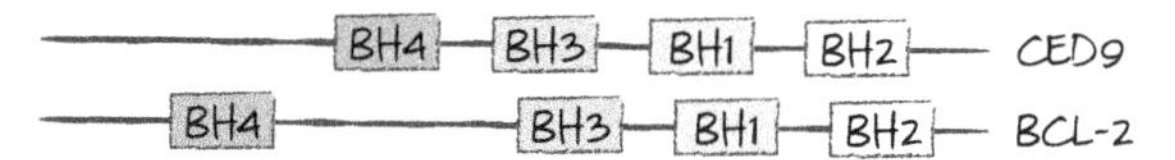

Figure 5.25. CED9 is a BCL-2 protein. The BH1–BH4 domains of CED9 and BCL2 are shown for comparison. Structural studies confirm that CED9 and BCL-2 are similar.

CED9 has a BH groove like that of the mammalian anti-apoptotic BCL-2 proteins, but this is not involved in binding to CED4. However, another protein, EGL1, binds to this groove and leads to a conformational change in CED9 that causes the release of CED4, promoting apoptosis. This is shown in Figure 5.26.

EGL1 is a BH3-only protein that is expressed in response to transcription factors that are activated developmentally or in response to DNA damage in germ cells in the adult (considered in more detail in Chapter 10). We can now complete the core signaling pathway for apoptosis in nematodes, as shown in Figure 5.27.

As we know, the conformational changes that mammalian BCL-2 proteins undergo during apoptosis require the presence of the outer mitochondrial membrane, and this leads to an intriguing speculation: perhaps the conformational change that is induced by EGL1 in CED9 to release CED4 is facilitated by the mitochondrial membrane itself. This would help to explain the location of CED9 on mitochondria, despite the absence of MOMP upstream of the activation of the APAF1 homolog CED4 in this animal.

In *Drosophila*, two BCL-2 proteins have been identified on the basis of sequence similarity to mammalian BCL-2 proteins. These are DeBCL (pronounced "debacle") and Buffy (named for the famous vampire slayer), also called DBorg1 and DBorg2, respectively. However, neither has been shown to have any role in apoptosis in *Drosophila*, which is intriguing, and their functions in flies are unknown. When we consider other roles for the vertebrate BCL-2 proteins later, it could be interesting to remember these.

Pro-apoptotic BCL-2 family effector proteins resembling BAX and BAK are also found among many invertebrate animal phyla, although not in nematodes or

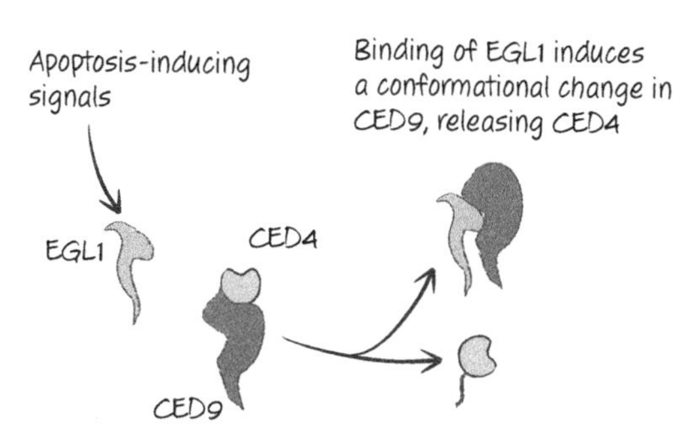

Figure 5.26. EGL1 functions to release CED4 from CED9.

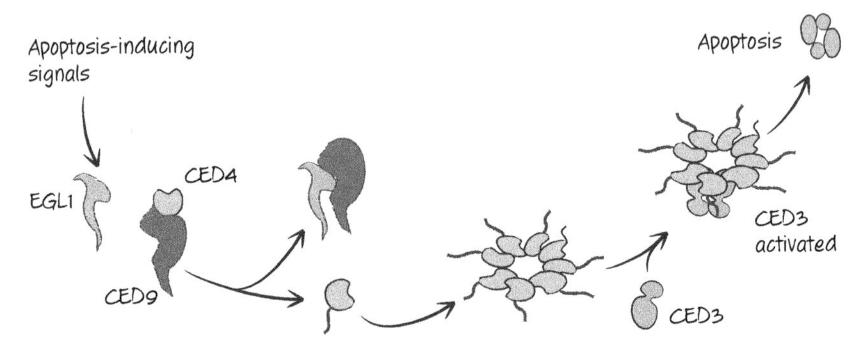

Figure 5.27. The action of participants in apoptotic signaling in nematodes.

arthropods. This further supports the idea that the mitochondrial pathway of apoptosis in animals might be ancestral, but lost in some lineages.

VIRAL BCL-2 PROTEINS

Because viruses have come up with ways to block apoptosis, it should not be surprising that many viruses that infect mammalian cells express anti-apoptotic proteins related to the BCL-2 proteins. Examples can be found in several types of virus, including adenoviruses, herpes viruses, and pox viruses (Fig. 5.28). In each case, the anti-apoptotic protein shares weak sequence similarity to BCL-2, but regions can be identified that correspond to the BH domains.

The structures of a few viral BCL-2 proteins have been solved, and these very closely resemble those of the cellular anti-apoptotic BCL-2 proteins (Fig. 5.29). In general, however, they lack the regions where regulatory events tend to occur in the cellular proteins, such as phosphorylation; perhaps this is not surprising.

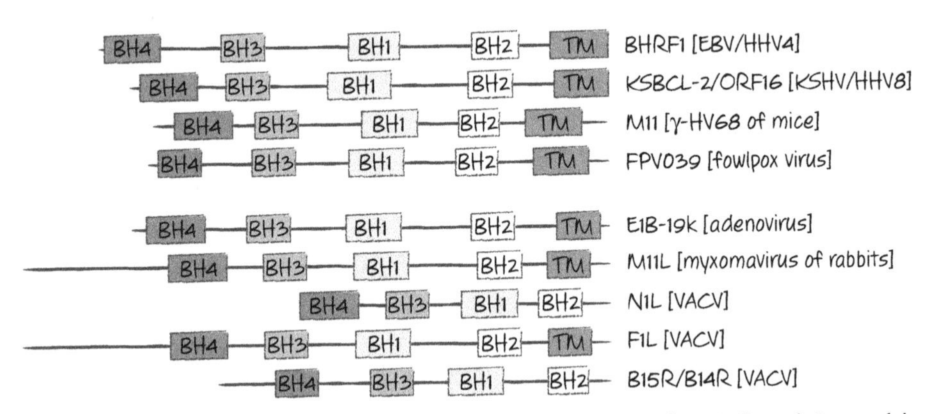

Figure 5.28. Viral BCL-2 proteins. All are from human viruses except where indicated. Some of the BH regions are based on structural considerations despite low amino acid homology.

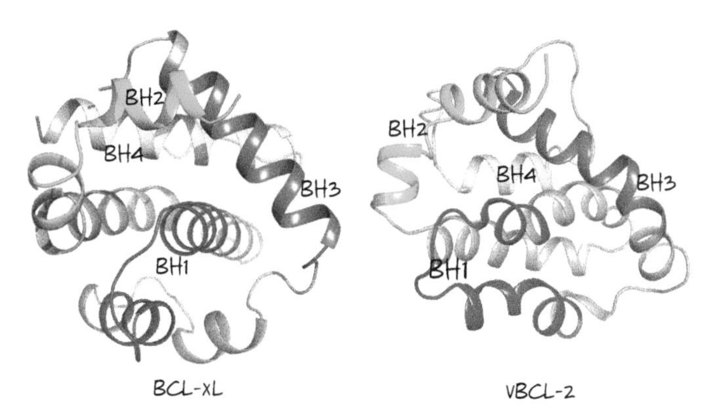

Figure 5.29. Structures of mammalian BCL-xL and Kaposi sarcoma virus vBCL-2 showing the arrangement of the BH1–BH4 regions.

Viruses also express proteins that function like anti-apoptotic BCL-2 proteins (e.g., they bind and neutralize active BAX and BAK) but share no sequence or structural similarity with cellular BCL-2. An example is the vMIA protein from human cytomegalovirus. Such proteins can teach us a lot, not only about viral infection, but also how the cellular BCL-2 proteins work.

BACTERIAL TOXINS AND THE RETURN OF THE JUST-SO STORY

In the last chapter, we discussed a just-so story—a speculation that apoptosis might have arisen with ancient endosymbiosis that resulted in mitochondria. A piece of that story involves BCL-2 proteins.

The structures of BCL-2 proteins BCL-2, BCL-xL, BCL-w, MCL-1, A1, BAX, BAK, BID, and vBCL-2 are remarkably similar to one another. Intriguingly, a similar structure is also found in some bacterial toxins, including diphtheria toxin β-chain and the colicins (Fig. 5.30). Each of these proteins undergoes a conformational change to bring a hydrophobic core into a membrane. The structural similarity might simply reflect this common function. But it is intriguing to think about this in the context of our just-so story.

Bacterial toxins of this type are often encoded in a discrete genetic element that actually makes two components—the toxin that creates holes in membranes and an immunity factor that makes the bacterium that produces it resistant to the toxin. When a bacterium harboring this element is stressed (e.g., by low levels of nutrients), it produces the toxin. This kills any bacteria that do not have the immunity factor, reducing competition for resources.

The original endosymbiosis, as we speculated, might have begun anything but cooperatively, with the bacterium being a parasite in the infected cell. In this scenario, the inner membrane of what would become the mitochondrion is the bacterial membrane

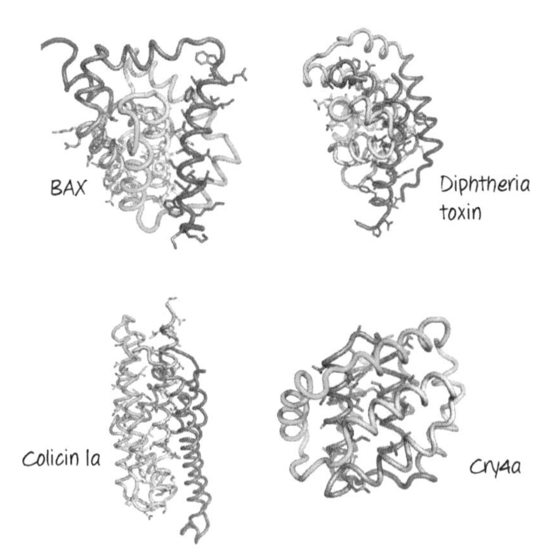

Figure 5.30. Structures of several pore-forming regions of bacterial toxins, compared with that of BAX. Several α-helical pore-forming toxins have a globular domain reminiscent of the BCL-2 core of BAX. The structures are colored blue to red from the amino to the carboxyl terminus, and the putative pore-forming helices are identified by their side-chain sticks.

(indeed, the inner membranes of modern mitochondria have lipid compositions that more closely resemble bacterial membranes than those of eukaryotes), whereas the outer membrane would have been derived from the infected cell (again, consistent with the composition of modern outer mitochondrial membranes). If stressed, the bacterium might synthesize a toxin that would target the next-nearest membrane—the mitochondrial outer membrane. So, was this the original pro-apoptotic effector?

With time, genetic control of the toxin might have transferred to the nucleus (as did nearly all of the original genes encoding mitochondrial elements, except for those encoding parts of the electron transport chain, ribosomal RNAs, and the tRNAs). Control of the integrity of what became the mitochondrial outer membrane now resided in nuclear genes. And the toxin, perhaps, became the BCL-2 family.

Finally, note that the mitochondrial inner membrane is not permeabilized by active BAX or BAK (we know this, because matrix proteins are not released with the proteins from the intermembrane space). Why not? Is there a distant descendant of the bacterial immunity factor that protects this membrane?

BCL-2-FAMILY PROTEINS HAVE OTHER ROLES IN CELLS

From the perspective of apoptosis, the major role of BCL-2 proteins is to control the integrity of the outer mitochondrial membrane. But these proteins appear to have additional functions in cells, and how these relate to their apoptotic effects and other aspects of cell physiology is emerging as an important area of research. If cell death is the work of the "night crew." then what are the "day jobs" of the BCL-2 proteins?

Given their limited sequence homologies, BH3-only proteins might have numerous functions beyond apoptosis. In at least one case, a day job for a BH3-only protein has been identified. That protein is BAD. In addition to neutralizing some anti-apoptotic BCL-2 proteins, BAD has a role in regulating glucose metabolism in mammals.

We consider here three other day jobs for BCL-2 proteins: control of calcium homeostasis, regulation of mitochondrial dynamics (fission and fusion), and removal of mitochondria during development of some cells. As we will see, these might all have connections to the control of cell death as well as other effects in cells, and of course, there are probably other non-apoptotic roles yet to be identified.

BCL-2 PROTEINS ACT AT THE ER TO REGULATE CALCIUM

Inositol (1,4,5)-trisphosphate (IP3) is a phospholipid-derived molecule produced in some signaling pathways. IP3 binds to a receptor on the ER (the IP3 receptor), causing an efflux of calcium that regulates many different enzymes in the cytosol.

BCL-2 proteins influence the function of the IP3 receptor. Cells that lack BAX and BAK or overexpress BCL-2 show defective calcium efflux in response to IP3. For example, T lymphocytes lacking BAX and BAK do not elevate calcium in response to T-cell receptor engagement (which induces IP3 production) and are defective for T-cell activation.

How BCL-2 proteins control the IP3 receptor is not entirely clear. BCL-2 binds to the IP3 receptor, but beyond this, the mechanism is obscure. Intriguingly, some BH3-only proteins can induce an increase in intracellular calcium, and it might be that this is through interaction with BCL-2 or another BCL-2 protein. However, we do not know how this contributes to apoptosis or other cellular effects.

MITOCHONDRIAL DYNAMICS ARE INFLUENCED BY BCL-2 PROTEINS

One of the most intriguing day jobs for BCL-2 proteins is in mitochondrial dynamics. Mitochondria do not simply sit around in cells as discrete organelles; instead, they are constantly undergoing active fission and fusion by complex mechanisms that are conserved among the eukaryotes.

The first clue to this function is what happens during MOMP. At about the same time as the mitochondrial outer membrane becomes permeable and proteins from the intermembrane space are released, mitochondria often appear to undergo extensive fragmentation. At first, it was thought that this fission might contribute directly to MOMP, but that does not seem to be the case because MOMP can occur in isolated mitochondria without fission.

A second clue is that, in the absence of BAX and BAK, mitochondria are extensively fragmented, and this turns out not to be due to excess fission but rather to a decrease in mitochondrial fusion. In contrast, increasing the expression of anti-apoptotic BCL-xL increases mitochondrial fusion. Therefore, in two different situations in which MOMP is prevented (increased BCL-xL and decreased BAX and BAK), mitochondria display either more or less fusion. How can these observations and those of mitochondrial fission upon MOMP be reconciled?

Here is one way. If BAX and BAK promote mitochondrial fusion independently of their role in MOMP, when they become engaged during apoptosis (effectively leaving their day jobs to join the night crew), fusion decreases. Because BCL-xL prevents this (e.g., by sequestering activators of BAX and BAK, letting them remain doing their day jobs), fusion is enhanced.

How all this might occur is not known, but BAX and BAK can associate with a protein called mitofusin-2 that participates in mitochondrial fusion. Remarkably, the *C. elegans* protein CED9 can also associate with mitofusin-2 in mammalian cells and similarly enhances mitochondrial fusion. CED9 cannot block MOMP in mammalian cells but stimulates mitochondrial fusion under these artificial conditions.

When apoptosis occurs in *C. elegans*, mitochondrial fragmentation is observed, even though there is no MOMP (as we know, MOMP is not upstream in the apoptotic pathway in these animals). EGL1, which disengages CED9 from CED4 (Fig. 5.22), might therefore also take CED9 away from its day job, that of promoting mitochondrial fusion.

Mitochondrial fragmentation is also observed in *Drosophila* cells undergoing apoptosis. Here, not only is MOMP not upstream in the apoptotic pathway, but the BCL-2 proteins DeBcl and Buffy do not seem to have a role in cell death. Whether they participate in the mitochondrial fragmentation is not known.

Do these changes in mitochondrial dynamics have anything to do with apoptosis? Some studies have indicated that a protein involved in mitochondrial fission, DRP-1, might help to promote MOMP, and cells in which this protein is blocked can become refractory to MOMP and apoptosis. However, this seems to be the case even when mitochondrial fission and fusion cannot occur, and so the role of this protein in MOMP remains somewhat obscure. However, as we saw above, the carboxyl terminus of BAX must insert into the outer mitochondrial membrane for subsequent activation and MOMP (e.g., Fig. 5.14). This turns out to depend on the curvature of the mitochondrial membrane, and this, in turn, is affected by mitochondrial dynamics. Smaller, fragmented mitochondria are more curved, and this facilitates the insertion of the BAX carboxy-terminal α9-helix. By promoting mitochondrial fragmentation, active BAX (and BAK) facilitate the insertion and activation of more BAX molecules.

Therefore, the interplay between mitochondrial dynamics and the function of the BCL-2 proteins might go both ways. Other conditions that favor mitochondrial fusion or fission (such as the availability of oxygen or nutrients and the engagement of metabolic pathways) can therefore influence apoptosis.

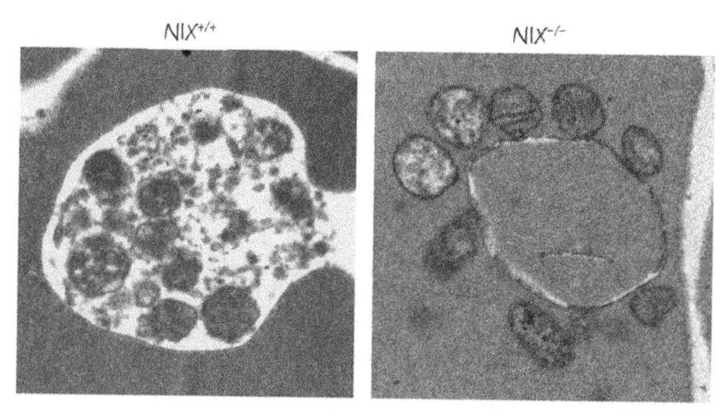

Figure 5.31. Mitochondria in NIX-deficient red blood cells. In developing wild-type red blood cells, mitochondria are removed by mitophagy (*left*), and these organelles are not present in the mature cells. In NIX-deficient mature red cells, mitochondria persist, many showing signs of damage (*right*).

BCL-2 PROTEINS CAN PARTICIPATE IN THE REMOVAL OF MITOCHONDRIA

Yet another function of the BCL-2 proteins also relates to mitochondria, but in this case it is the way in which excess or damaged mitochondria are removed. This occurs by autophagy, the "self-eating" process that allows cells to survive nutrient deprivation and remove damaged or unwanted organelles. We have much more to say about autophagy in Chapter 8. Autophagy of the mitochondria (mitophagy) involves the creation of a membrane vesicle around the organelle and fusion of the vesicle with a lysosome results in the digestion of the contents of the vesicle.

In at least some cases, the signals that bring the autophagy machinery to the mitochondria are members of the BCL-2 family—in particular, two closely related BH3-only proteins, BNIP3 and NIX. Neither BNIP3 nor NIX seems to have a significant role in apoptosis; instead, they appear to function in mitophagy. In mice lacking NIX, mature red blood cells, which normally lack mitochondria because these have been removed by mitophagy, are found to harbor these organelles (Fig. 5.31). Another mechanism that promotes mitophagy is discussed in Chapter 8.

Other studies have suggested that, during hypoxia (reduced oxygen levels, a situation in which having fewer mitochondria is desirable), a transcription factor called HIF-1 is activated that causes expression of BNIP3. The latter then appears to promote the autophagic removal of mitochondria.

How these BH3-only proteins cause mitophagy is not completely clear, but the process might involve binding to BCL-2 and displacing a protein that is bound to it, beclin-1. Beclin-1 is part of the machinery that initiates autophagy, and its release from BCL-2 on mitochondria might attract components of the autophagy pathway to these organelles (Fig. 5.32). It is particularly noteworthy that beclin-1 contains a BH3 region and binds to the BH groove in BCL-xL (Fig. 5.33).

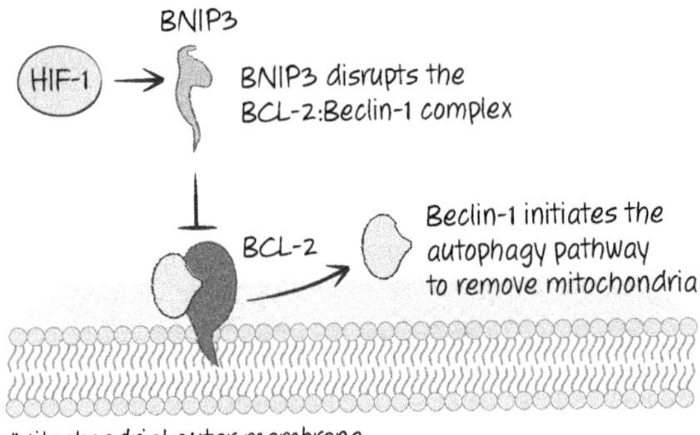

Figure 5.32. Hypoxia leads to stabilization of hypoxia-induced factor HIF and removal of mitochondria by autophagy.

We return to the role of beclin-1 and BCL-2 in autophagy in Chapter 8. For now, it is sufficient to realize that the day jobs of BCL-2 proteins can involve the type of interactions involved in apoptosis (as in this example) or other activities and interactions.

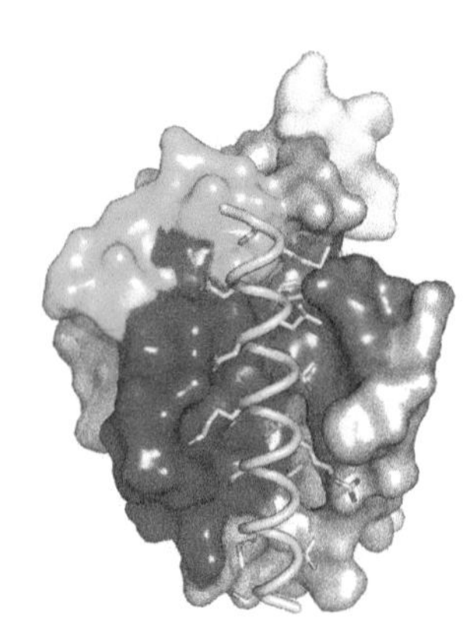

Figure 5.33. Beclin-1 BH3 region bound to BCL-xL. The BH regions of BCL-xL are colored as in Figure 5.4, and the BH3 peptide is pale blue.

CHAPTER 6

The Death Receptor Pathway of Apoptosis

DEATH RECEPTORS ARE A SUBSET OF THE TUMOR NECROSIS FACTOR RECEPTOR FAMILY

So far, we have considered a pathway of apoptosis, the mitochondrial pathway, which is activated in response to cellular stress and some developmental cues. In this chapter, we discuss a different pathway of apoptosis that is triggered by the interaction of extracellular ligands with receptors on the cell surface.

On the surfaces of some mammalian cells (and probably those of other vertebrates as well) are receptors that, when bound by their ligands, engage a pathway of caspase activation and apoptosis that can be distinct from the mitochondrial pathway that we have discussed. These surface molecules are referred to as "death receptors," and their mode of caspase activation is called the death receptor pathway of apoptosis.

The death receptors and the proteins that activate them ("death ligands") are all closely related molecules. The death ligands are members of a larger family of proteins, the tumor necrosis factor (TNF) family, and the receptors for these ligands are members of the TNF receptor (TNFR) family.[1]

The major death receptors are tumor necrosis factor receptor-1 (TNFR1), CD95 (also called Fas and APO-1), the TRAIL receptors (TRAIL receptor-1, also called DR4, and TRAIL receptor-2, also called DR5 in humans; rodents have only one, which resembles DR5), DR3 (death receptor 3), and DR6 (death receptor 6). These are shown in Figure 6.1. The death ligands are also shown and include TNF, CD95-ligand

[1] The TNF and TNFR superfamilies have been given the designations TNFSF and TNFRSF, respectively, followed by a number (those for death ligands and receptors are shown in Fig. 6.1). We have chosen to employ the most widely used names in the text, rather than these accepted designations.

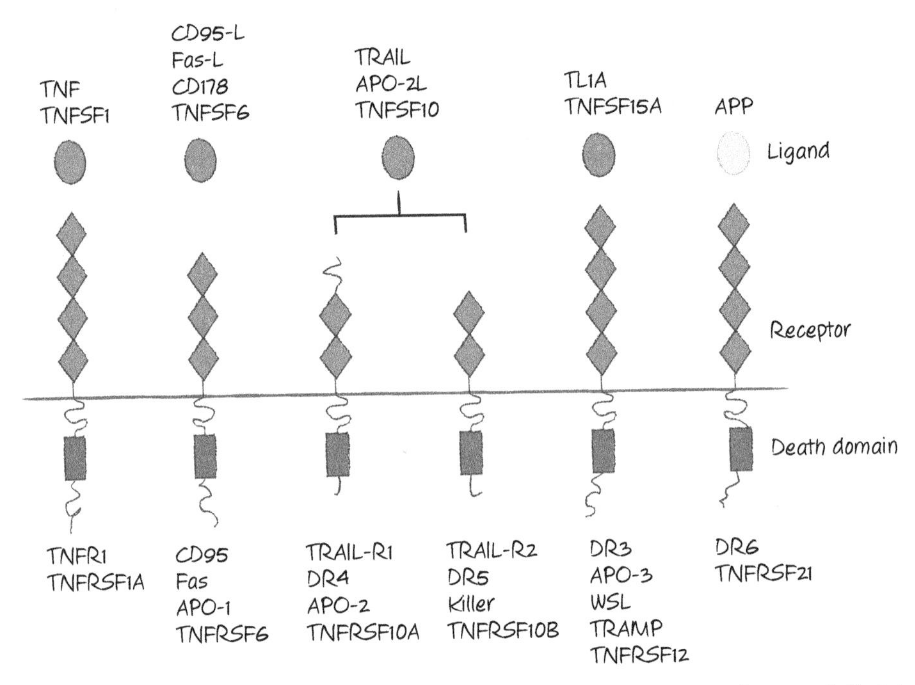

Figure 6.1. The death receptors and their ligands. Several names for each are shown. Of all of the ligands, only APP is not a member of the tumor necrosis factor (TNF) family.

(CD95-L or Fas-L), and TRAIL (TNF-related apoptosis-inducing ligand, also called APO-2L). Although a ligand for DR3 has been identified, this does not readily trigger apoptosis, and therefore there could be other ways DR3 is engaged or it might not normally function in apoptosis induction. The apoptosis-inducing ligand for DR6 has been identified as the cell-bound amyloid precursor protein, APP. Unlike the other ligands, this is not a member of the TNF family, and its function as a DR6 ligand is somewhat controversial.

All of the TNF-family ligands, including the death ligands, are trimers (an example is shown in Fig. 6.2). This is also true of the receptors. It is often mistakenly suggested that the unligated death receptors are monomeric and assemble into trimers when bound, but there is very good evidence that the receptors are already trimeric on the cell surface before engaging their ligands. On binding a death ligand, the trimers form higher-order complexes and expose a region located in the intracellular portion of the receptor. This region contains a "death domain" (DD; see Chapter 3). An exception is the interaction between APP and DR6. In this case, the receptor appears to be monomeric, and it is dimerized by interaction with its ligand.

DDs are found in the death receptors but not in any of the other TNFR-family members. Recall that, structurally, they show the death fold also adopted by the

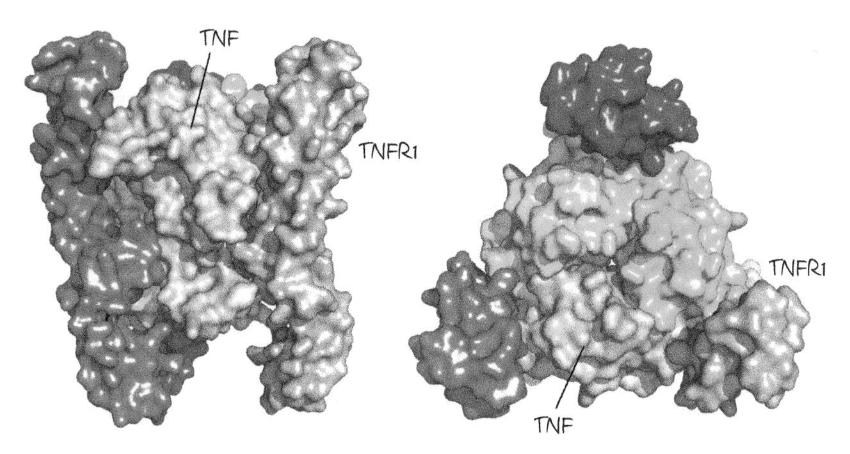

Figure 6.2. Death receptors and death ligands are trimeric. Shown are two views of the structure of tumor necrosis factor (TNF; green) bound to the extracellular regions of the TNF receptor (TNFR1, blue).

death effector domains (DEDs) and caspase-recruitment domains (CARDs) found in caspases and their adapters (Chapter 3) (Fig. 6.3). The death domains share no sequence homology with the other death fold domains, but they interact with proteins that share the same domain (giving rise to DD–DD interactions).

Different death receptors bind to different DD-containing adapter proteins as the next step in the signaling pathway. We begin with the simplest version of this pathway, seen in the case of the death receptor CD95.

HOW CD95 TRIGGERS APOPTOSIS

When CD95 is bound by its ligand CD95-L, or by some antibodies against CD95 that mimic this, the resulting conformational change in the intracellular DD allows it to

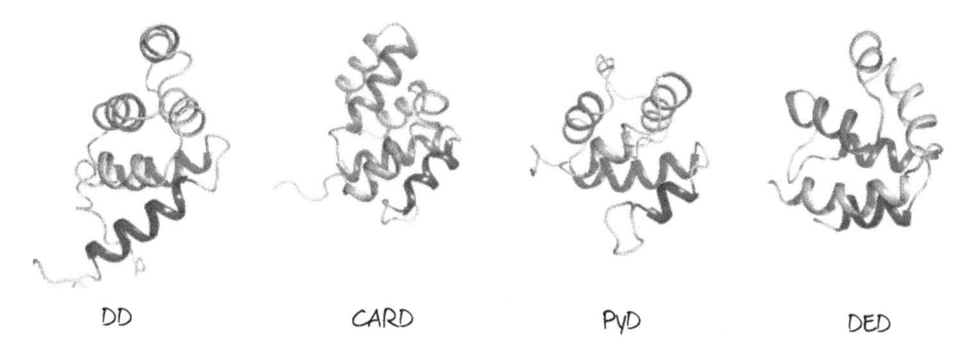

Figure 6.3. Ribbon diagrams of the representative structures of distinct death folds. DD, death domain; CARD, caspase-recruitment domain; PyD, pyrin domain; DED, death effector domain.

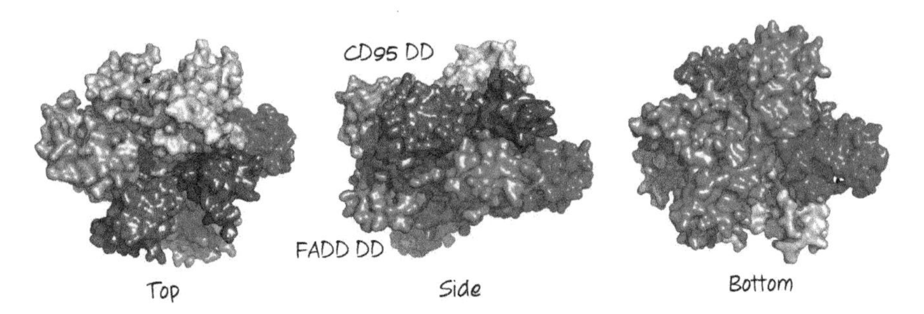

Figure 6.4. The structure of the CD95 receptor death domain (DD) binding to the DD of the FAS-associated death domain protein (FADD) viewed from top, side, and underside.

interact with a small cytosolic protein called FADD (FAS-associated death domain protein). FADD contains a DD through which it binds to CD95 through a DD–DD interaction. The DDs bind to each other at three different interfaces, producing a two-layer structure (Fig. 6.4).

These assemble into arrays, as shown in Figure 6.5. This explains why at least two trimers of CD95 must be brought together to produce the complex that leads to apoptosis.

The binding of the CD95 DD to the FADD DD causes FADD to expose a different death fold, termed a DED. In turn, the exposed DED on FADD now binds to the prodomain of the initiator caspase, caspase-8, through one of the two DEDs in this caspase. In this way, ligated CD95 binds to FADD, which exposes its DED to recruit caspase-8. This results in the oligomerization of caspase-8. This activates the caspase, which then cleaves itself, stabilizing caspase-8 as dimers, as we saw in Chapter 3.

Actually, this is a bit more complicated. Caspase-8 has two DED domains (Fig. 6.6) and only one binds to FADD (DED1). The second one (DED2) is then exposed,

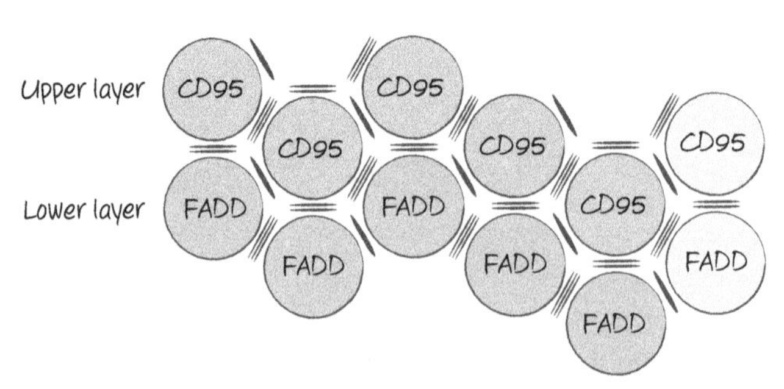

Figure 6.5. Model for the clustering of the CD95 receptor and FADD, based on the structure in Figure 6.4. The three different types of interface between different death domains (DDs) are shown by different lines.

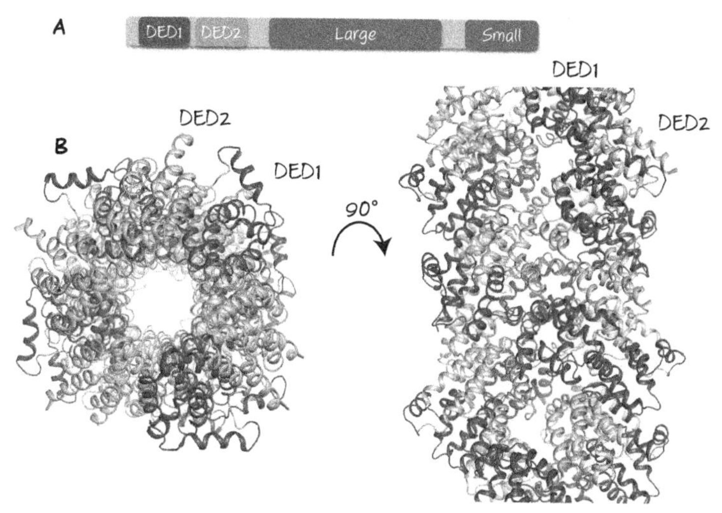

Figure 6.6. Caspase-8 oligomerizes into filaments. (*A*) Domains in caspase-8. DED, death effector domain; Large, large subunit; Small, small subunit. (*B*) Cryoelectron microscopy structure of filament formed when caspase-8 prodomains are brought into proximity by FADD. Only death effector domain (DED) regions (1 and 2, colored as in *A*) are shown. (*Left*) "Top-down" view. (*Right*) Side view.

allowing it to bind to the DED1 of another caspase-8 monomer, which exposes its own DED2. This results in a chain reaction, resulting in an extended filament of caspase-8, unless something else stops it (Fig. 6.6). We consider this "something else" below. As a result, one FADD protein recruits a complex of up to eight caspase-8 molecules. Caspase-8 can now cleave and activate the executioner caspases-3 and -7, which in turn orchestrate apoptosis (Fig. 6.7).

The complex comprising ligated CD95, FADD, and caspase-8 is called the CD95 death-inducing signaling complex (DISC). The DISC forms within seconds of CD95 ligation, and apoptosis can proceed within tens of minutes, presumably limited only by the rate of executioner caspase activation and substrate cleavage. However, in some cells, as we will see, this pathway can be slower and somewhat more complex.

In humans, the DISC can also include caspase-10, which appears to be similarly activated by binding to FADD through DED–DED interactions (note that caspase-10 is not found in rodents). However, at present, there is no compelling evidence that caspase-10 can replace caspase-8 for CD95-induced apoptosis. Human cells lacking caspase-8 are resistant to CD95 ligation (despite the presence of caspase-10).

Not all cells expressing CD95 are sensitive to CD95-induced apoptosis. Another protein, FLICE-like inhibitory protein (FLIP), which is related to caspase-8 but lacks an active cysteine, can also bind to FADD in the DISC through a DED–DED interaction (see Chapter 3). If caspase-8 and FLIP are dimerized, caspase-8 becomes activated and cleaves FLIP, but the caspase-8 molecule does not require cleavage for enzymatic

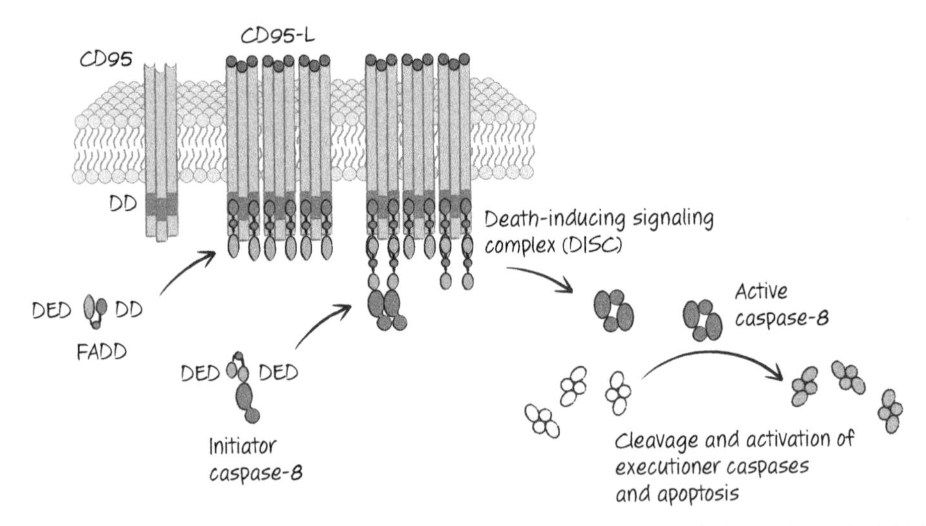

Figure 6.7. The CD95-induced apoptotic pathway, showing the principal players from the initial binding of the CD95 ligand (CD95-L) to the clustered receptors at the cell plasma membrane to formation of the death-inducing signaling complex (DISC), leading to eventual activation of executioner caspases.

activity of this complex. Nevertheless, apoptosis does not ensue (Fig. 6.8). The reason for this is that although FLIP, like caspase-8, contains two DEDs, the second does not bind to the DED1 of caspase-8 (or of FLIP). FLIP, then, is the "something else" that is able to prevent formation of the caspase-8 filament, thereby restricting its activity and preventing apoptosis.

The requirement for caspase-8 cleavage for apoptosis to occur is supported by the observation that replacing this caspase with a noncleavable form fails to restore

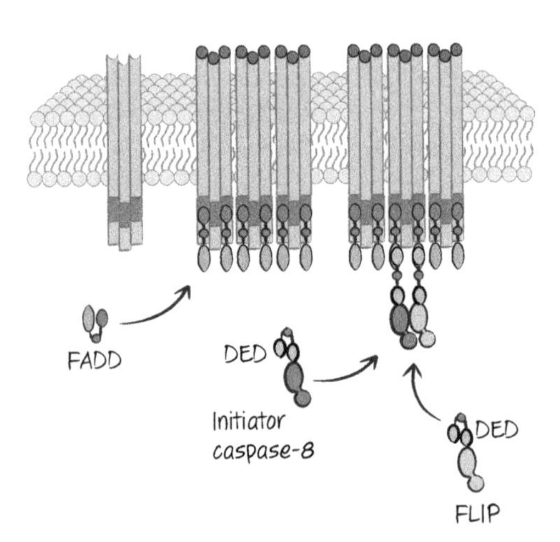

Figure 6.8. FLIP inhibits apoptotic signaling. When FLICE-like inhibitory protein (FLIP) is present, the formation of the death-inducing signaling complex (DISC) recruits it, and this prevents caspase-8 from forming an active filament that would lead to apoptosis (as illustrated in Fig. 6.6). DED, death effector domain; FADD, FAS-associated DD protein.

CD95-induced apoptosis, even though the noncleavable caspase-8 dimer is enzymatically active. Cleavage of FLIP might generate a signal with other roles in the cell; alternatively, the caspase-8–FLIP dimer might interact with some substrates but not with others (such as the executioner caspases, which are not activated when FLIP is expressed). We return to FLIP and its interaction with caspase-8 when we consider another form of cell death in Chapter 8.

Some herpes viruses express a short form of FLIP (v-FLIP) that blocks CD95-induced apoptosis by binding to FADD and preventing caspase-8 filament formation and activation. There is also a short isoform of mammalian FLIP (called $FLIP_S$), which similarly prevents caspase-8 activation (but does not form a proteolytically active dimer).

DEFECTS IN CD95 SIGNALING CAUSE A LYMPHOACCUMULATIVE DISEASE

Inactivating mutations in CD95, its ligand, or caspase-8 can cause the childhood disease acute lymphoproliferative syndrome (ALPS). This is marked by accumulation of T and B lymphocytes, including a massive increase in an unusual T-cell subset not found in normal individuals (for aficionados, this subset is $CD3^+$, $CD4^-$, $CD8^-$, and $B220^+$), presumably because apoptosis that would normally control the numbers of these cells does not occur. Autoimmunity and lymphoid tumors are not uncommon in people with ALPS, and the disease is usually lethal if not treated.

Mice with naturally occurring mutations in CD95 (lymphoproliferative, "*lpr*") or CD95-L (generalized lymphoproliferative disorder, "*gld*") show the same lymphocyte accumulation seen in ALPS, including accrual of the peculiar T-cell type and often become autoimmune (Fig. 6.9). Genetic deletion of CD95 or CD95-L has the same effects, but this is not seen when the deletion of CD95 is restricted to only T cells; it appears that loss of CD95 in another cell type, dendritic cells, is also involved in the disease. Such observations tell us that interactions between CD95 and its ligand are important for homeostasis in the immune system.[2]

TRAIL AND APOPTOSIS

TRAIL is a TNF-family ligand produced by T cells and other cell types. Like CD95-L, it is a trimer that can trigger apoptosis in target cells bearing the appropriate death receptors. In addition, like its relative, not all cells bearing receptors for TRAIL are

[2] Intriguingly, background genes influence the susceptibility to the disease, which manifests differently in different mouse strains. This is also true in people. Families that carry dominant-negative forms of CD95 (which can act in a partially dominant manner) can sometimes show ALPS in some children, whereas others are generally normal.

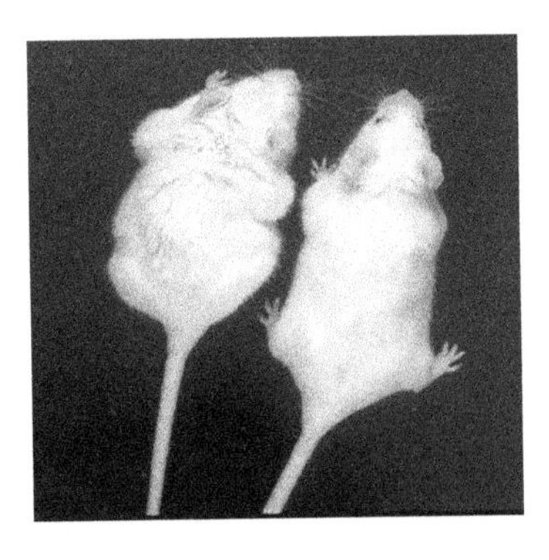

Figure 6.9. Lymphoproliferative *lpr* disease. The mouse on the *left* carries the *lpr* mutation in the *Cd95* (*Fas*) gene, resulting in massive enlargement of the lymphoid organs. The control mouse on the *right* is wild type.

sensitive to TRAIL-induced apoptosis, but in this case the mechanisms of resistance are less well understood (although one mechanism probably involves FLIP).

The mechanism of apoptosis induced by ligation of a TRAIL receptor appears similar to that induced by CD95; ligation of the receptor recruits FADD to the DD on the intracellular region of the TRAIL receptor, and this binds to and dimerizes caspase-8 to activate it. Cells lacking either FADD or caspase-8 are resistant to TRAIL-induced apoptosis.

At least one of the TRAIL receptors (DR5) can engage apoptotic signaling in a manner that is independent of its ligand, TRAIL. Endoplasmic reticulum (ER) stress can induce DR5 gene expression and cause DR5 protein in the ER to spontaneously oligomerize, resulting in caspase-8 activation and apoptosis. This is not the only way in which ER stress can cause apoptosis (other mechanisms involve engagement of the Bcl-2 family proteins, such as BIM, among others; see Chapter 5).

APOPTOTIC SIGNALING BY THE TNFR1 IS COMPLEX

Two different, but related, ligands bind to the death receptor TNFR1. These are TNF itself and another TNF-family ligand called lymphotoxin.[3] Ligation of TNFR1 by either causes it to expose its DD and interaction sites for another type of signaling molecule, TNF-receptor-associated factor-2 (TRAF-2). The DD of TNFR1 does not bind to FADD but, instead, to another DD-containing adapter protein, TRADD (tumor necrosis factor receptor type 1–associated death domain). TRADD helps to stabilize the binding of TRAF-2 and also recruits other molecules to the complex, RIPK1, cIAP, and cIAP2.

RIPK1 is a kinase, but here it is not the kinase activity that concerns us; the following steps do not depend on its enzymatic function. (We return to RIPK1 kinase activity in another form of cell death in Chapter 8.)

[3] Both of these also bind to another TNFR-family member that is not a death receptor; TNF binds to TNFR2, and lymphotoxin binds to the lymphotoxin receptor. However, because these are not death receptors, we do not further discuss them nor other non-death-receptor members of this extended receptor family.

The cIAPs are members of the inhibitor of apoptosis protein family (IAP family; see Chapter 3), but they do not inhibit caspases. Instead, they act as E3-ubiquitin ligases to polyubiquitylate RIPK1. This involves lysine 63 of the ubiquitin chain and so has a function different from degradation (characteristic of lysine 48 linkages). The lysine-63-linked polyubiquitin on RIPK1 recruits other proteins, including a complex called LUBAC, which then creates a linear ubiquitylation on the protein. This, in turn, recruits NEMO (also known as I-κB kinase-γ, IKKγ). This recruits and activates the rest of the IKK complex, which in turn activates the transcription factor NF-κB (Fig. 6.10).

NF-κB induces the transcription of a number of genes involved in cell survival, including FLIP. When NF-κB is activated by TNFR1 ligation, apoptosis does not occur, and instead the cell responds in other ways—for example, participating in inflammatory responses by producing cytokines.

There are additional signaling consequences of this complex. For example, ligation of TNFR1 induces the activation of c-Jun amino-terminal kinase (JNK). This phosphorylates the protein Jun, which forms part of another transcription factor, AP-1. TRAIL ligation can also activate JNK. JNK has other targets in the cell, and its activation can promote apoptosis in some cell types, as discussed below.

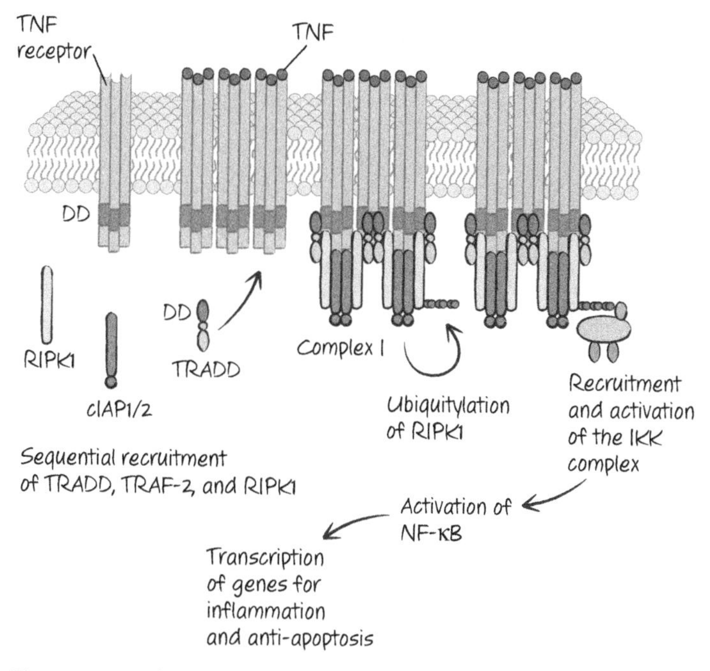

Figure 6.10. Tumor necrosis factor (TNF) signaling for NF-κB, antiapoptosis, and inflammation. Other proteins and modifications involved in the process are not shown.

If NF-κB is not functional or if survival genes induced by NF-κB are not active, a second signaling step can lead to apoptosis. This happens in some cell types or under conditions in which cell signaling is experimentally altered. An enzyme that removes ubiquitin chains (a deubiquitinase, or DUB), called CYLD, removes ubiquitin from RIPK1. The modified TRADD–RIPK1 complex (called TNFR1 signaling complex I) disengages from the receptor and becomes cytosolic. The DD of TRADD, previously associated with TNFR1, is now exposed, and this recruits FADD. FADD, in turn, binds and activates caspase-8 (forming TNFR1-signaling complex IIa). The caspase-8 then precipitates apoptosis, as we have seen (Fig. 6.11). Alternatively, RIPK1, which also has a DD, can recruit FADD (TNFR1-signaling complex IIb).

Mice that display defective NF-κB function (because parts of the NF-κB pathway have been knocked out) die during development because of extensive apoptosis in the liver. However, if these animals also lack TNFR1, they survive beyond birth. This tells us that it is apoptotic signaling from TNFR1 that is responsible for liver failure in NF-κB-defective embryos.

The rescue of these mice by TNFR1 deficiency tells us that other death ligands, such as TRAIL or CD95-L, do not contribute significantly to the lethal effects of loss of NF-κB signaling during development. Note, however, that activation of NF-κB can

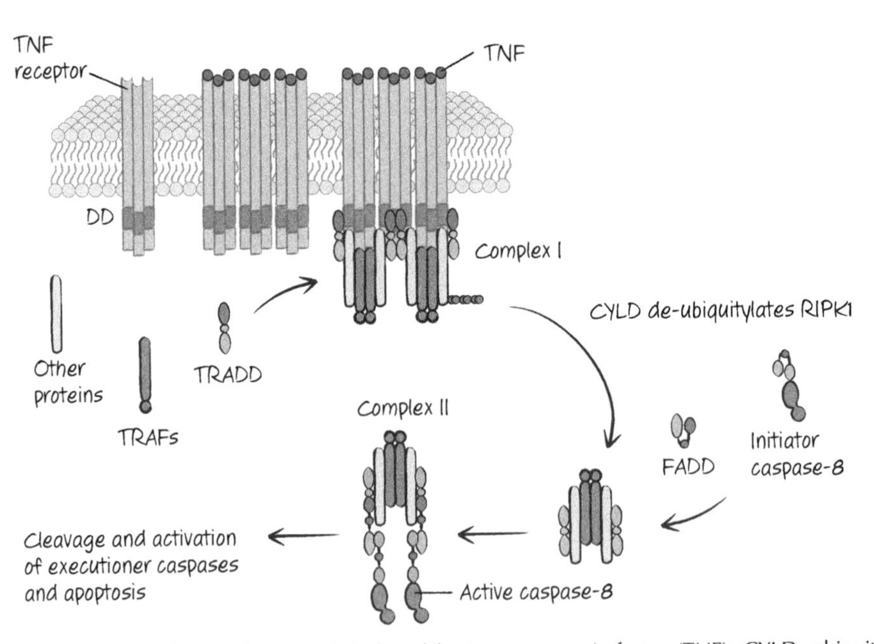

Figure 6.11. The pathway of apoptosis induced by tumor necrosis factor (TNF). CYLD, ubiquitin carboxy-terminal hydrolase CYLD; DD, death domain; FADD, FAS-associated death domain protein; RIPK1, receptor-interacting serine/threonine-protein kinase 1; TRAF, TNF-receptor-associated factor; TRADD, tumor necrosis factor receptor type 1–associated death domain protein.

produce resistance to apoptosis induced by ligation of other death receptors by inducing expression of FLIP and other proteins.

There is evidence that other death receptors, including CD95 and the TRAIL receptors, might also form cytosolic signaling complexes following ligation that resemble complex II in TNFR signaling. At this point, it is not clear that this is necessary for caspase-8 activation and apoptosis in these cases. Furthermore, these and other death receptors can also activate JNK or NF-κB in some cells, although the precise mechanisms are less well defined.

DEATH RECEPTOR SIGNALING CAN ENGAGE THE MITOCHONDRIAL PATHWAY OF APOPTOSIS

In Chapter 5, we discussed how the BH3-only protein BID is a sensor for proteases, because cleavage of the protein can activate it to engage the mitochondrial pathway of apoptosis by activating BAX and BAK. When death receptors are ligated and activate caspase-8, this causes mitochondrial outer membrane permeabilization (MOMP) and engages the mitochondrial pathway of apoptosis. This is because caspase-8 is very efficient at cleaving BID and activating this BH3-only protein (Fig. 6.12).

In some cells, inhibition of MOMP by antiapoptotic BCL-2 proteins has little or no effect on death receptor–mediated apoptosis. These have been called type I cells in terms of their response to death ligands.[4] Lymphocytes and lymphoid tumors are examples of type I cells. In other cells, inhibition of MOMP effectively blocks apoptosis, and these are referred to as type II cells. Hepatocytes are an example of type II cells, as we will see.

The crucial difference between type I and type II cells in the response to death receptor ligation could be the expression of XIAP. This protein blocks the activities of caspase-9 and the executioner caspases-3 and -7 (see Chapter 3) but does not inhibit caspase-8. When MOMP is triggered by cleaved BID, antagonists of XIAP, such as Smac and Omi, are released from the mitochondrial intermembrane space and disrupt XIAP function, allowing apoptosis to proceed. This effect can be seen in cells lacking APAF1 (which therefore cannot activate caspases as a result of cytochrome *c* release), and this tells us that the function of MOMP to block XIAP can be crucial for caspase-8 to activate caspases-3 and -7 and promote apoptosis by this pathway. This scheme is illustrated in Figure 6.13.

One example is the cells of the liver—the hepatocytes. Injection of mice with CD95-ligand or with antibodies that activate CD95 causes a rapidly lethal

[4] It is unfortunate (and confusing) that the forms of cell death are referred to as "type 1," "type II," and "type III," whereas the cells that respond differently to death receptor signaling are also referred to as "type 1" and "type 2" cells. In this chapter, any reference to "type 1 cells" or "type II cells" is to the latter.

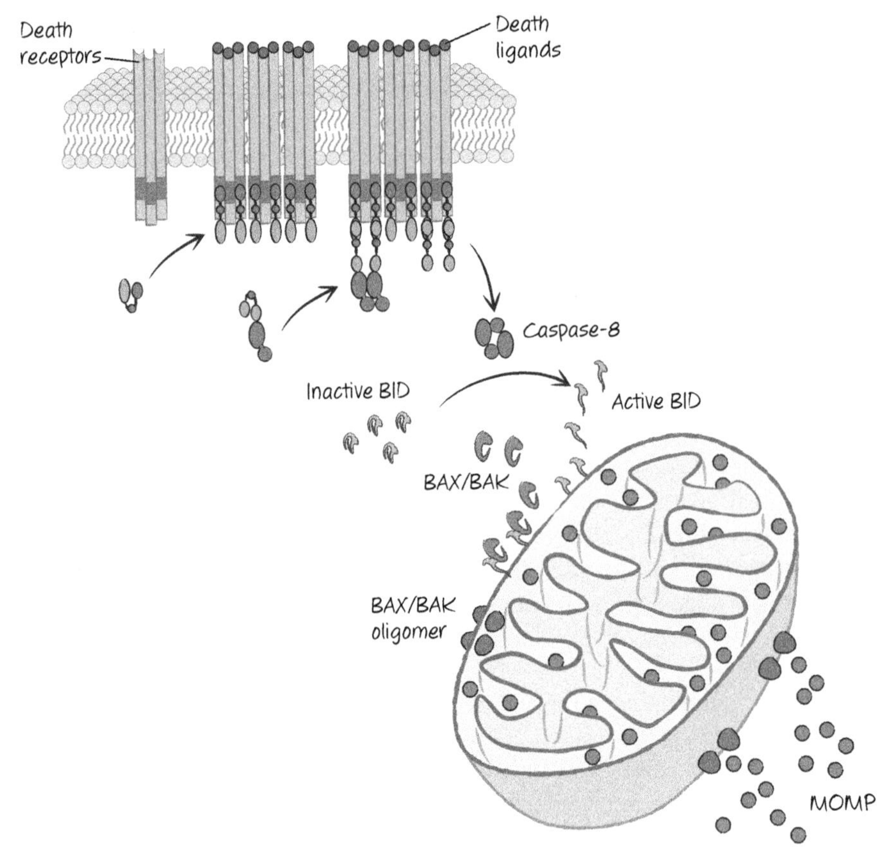

Figure 6.12. The pathway of death receptor signaling giving rise to mitochondrial outer membrane permeabilization (MOMP) through caspase-8-mediated cleavage of BID. TNF, tumor necrosis factor; FADD, FAS-associated DD protein; TRAF, TNF-receptor-associated factor; TRADD, tumor necrosis factor receptor type 1–associated DD protein.

destruction of the liver. Expression of BCL-2 in hepatocytes prevents this lethal effect. Mice that lack BID are also resistant to these treatments. Therefore, these cells are type II cells that require engagement of the mitochondrial pathway following CD95 ligation to cause apoptosis. However, mice lacking both BID and XIAP are sensitive.

CD95 ligation on hepatocytes therefore engages caspase-8, but, although caspase-8 can activate the executioner caspases, it seems that XIAP blocks them and preserves survival of both the cells and the animals. However, when caspase-8 cleaves and activates BID, the resulting MOMP releases proteins such as Smac and Omi that neutralize XIAP, and death ensues. At least in the case of hepatocytes, it is the expression and function of XIAP that make these type II cells; and MOMP must be engaged to inhibit XIAP in order for apoptosis to proceed.

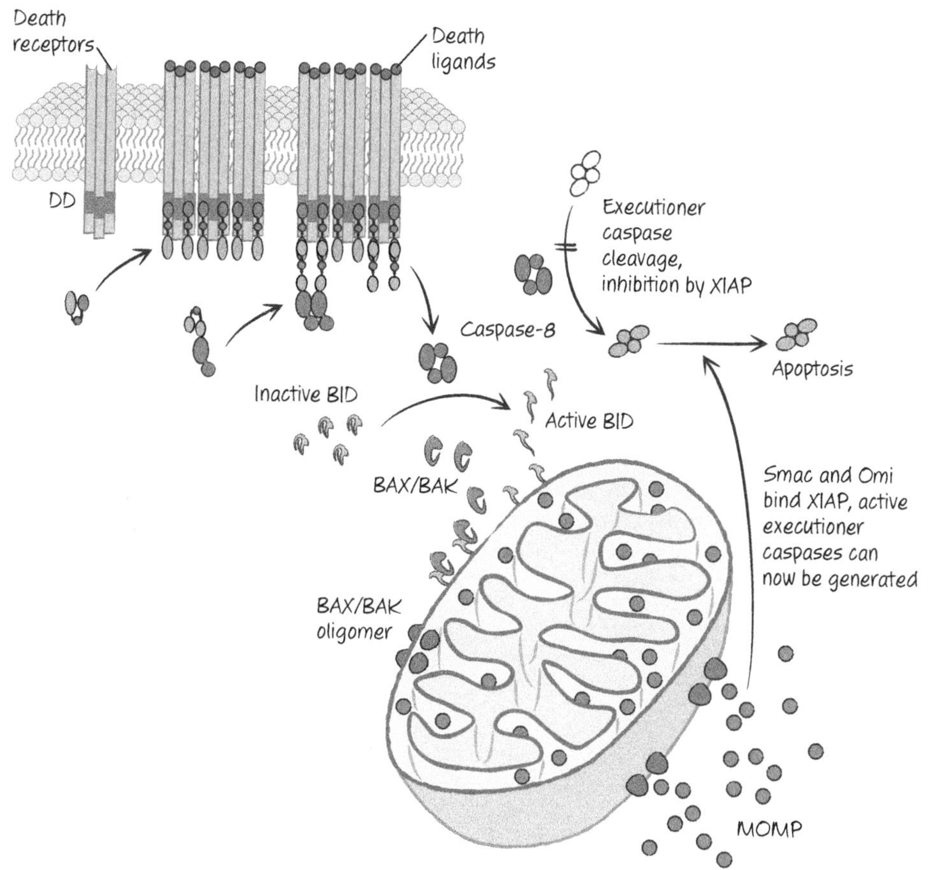

Figure 6.13. Pathway of death receptor–induced apoptosis in type II cells. BID-induced mitochondrial outer membrane permeabilization (MOMP) can also engage apoptosome formation and caspase activation, but this does not appear to be necessary for apoptosis by this mechanism. DD, death domain.

There could be another way in which some death receptors engage apoptosis. As mentioned above, signaling from the TNFR1 and TRAIL receptors can activate JNK. As we saw in Chapter 5, JNK can phosphorylate the BH3-only protein BIM, promoting its activity. Signaling from some death receptors might therefore trigger apoptosis in this way, and there is evidence to support this idea. If this is the case, we have to reconcile this with the finding that caspase-8 is required for apoptosis induced by death receptors. Unfortunately, at present, we know of no way for caspase-8 to participate in the activation of JNK or BIM. This could change, however. Recently, another function for caspase-8 was identified that is independent of its caspase activity, instead acting as a scaffold for additional signaling events following death receptor ligation. If correct, this might provide some insights into how caspase-8 might be involved in apoptosis via JNK and BIM.

DEATH RECEPTORS IN OTHER ANIMALS

Although there is a TNFR-family protein in *Drosophila*, its ligation does not appear to activate caspases or apoptosis in any situation. *Drosophila* also have a homolog of FADD (called dFADD) and a homolog of caspase-8 containing DEDs (Dredd). If these proteins are not involved in apoptosis, what do they do?

It could be that the role of this pathway in *Drosophila* (and presumably other insects) is in host defense. Dredd and dFADD participate in the production of antibacterial peptides in the fly. Intriguingly, this is because Dredd cleaves and thereby activates a protein related to NF-κB called Relish. This has led to the idea that caspase-8 in mammals might have a similar role in a NF-κB pathway, but currently this is controversial.

In contrast to *Drosophila*, a TNFR-like receptor has been identified in corals, and remarkably this interacts with human TNF to trigger apoptosis, both in coral cells and in human cells in which it is experimentally expressed. It might be that the death receptor pathway, like the mitochondrial pathway, is more common in animals than we once thought.

CHAPTER 7

Inflammasomes and Other Caspase-Activation Platforms

OTHER WAYS TO CELL DEATH

Although most apoptosis in mammals occurs by means of the mitochondrial pathway or death receptor ligation, caspase activation can occur by other mechanisms as well. Some of these lead to cell death, and some have other functions.

Apart from the executioner caspases, caspases are activated by the mechanism known as induced proximity. As we have seen, this allows the formation of dimers or higher-order oligomers, which is generally effected by the binding of one or more adapter proteins to the prodomains of the caspase monomers. These adapter proteins help to define the caspase-activation pathway involved. In this chapter, we discuss the activation and functions of other caspases that can participate in cell death but can also have other functions as well—caspase-1, caspase-2, caspase-4, and caspase-5 (as well as caspase-11 in mice). These caspases are not activated by the pathways that we have described in Chapters 4 and 6 but, rather, by other activation platforms. And although some of these can induce apoptosis, others cause another form of cell death—pyroptosis (a form of necrosis).

THE "ADAPTERLESS" ACTIVATION OF CASPASES-4, -5, AND -11

So far, we have discussed two ways in which caspases are activated: cleavage in the case of executioner caspases and induced proximity by the binding of initiator caspases to adapter molecules. But the activation of caspases-4 and -5 (in humans), and caspase-11 (in rodents), uses a different approach. Like the initiator caspases, these caspases exist as monomers expressed in myeloid cells (such as macrophages and dendritic cells) and most epithelial cells. And, like initiator caspases, they are activated by induced proximity. However, it appears that this can occur without a requirement for an adapter protein.

Gram-negative bacteria contain lipopolysaccharide (LPS) in their outer coats, which is a molecule composed of a lipid and carbohydrates and is sensed by the immune system as an indication of infection. If such bacteria invade the cytoplasm of a cell, the lipid portion of LPS binds to the caspase-recruitment domain (CARD) of caspases-4 and -5, or caspase-11 in rodents, forcing the caspase to form dimers and thus activating the enzyme. This appears to involve direct binding; it is the LPS that serves as the platform on which the caspase activates. In this sense, these caspases can be thought of as intracellular LPS sensors. At this point, we do not know whether there are other bacterial products that can activate these (or other) caspases through interaction with their prodomains, but this remains a possibility.

The activities of caspases-4, -5, and -11 do not cleave and activate executioner caspases nor promote apoptosis in cells. Instead, they cause cell death by cleaving a specific substrate, gasdermin D. Gasdermin D is auto-inhibited, and this inhibition is disrupted by cleavage. Active gasdermin D then moves to the plasma membrane, where it creates pores, causing the cell to swell and ultimately burst (Fig. 7.1). It is the action of gasdermin D that promotes this necrotic cell death, and mutation of the cleavage site of gasdermin D prevents cell death by active caspase-11. Mice lacking either caspase-11 or gasdermin D are resistant to LPS shock, a toxic condition caused by intravenous injection of LPS, and to bacterial sepsis. There are several gasdermin proteins, all of which have the pore-forming potential, although only gasdermin D is activated through cleavage by caspase-4, -5, or -11. The functions of the other gasdermin molecules are largely unknown, but it is possible that these proteins participate in other pathways of cell death. In Chapter 8, we will mention a molecule related to gasdermin D, which functions in a form of necrosis.

Although LPS can activate these caspases without a requirement for an adapter protein, this does not necessarily mean that no such adapter exists. Although at present the existence of an adapter protein for caspases-4, -5, and -11 is speculative, there are conditions that appear to activate these caspases without intracellular LPS. One example is the cholera toxin b-chain, which binds to the cell surface and triggers the activation of these caspases.

The activation of caspases-4, -5, or -11 is often referred to as the "noncanonical inflammasome." Next, we consider the "canonical inflammasomes," which activate caspase-1.

CASPASE-1 IS ACTIVATED BY INFLAMMASOMES

As outlined in Chapter 3, caspase-1 is involved in the processing of interleukin-1 and interleukin-18, secretion of these and other proteins, and cell death by pyroptosis. The activation of caspase-1 occurs in complexes called inflammasomes, and, unlike

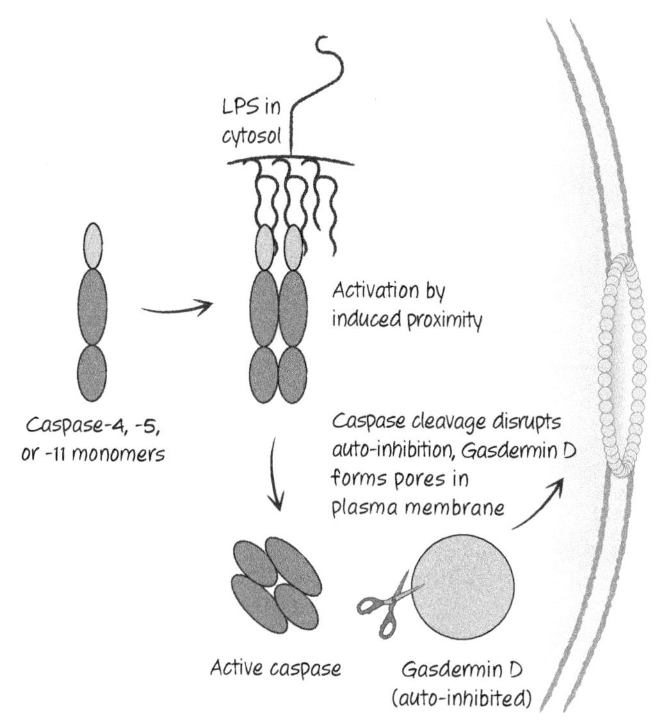

Figure 7.1. Cytosolic lipopolysaccharide (LPS) directly induces caspases and pyroptosis. This appears to be restricted to caspase-4 and caspase-5 in humans, and caspase-11 in mice. Other caspases we have discussed are not activated in this direct manner.

the other caspase-activation platforms that we have discussed (the apoptosome and the death-inducing signaling complex (DISC) for caspases-9 and -8, respectively), inflammasomes can comprise different adapter molecules. In most inflammasomes, however, a common feature is a molecule that binds to the CARD in the prodomain of a caspase-1 monomer (Fig. 7.2).

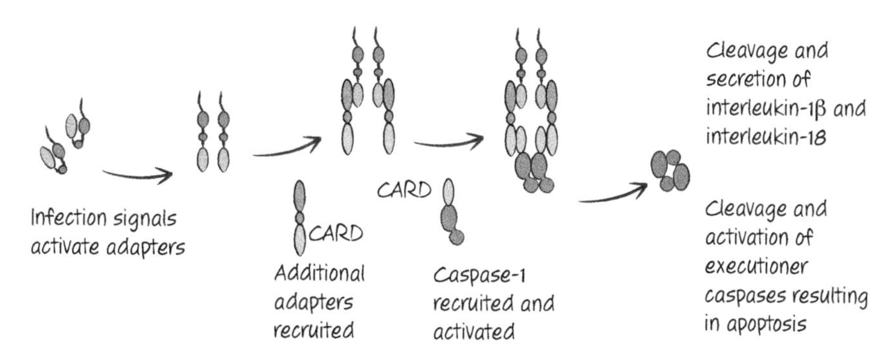

Figure 7.2. The basic inflammasome and activation of caspase-1.

Caspase-1 is expressed in myeloid cells, such as macrophages and dendritic cells, as well as epithelial cells, and inflammasomes that activate this caspase assemble in these cells in response to a wide range of signals, most of which are associated with infectious agents. These signals can be proteins, lipids, DNA, or RNA from pathogens, collectively referred to as pathogen-associated molecular patterns (PAMPs). Some inorganic materials, such as some crystals, also induce formation of inflammasomes, as do (probably) some materials released from necrotic cells. Another general term is sometimes used for the latter—damage-associated molecular patterns (DAMPs).[1] How PAMPs/DAMPs trigger inflammasome formation and how different inflammasomes are engaged are at the heart of a rapidly emerging area of inflammation research.

SEVERAL INFLAMMASOMES FOR ACTIVATION OF CASPASE-1 INVOLVE ASC

A small CARD-containing protein called ASC ("apoptosis-associated speck-like protein containing a CARD") can oligomerize and bind caspase-1 through CARD–CARD interactions to activate the protease by induced proximity. In addition to its CARD domain, ASC contains another death fold, a pyrin domain (PyD). If ASC is experimentally overexpressed, it oligomerizes to form large aggregates (appearing as specks in the cytoplasm). These aggregates of ASC are sufficient to bind and activate caspase-1 (Fig. 7.3). As we will see, however, this artificial situation is not how ASC functions to promote caspase-1 activation. Instead, there are intracellular receptors that interact with ASC to promote the formation of inflammasomes, and the general structure of these has been elucidated in cryoelectron microscopy studies. All of these inflammasomes form disc-like structures when activated, and these, in turn, recruit

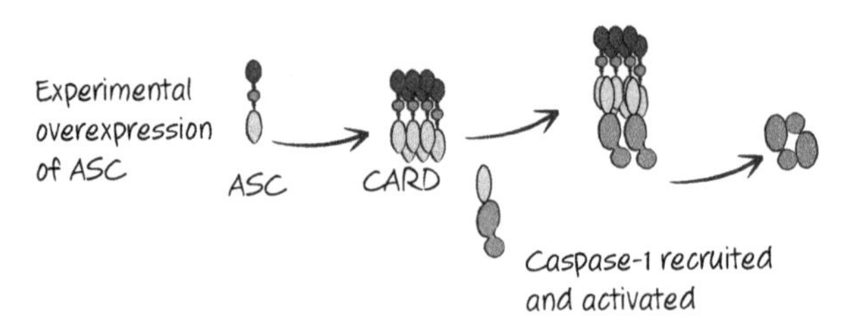

Figure 7.3. Overexpression of ASC can activate caspase-1. This is not a physiological mechanism of caspase-1 activation, but the observation is informative.

[1] DAMPs are sometimes called danger-associated molecular patterns, based on the "danger hypothesis." Because danger is rather tautological, defined as anything that elicits a response, we prefer the term "damage," but this has its own problems.

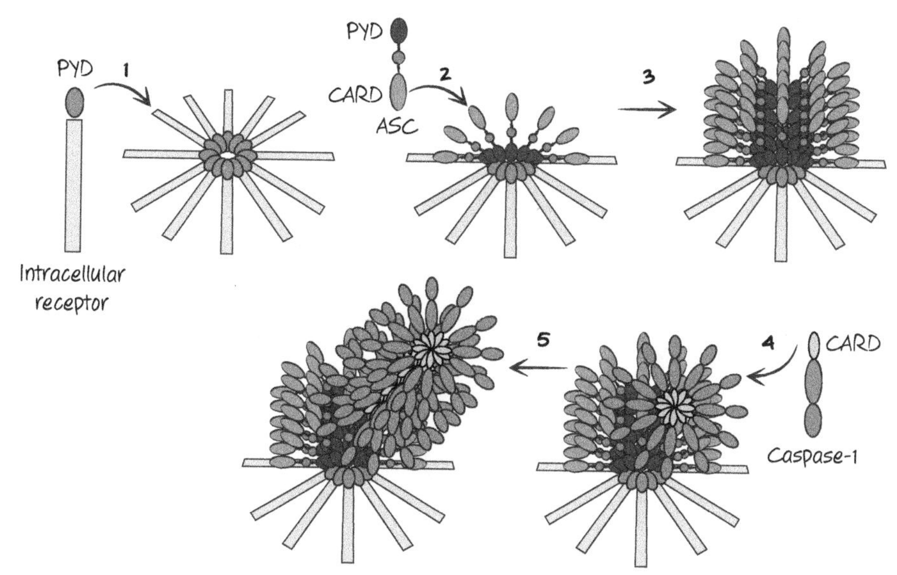

Figure 7.4. General structure of inflammasomes. 1. The intracellular receptor, containing a pyrin domain (PYD), on interaction with its ligand forms a disc-like structure. 2. The center of this structure interacts with ASC through PYD–PYD interactions. 3. The PYD domains of ASC interact with additional ASC PYD domains to form a cylindrical structure. 4. The CARD domains of caspase-1 monomers bind to the CARD of ASC. 5. Bound caspase-1 recruits additional caspase-1 molecules through CARD–CARD interactions, forming a fibril. A "mature" inflammasome is shown in Figure 7.5.

ASC, which itself forms a cylindrical structure. When the caspase-1 CARD binds to the exposed ASC CARD, the caspase-1 CARD then binds further caspase-1 CARD domains. Filaments of oligomerized caspase-1 then extend from this ASC cylinder (Figs. 7.4 and 7.5). Therefore, the structures of inflammasomes are distinct from the structures of other caspase-activation platforms, such as the apoptosome (Chapter 4).

TLRs INDUCE OTHER INFLAMMASOME COMPONENTS AND CASPASE-1 SUBSTRATES

A set of receptors related to the *Drosophila* Toll protein is involved in recognizing many PAMPs. These Toll-like receptors (TLRs) recognize components of bacterial cell walls, bacterial RNA, and fungal components. When they interact with PAMPs, TLRs activate nuclear factor-κB (NF-κB) and interferons, which in turn induce the expression of caspases-1, -4, -5 (and caspase-11 in rodents), the interleukin targets of caspase-1, as well as additional inflammasome components that can bind to and oligomerize ASC to facilitate caspase-1 activation. Because of this, cells with potential inflammasome activity often have to be "primed" with agents that activate a TLR. In

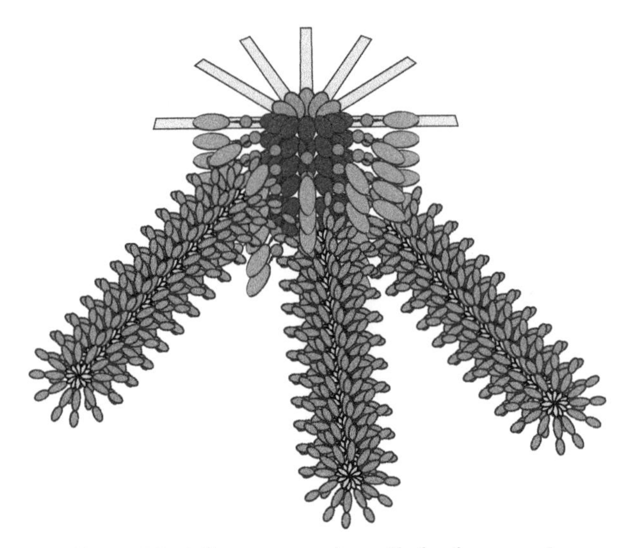

Figure 7.5. Inflammasomes have fibrils of caspase-1.

Figure 7.6, some of the TLRs in human cells are illustrated, together with some of the PAMPs that they recognize.

There is a logic to this arrangement. The production of inflammatory cytokines and the death of cells by pyroptosis can cause tissue damage. By restricting the system to conditions in which an infection is sensed (by TLRs), the activation of the inflammasome is limited.

Unfortunately, however, there are many conditions that can prime myeloid and epithelial cells and also activate caspase-1 in the absence of an overt infection. We return later to these in considering the roles for inflammasomes in promoting disease.

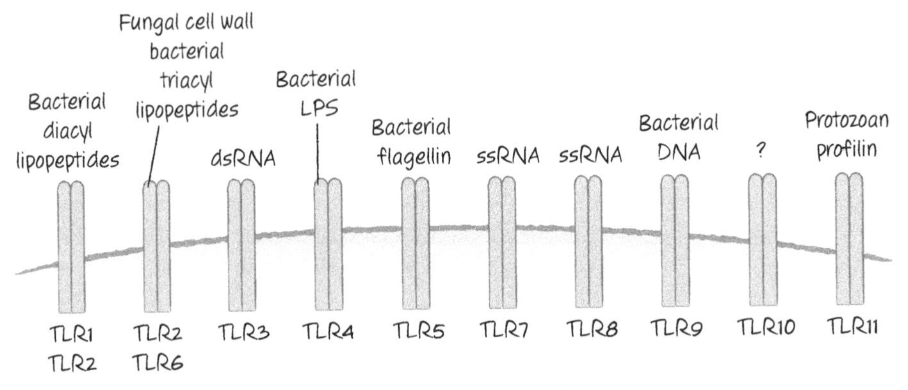

Figure 7.6. Some TLRs and the PAMPs that activate them.

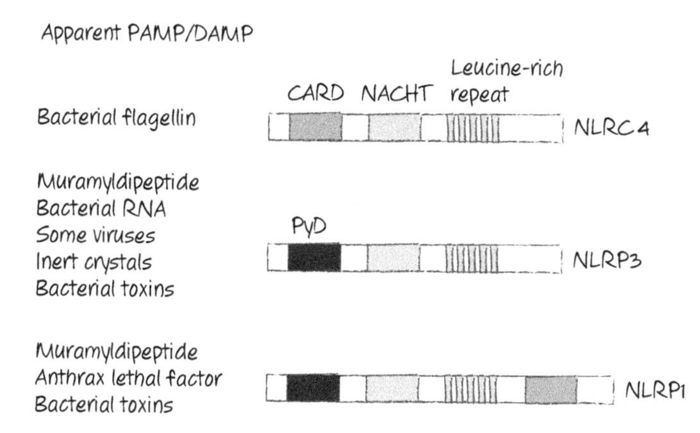

Figure 7.7. Some NLRs that function in inflammasomes. The PAMPs and DAMPs that induce them are listed. Toxins from bacteria can also alter potassium levels, thereby indirectly facilitating inflammasome formation.

NLRs PARTICIPATE IN INFLAMMASOMES

In addition to the TLRs, there is a large set of intracellular molecules that appear to recognize PAMPs and DAMPs—the Nod-like receptors (NLRs, so named for the first one found, Nod-1). The NLRs are found throughout the animal kingdom and even in plants, where they are involved in host defense against infection. Collectively, TLRs and NLRs are often referred to as "pattern-recognition receptors" (PRRs).

Several NLRs contain death folds, including CARDs and PyDs, and some of these have been shown to participate in the generation of inflammasomes. The most important of these to date are NLRP3 ("NACHT, LRR, and PYD domains-containing protein 3," also called NALP3 and cryopyrin) and NLRC4 ("NLR family CARD domain-containing protein 4," also called CARD12 and IPAF). All of the NLRs (including those in plants) also have a NACHT domain (see Chapter 4) and a long tail. In many NLRs, this tail is a leucine-rich region (LRR) believed to be specific for a type of PAMP or DAMP that appears in the cell. Some of the mammalian NLRs and what they appear to recognize are listed in Figure 7.7.

Whether NLRs respond directly to a DAMP or PAMP has not been formally proven. It remains possible (and in some cases likely) that other molecules participate in this recognition event, as is the case for NLRC4 (discussed below).

THE NLRP3 INFLAMMASOME

NLRP3 contains a PyD that binds to the PyD of ASC. In addition, NLRP3 has a NACHT domain and an LRR. NLRP3 and ASC are required for the formation of inflammasomes in response to a number of PAMPs, including those from *Staphylococcus aureus*, *Listeria*

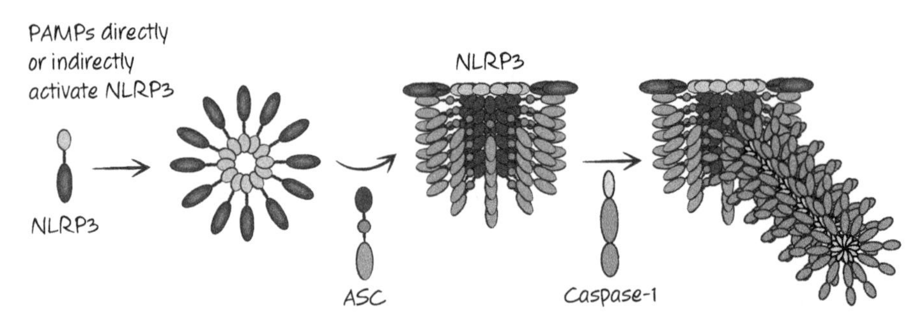

Figure 7.8. Activation of the NLRP3 inflammasome.

monocytogenes, and bacterial RNAs. This does not depend on TLR signaling. However, TLR signals induce the expression of NLRP3 (and other inflammasome components, as we have seen) to increase sensitivity. Once activated, NLRP3 promotes inflammasome formation and caspase-1 activation (Fig. 7.8).

Some noninfectious materials also act to engage the NLRP3 inflammasome to activate caspase-1. These include asbestos, crystals of uric acid (the cause of gout), and crystals of calcium pyrophosphate dihydrate (CPPD, the cause of pseudogout), as well as crystals of cholesterol (Fig. 7.9).

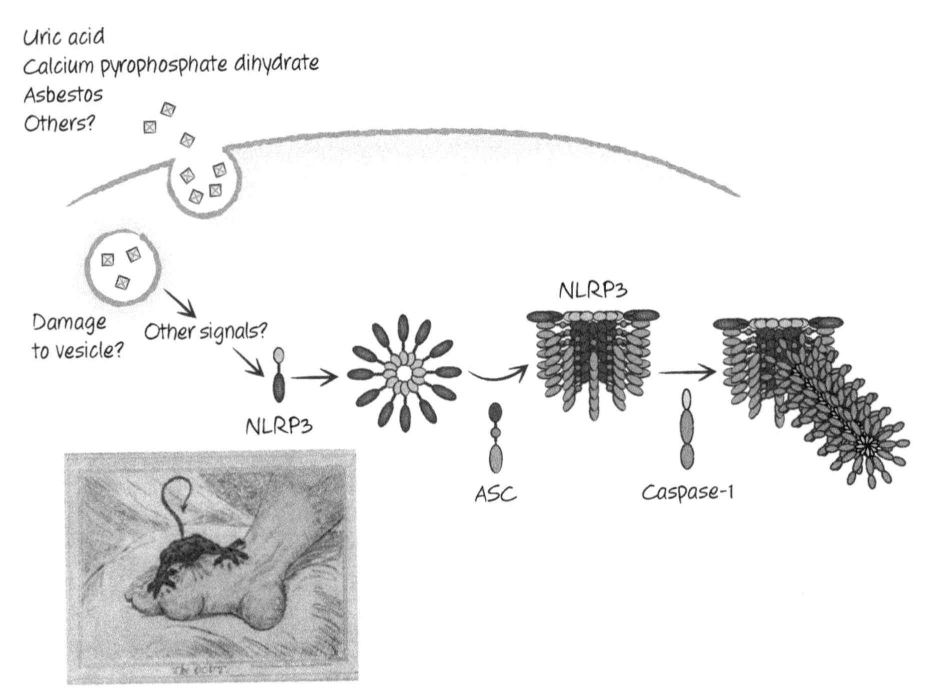

Figure 7.9. Inert crystals induce the NLRP3 inflammasome in gout and other diseases.

It is unlikely that all of these interact directly with NLRP3 to trigger its interaction with ASC and caspase-1. There might be common denominators. One of these appears to be the mitochondria. On any of these treatments, mitochondria seem to become stressed, producing reactive oxygen by the action of the mitochondrial electron-transport chain. Indeed, an inhibitor of mitochondrial complex I—rotenone—can induce the activity of the NLRP3 inflammasome through the production of reactive oxygen. Furthermore, cells that lack an electron-transport chain because of loss of mitochondrial DNA do not activate the NLRP3 inflammasome in response to these agents. However, it is not clear whether, or how, reactive oxygen activates the NLRP3 inflammasome. Some studies have suggested that it is not reactive oxygen, but instead oxidized mitochondrial DNA, that performs this function, but again the mechanism underlying this remains obscure.

Another potential common denominator is potassium. The high levels of potassium in cells inhibit the NLRP3 inflammasome. Activation of potassium channels in the plasma membrane allows an efflux of potassium, permitting inflammasome activation.

One such potassium channel is the P2X purinoceptor 7 (P_2X_7). This is activated by extracellular ATP, which might act as a damage signal that cells (pathogens or host cells) are lysing in the vicinity. Extracellular ATP activates the NLRP3 inflammasome in primed macrophages. Bacteria produce a variety of toxins that bind to the cell surface and also cause potassium efflux. Again, such toxins are potent activators of the NLRP3 inflammasome.

One other way in which potassium levels in a cell can decrease to promote the activation of the NLRP3 inflammasome is when caspase-4, -5, or -11 is activated by intracellular LPS (see above). When gasdermin D is cleaved, and before the cell dies, this effector causes potassium efflux, perhaps as a consequence of disruption of the plasma membrane. Therefore, although these caspases do not process interleukin-1β or interleukin-18 directly, they can promote the activation of the NLRP3 inflammasome to activate caspase-1, which has this cytokine-processing ability.

However, all of this is complicated by the fact that potassium levels also affect mitochondrial function, and therefore it remains possible that it is the mitochondria, and not only potassium levels, that somehow activate the NLRP3 inflammasome. The production of reactive oxygen species from mitochondria has been implicated in NLRP3 activation, although the mechanism for this is obscure. Until we fully understand the roles for potassium and mitochondria in NLRP3 activation, the common denominators that link all of the agents that can induce this inflammasome remain murky.

THE NLRC4 INFLAMMASOME

Another important NLR for inflammasome formation is NLRC4, which is required for the response to bacterial flagellin from *Salmonella typhimurium* and *Legionella*

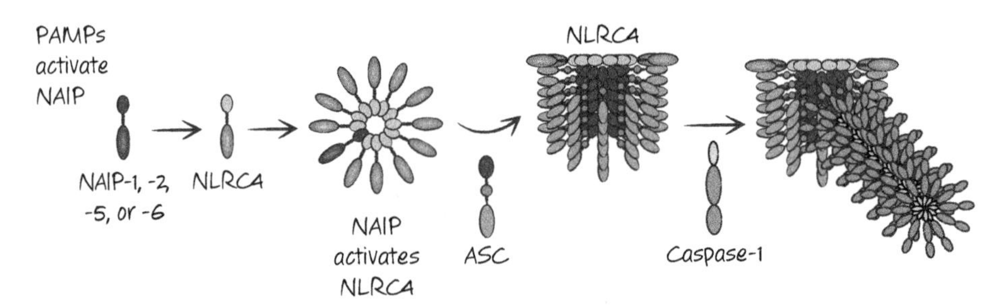

Figure 7.10. Activation of the NLRC4 inflammasome.

pneumophila and to *Shigella flexneri* infection, as well as the response to pathogenic *Escherichia coli*.

It turns out that NLRC4 does not recognize these PAMPs directly. Another set of molecules, comprising NAIP1, 2, 5, and 6, is important in the NLRC4 response. These proteins do not have a death fold. Instead, each has a region with three BIR domains. It therefore resembles an IAP protein (see Chapter 3) but without caspase-inhibitory activity (as far as we know). NAIP5 and NAIP6 respond to bacterial flagellin, whereas NAIP1 and NAIP2 respond to components of the bacterial type III secretion system, which functions to inject bacterial components into the cell. One molecule of any of these NAIP proteins, on activation, binds to an NLRC4 molecule, inducing a conformational change in the latter. This activated NLRC4 can then activate another NLRC4 molecule, thus creating a chain reaction that results in the formation of a disc-like structure containing one NAIP protein and several NLRC4 proteins. The NLRC4 complex then recruits ASC, which in turn activates caspase-1 (Fig. 7.10).

NLRC4 has a CARD domain. Following activation, this interacts with ASC through a CARD–CARD interaction. It is likely that this then promotes the PYD–PYD interactions of ASC to form the ASC cylinder, which in turn recruits the CARD of caspase-1. Alternatively, it is possible that the CARD of NLRC4 binds directly to the CARD of caspase-1 in some cases as the activation of caspase-1 by NLRC4 is enhanced by, but does not always require, the presence of ASC. Therefore, if NLRC4 oligomerizes, it can bring together and activate caspase-1 monomers, through direct or indirect interactions.

OTHER INFLAMMASOMES

The lethal bacterial toxin of anthrax activates caspase-1 through another inflammasome, which involves the NLR NLRP1b (also called NALP1b). NLRP1b is different in humans and rodents, and the way in which it activates caspase-1 is somewhat controversial. It has a CARD, and, although some studies indicate that the activation of caspase-1 by NLRP1 requires ASC, others show that this protein can interact directly

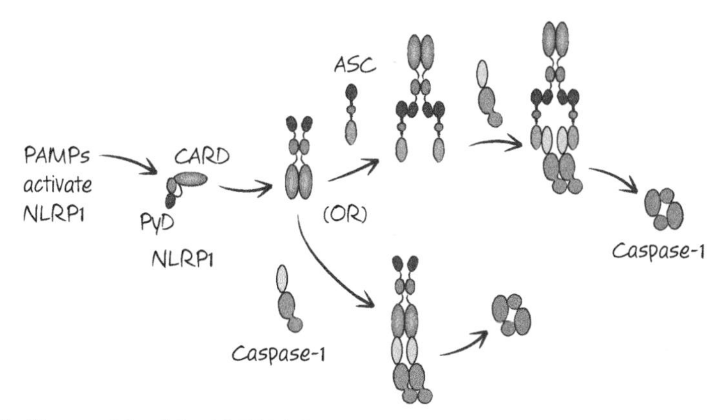

Figure 7.11. Two models of the NLRP1 inflammasome. Both may be correct in different settings.

with caspase-1 through a CARD–CARD interaction.[2] These two possibilities are shown in the simplified scheme in Figure 7.11.

Another inflammasome comprises yet another NLR, NLRP2 (also called NALP2), and ASC. It is not known what NLRP2 recognizes to trigger caspase-1 activation.

There is a type of inflammasome that does not involve an NLR. Double-stranded DNA in the cytosol binds to a protein called AIM2. AIM2 has a PyD that binds to the PyD of ASC. Therefore, when DNA appears in the cytosol, clusters of AIM2 form and, in turn, activate caspase-1 (Fig. 7.12).

AIM2 is a sensor for infection by both RNA and DNA viruses, although it also binds to any other double-stranded DNA, including that from the host (provided it appears in the cytosol). For this reason, the AIM2 inflammasome might have a role in the autoimmune response to double-stranded DNA in systemic lupus erythematosus and related diseases (although other intracellular DNA sensors, not relevant to our discussion here, have also been implicated).

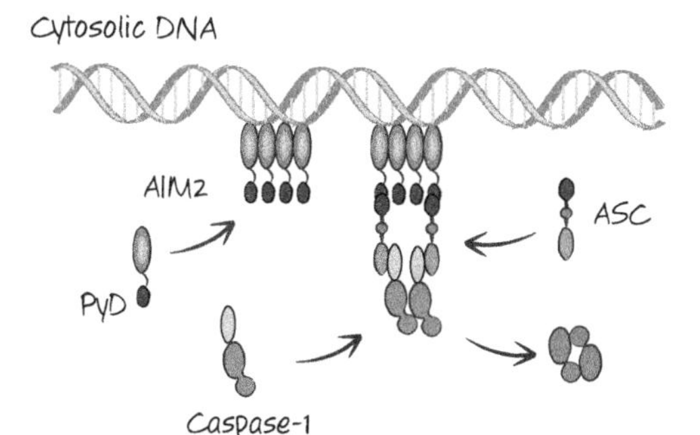

Figure 7.12. The AIM2 inflammasome. This is a simplified scheme, and AIM2 facilitates the formation of caspase-1 fibrils, as we saw for other inflammasomes.

[2] It is possible that this is a difference between the human and the murine protein.

CELL DEATH BY CASPASE-1 AND NONCANONICAL SECRETION

As we have discussed, the activation of caspase-1 can lead to apoptotic cell death, but this is often not the case. Like caspases-4, -5, and -11, caspase-1 can cleave and activate gasdermin D to cause necrotic pyroptosis. If gasdermin D is absent, caspase-1 can cause cell death by cleaving and thereby activating executioner caspases, caspase-7 being more important than caspase-3 in this case. Caspase-1 can also activate BID and thereby engage the mitochondrial pathway of apoptosis (see Chapter 5). Bacteria that infect cells often induce cell death through activation of caspase-1, and, in several cases, the inflammasome responsible for this effect has been identified (as indicated above). For example, *Shigella flexneri* kills macrophages in a manner that depends on NLRC4, ASC, and caspase-1.

When caspase-1 is activated, it processes interleukin-1β and interleukin-18, activating these cytokines. These are secreted, along with the unprocessed forms, as well as other inflammatory mediators, and this secretion occurs in a manner that is distinct from that of conventional secretion.[3]

A number of mechanisms have been proposed to explain this noncanonical secretion. However, there is an emerging idea that the release of these and other molecules from cells with active caspase-1 is actually due to the breakdown of the plasma membrane by gasdermin D during pyroptosis. This controversy is difficult to resolve one way or the other as the assays for the bioactive molecules and for cell death are different; if only a few cells die by pyroptosis and release potent factors, it can appear that the release is independent of cell death.

INFLAMMASOMES IN DISEASE

We need our inflammasomes to respond to many types of infections, but, when triggered inappropriately, inflammasomes can cause disease. A wide variety of inflammatory diseases are associated with activation of inflammasomes, especially NLRP3. We have already mentioned its role in gout and pseudogout, where inert crystals activate this inflammasome. The ability of crystals of cholesterol to activate the NLRP3 inflammasome also suggests a role in atherosclerosis, although this is not proven. Diets that are rich in fats (so-called "Western diets") promote inflammation by inflammasome activation, and this inflammation can have a positive-feedback effect in adipose tissue to cause obesity. Inflammatory conditions associated with aging also have an NLRP3 component, and strikingly animals lacking NLRP3 do not show common pathologies of aging (although they are more susceptible to a wide range of infections, if exposed).

[3] Canonical secretion involves the transport of the protein into the endoplasmic reticulum, where it is then packaged into secretory vesicles that fuse with the plasma membrane to release the proteins. Interleukin-1β, interleukin-18, and other molecules released following caspase-1 activation are not secreted in this manner.

Familial mutations in NLRP3 have been identified that allow it to be activated spontaneously. Patients with such activating mutations often display recurrent fevers (without infection), pain in their joints and extremities, rashes, severe abdominal pain, and conjunctivitis, all of which can be catastrophic.

Many of the consequences of infection are due to inflammation, and, in turn, are due to activation of inflammasomes. The lethal effects of bacterial sepsis by a pathogenic *E. coli* was shown to depend on NLRC4; for example, interleukin-1, produced by inflammasome activation during infections, can promote damaging inflammation and can act on the vagus nerve to alter feeding behavior, causing anorexia.

But not all inflammasome activation is harmful. Inflammasome activation, in addition to promoting immunity, can also have other beneficial effects for health. For example, interleukin-18 can act to promote healing of the intestine and lungs. The production of this cytokine from macrophages in adipose tissue can prevent muscle wasting, a severe consequence of infection. Therefore, inflammasomes, like many things, present a yin–yang situation: although their activation can cause inflammation and damage, it can also promote immunity and repair.

NLR ACTIVATION AND THE RETURN OF THE JUST-SO STORY

The similarity between APAF1 and the NLRs hopefully has not escaped the reader, and indeed we can include it among the NLRs on the basis of sequence similarity (Fig. 7.13).

This could mean that we can draw an analogy with APAF1 to explain how NLRs are activated by a ligand. If the LRR regions act like the WD region of APAF1, binding to a ligand might cause conformational changes that expose the nucleotide-binding site in the NACHT domain. Binding of a nucleotide would then, as in APAF1, induce further conformational changes that expose an oligomerization domain, and the protein–protein interaction site (e.g., CARD or PyD) would interact with ASC. At present, however, this is all speculation.

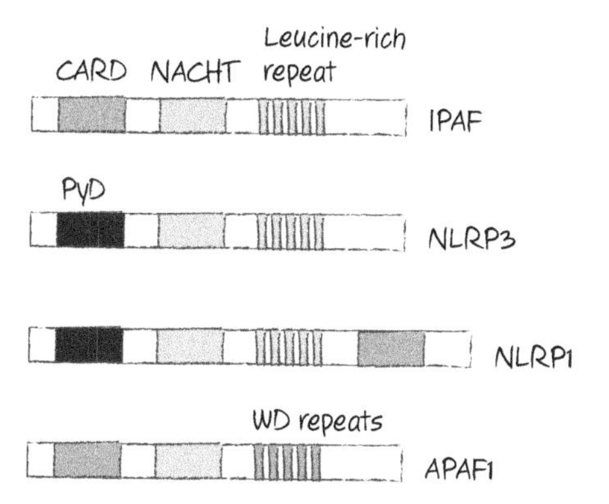

Figure 7.13. APAF1 is in the Nod-like receptor (NLR) family.

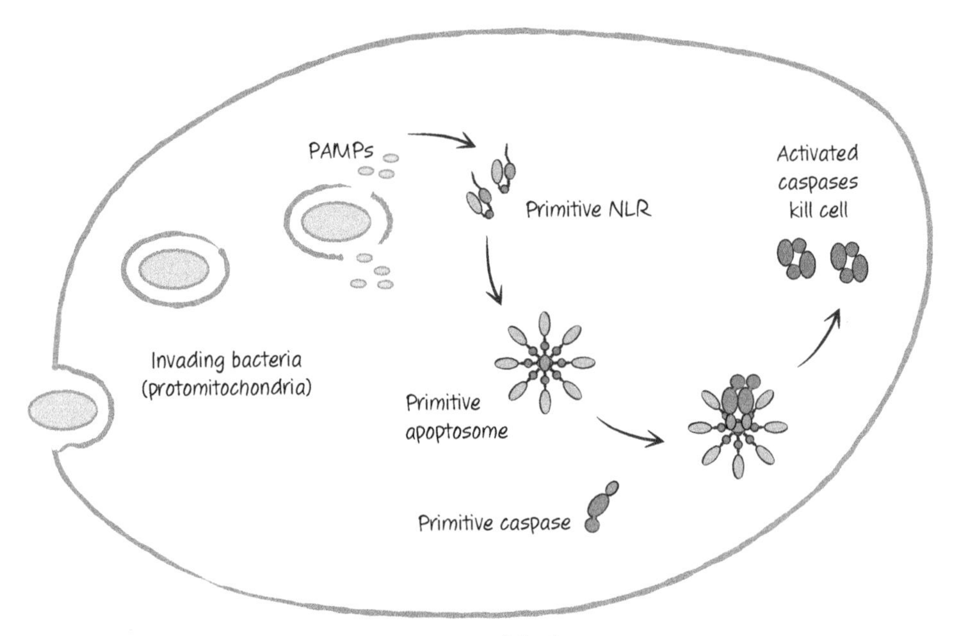

Figure 7.14. Return of the just-so story.

The relationship between APAF1 and the NLRs is tantalizing and brings us back to our just-so story about the evolution of apoptosis. It is easy to imagine that the ancient cell infected by the "mitochondrion to be" had a defense system in place to recognize such infections, using some type of NLR. Perhaps this NLR used bacterial cytochrome c as a kind of PAMP because this protein did not exist in the cell at that time. The recognition of the PAMP could have engaged a cell death mechanism to prevent the spread of the parasite to other cells. In this fantasy, the NLR eventually became APAF1, and the cell death mechanism became the mitochondrial pathway of apoptosis (Fig. 7.14).

THE ACTIVATION OF CASPASE-2

Caspase-2 is a bit of a puzzle. It is the most highly conserved of the caspases throughout the animals and has been implicated in cell death caused by heat shock, cytoskeletal disruption, metabolic perturbation, and, perhaps, DNA damage, among other situations. However, in none of these cases has caspase-2 been shown unambiguously to be required for cell death.

Caspase-2 has a long prodomain containing a CARD and is activated by induced proximity (see Chapter 3). Unlike the other caspases, however, it is a rather poor activator of executioner caspases. Instead, it might require BID to cause cell death by engaging the mitochondrial pathway of apoptosis.

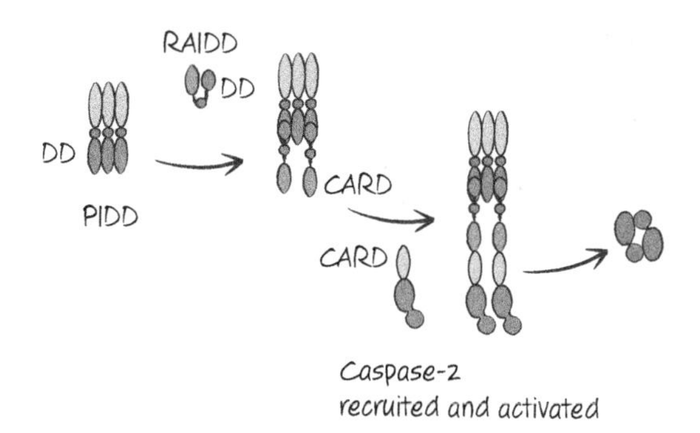

Figure 7.15. The PIDDosome activates caspase-2. DD, death domain.

These considerations raise an intriguing question: Is the primary function of caspase-2 to cause apoptosis? Caspase-2 might have another function in cells that has not been identified. But it does seem to be important. Mice lacking caspase-2 are developmentally normal but age prematurely and are more prone to cancer in at least two model systems. Why this might be is considered in more detail below (and in Chapter 11).

The activation of caspase-2 involves an adapter protein called RAIDD, which binds to caspase-2 by a CARD–CARD interaction. RAIDD does not appear itself to oligomerize but instead might require additional molecules. One such molecule is PIDD. PIDD has a death domain (DD), and it binds to RAIDD via a DD–DD interaction. This is illustrated in Figure 7.15.

The caspase-2 activation platform is called a PIDDosome. The PIDDosome has been resolved at a structural level, which shows it to have multiple PIDD and RAIDD subunits (Fig. 7.16). This is similar to the structure of the CD95-DD–FADD-DD structure involved in the activation of caspase-8 (see Chapter 6).

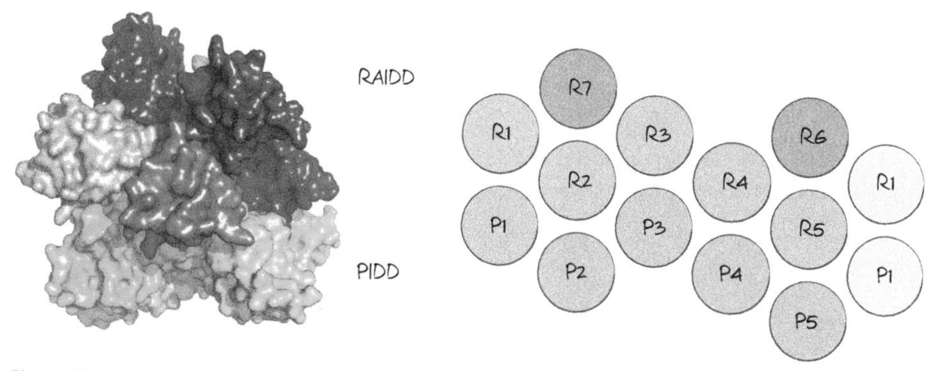

Figure 7.16. PIDDosome structure. The interactions of PIDD (P) and RAIDD (R) DDs are shown on the *right*.

PIDD is an interesting protein that undergoes extensive processing. It has a larger form that does not seem to be involved in activation of caspase-2 but instead activates the transcription factor NF-κB. However, this form of PIDD processes itself to a smaller form that can engage RAIDD. This processing is by an intein mechanism, in which sequences in the protein interact in a manner very similar to that of the function of proteases. This results in autocleavage of the protein to its smaller form. How the relative levels of the different species of PIDD are controlled is not clear.

This dual activity of PIDD is reminiscent of the ability of some of the death receptors to activate caspase-8 (and apoptosis) as well as NF-κB. It is not known whether NF-κB, induced by PIDD, can induce the expression of genes that influence the function of caspase-2, but the idea is intriguing.

Caspase-2 might function to monitor a cell-cycle process. During the cell cycle, the single centrosome in the cell replicates, and these move to opposite poles to orchestrate chromosome separation during mitosis. Only one of these centrosomes is "mature"—that is, it has associated proteins that only appear on the other centrosome after cell division. PIDD is localized to the mature centrosome (Fig. 7.17), and,

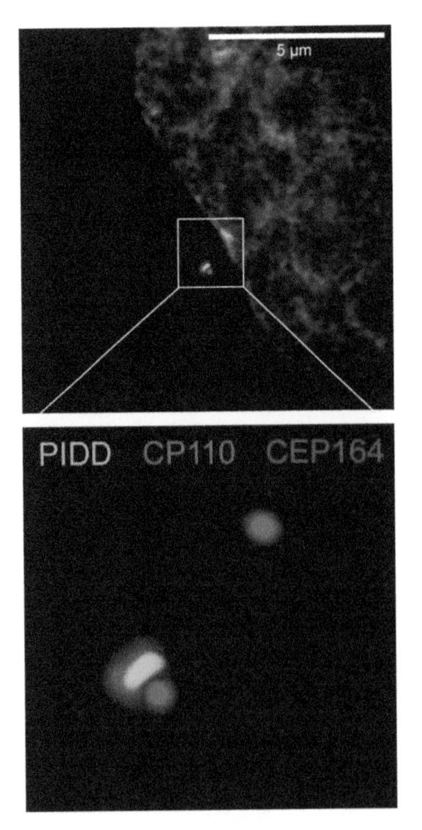

Figure 7.17. Fluorescence micrograph showing how PIDD localizes to mature centrosomes. PIDD (green) associates with the mature centrosome (which also contains CEP164, blue). The nascent (*upper* red spot) and mature centrosomes (*lower* complex of spots) both contain CP110 (red). The *lower* image is a magnified view of the area enclosed by the white square in the *upper* image.

under conditions in which extra mature centrosomes appear (such as failed cell division), caspase-2 is activated. This requires both PIDD and RAIDD (Fig. 7.18). As mentioned above, mice lacking caspase-2 show premature aging and an increased susceptibility to cancer in some model systems. Interestingly, cancers in caspase-2-deficient animals show less chromosomal stability than their caspase-2-sufficient counterparts—that is, they show increased abnormalities in chromosome numbers. These effects would be explained if caspase-2 activation resulted in apoptosis of cells with abnormal centrosomes. However, it seems that caspase-2 does not necessarily cause cell death in such cells, but can instead cause a cell-cycle arrest. This is because caspase-2 can stabilize the tumor-suppressor protein p53, which in turn can cause either apoptosis or cell-cycle arrest. We will return to this function of caspase-2 in Chapter 11, in our discussion of the regulation of p53.

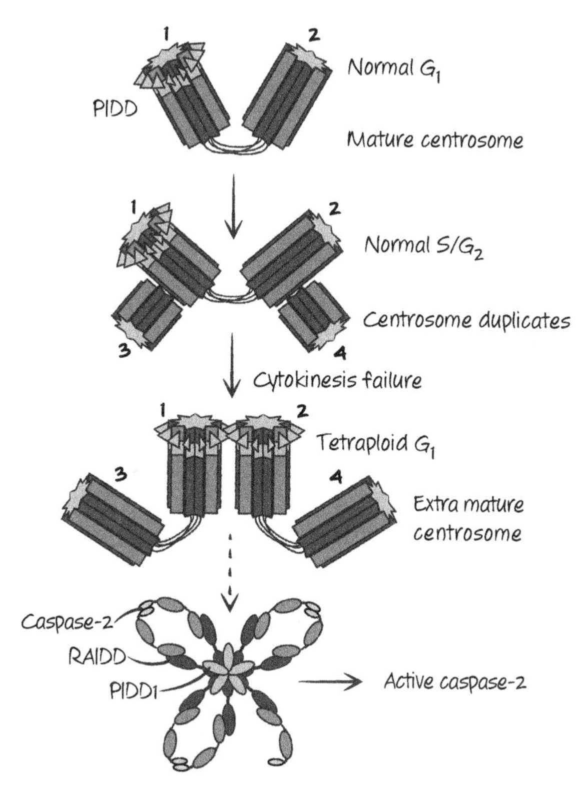

Figure 7.18. Aberrant extra mature centrosomes activate caspase-2. During the S/G_2 phases of the cell cycle, the centrosome duplicates, but PIDD remains associated with only the mature centrosome. If cytokinesis fails, leading to the formation of a tetraploid cell in the following G_1 phase, two mature centrosomes appear in the cell, leading to PIDD–PIDD interaction, recruitment of RAIDD, and activation of caspase-2.

During the postnatal development of the liver, hepatocytes can double (or even quadruple) their chromosome numbers because of a failure in cytokinesis (the final step in mitosis). This effect is greatly enhanced in animals lacking PIDD or caspase-2. Again, this appears to be caused by the ability of caspase-2 to activate p53, which prevents further proliferation of hepatocytes with failed cytokinesis (and extra mature centrosomes).

C H A P T E R 8

Nonapoptotic Cell Death Pathways

DIFFERENT WAYS TO DIE

So far, we have mostly focused on apoptosis, also known as type I cell death.[1] Along the way, we touched on other forms of cell death, in particular those most related to apoptosis—pyroptosis, caused by the action of caspase-1, -4, -5, or -11, and caspase-independent cell death, caused by mitochondrial outer membrane permeabilization (MOMP) when subsequent caspase activation is blocked. Two other major classes of cell death mainly concern us here: autophagic cell death (also called type II cell death) and necrosis (also called type III cell death). There is another type of cell death that may or may not fall into any of these categories—termed "mitotic catastrophe"—that we also consider in this chapter. Note that, although one pathway or type of cell death can appear to dominate in a particular setting, this might only be because it happens to be faster.

ACCIDENTS WILL HAPPEN

Cells are highly complex, and any of a wide variety of accidental events can compromise them, leading to cell death. There have been attempts to classify cell death on the basis of whether it is "accidental" or "active." But where do we draw the line? If a cell is ruptured, it dies immediately, and this death appears distinct from apoptosis. But a milder insult can result in the cell engaging apoptosis or another cell death pathway. Is this "intentional"? In the end, the distinction can be too artificial to be useful. Furthermore, below we consider necrosis, a form of death that can be "accidental" but, as we will see, is not always so.

---

[1] As mentioned in Chapter 6, it is unfortunate that "type I" can refer to a form of cell death as well as a way in which cells respond to death receptor signaling, leading to apoptosis. In this chapter, all reference to "type I, II, or III" is to a form of cell death, not to the death receptor signaling response, per se. It is easy to get confused!

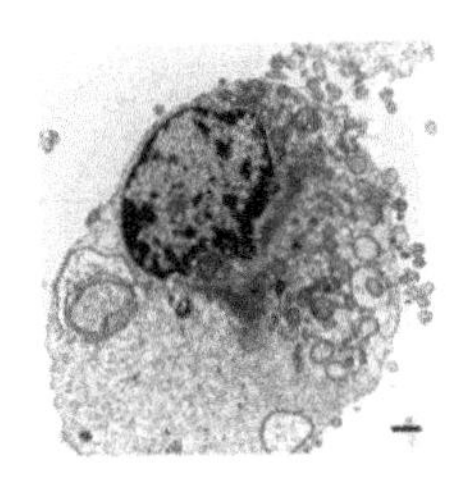

Figure 8.1. A fibroblast cell dying by necrosis.

NECROSIS (TYPE III CELL DEATH)

During necrosis, an influx of water causes organelles to enlarge and the cell swells, ultimately lysing. Unlike cells undergoing apoptosis, chromatin does not condense and the plasma membrane does not form blebs (Fig. 8.1).

Much of the energy in a cell is devoted to powering ion pumps that sustain gradients across the plasma membrane. If ATP levels become depleted because of a lack of nutrients or the actions of toxins, the cell can swell and rupture, undergoing necrotic cell death. But this and more direct forms of damage are not the only way necrosis comes about.

One way that necrosis can occur is as a consequence of apoptosis. Normally, if a cell dies by apoptosis, it is rapidly removed and degraded by phagocytic cells (discussed in detail in Chapter 9). If for any reason this clearance does not occur (e.g., in tissue culture or because of defects in the clearance mechanisms), the integrity of the plasma membrane of the apoptotic cell is ultimately lost and necrosis ensues. This is referred to as secondary necrosis and is a confounding variable in many studies of cell death. Cells that undergo secondary necrosis have features of both apoptotic cells (e.g., condensed chromatin) and necrotic cells (e.g., a ruptured plasma membrane) (Fig. 8.2). In many cases, necrosis proceeds independently of apoptosis, or the two forms of cell death can be mixed and difficult to tease apart. As we will see, the key to understanding what is happening is the molecular mechanism underlying *how* the cell died.

As it turns out, however, secondary necrosis of apoptotic cells can also be an active process. One of the gasdermins (see Chapter 7), DNFA5 (also called gasdermin

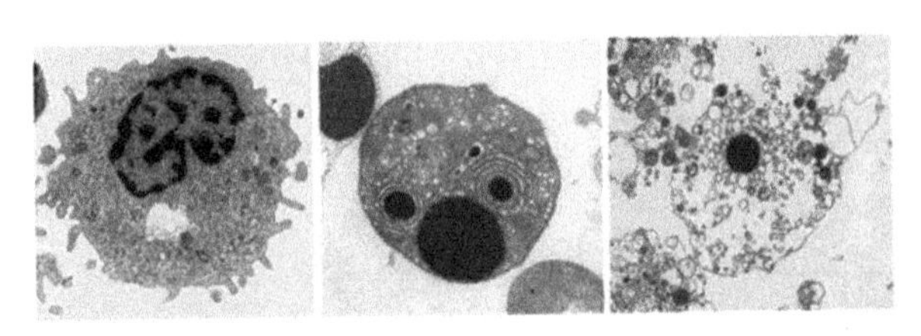

Figure 8.2. Apoptosis and secondary necrosis. Living cell (*left*), apoptotic cell (*center*), and secondary necrosis (*right*). Note that, in secondary necrosis, the nucleus is condensed, as in apoptosis.

E), is cleaved and activated by active caspase-3 or caspase-7, much as gasdermin D is activated by inflammatory caspases. Like active gasdermin D, active DNFA5 targets the plasma membrane, promoting secondary necrosis. Apoptotic cells that lack DNFA5 remain intact for some time, often breaking into small, membrane-bound pieces (although again if these are not cleared, the plasma membrane integrity will eventually be lost). In contrast, apoptotic cells that contain DNFA5 rapidly undergo secondary necrosis. Nevertheless, as in the situation in which clearance does not occur, these dying cells show features of both apoptosis and necrosis.

NECROSIS CAN BE AN ACTIVE PROCESS

In the above example of secondary necrosis, the death of a cell undergoing necrosis can be an active process. That is, a cell can participate in its death by virtue of molecularly controlled processes. Several forms of regulated (or "programmed") necrosis exist, and we discuss those that are best understood in this chapter. In considering these, it might be helpful to think about the concepts of suicide and sabotage we discussed in Chapter 1; it could well be that some of these processes did not evolve for the "purpose" of killing the cell but rather represent a normal cellular activity gone awry. In at least one case, however, regulated necrosis appears to have arisen as a bona fide cell suicide. This is the process of necroptosis.

DEATH RECEPTORS CAN CAUSE NECROPTOSIS

As we saw in Chapter 6, death receptors of the tumor necrosis factor receptor (TNFR) family can trigger an apoptotic pathway. However, in some cases, they can also induce necrosis, although this form of cell death has some features of type II (autophagic) cell death (discussed later in this chapter). The necrotic form of cell death induced by death receptors is referred to as necroptosis.[2]

When TNF engages its death receptor (TNFR1), this activates a complex of TRADD–TRAF–RIPK1 that dissociates from the receptor and recruits the adapter FADD and caspase-8 (see Chapter 6), and apoptosis proceeds. However, in many cells, if caspase-8 is inhibited, death ensues nevertheless. This death is necrotic and requires RIPK1. Unlike the activation of NF-κB by this receptor, wherein RIPK1 performs a scaffolding function (see Chapter 6), necroptosis induced through TNFR1 requires the serine/threonine (Ser/Thr) kinase activity of RIPK1 (Fig. 8.3).

Inhibitors called necrostatins block the kinase activity of RIPK1 and prevent necrosis induced by TNF. They can also reduce cell death caused by ischemia–reperfusion injury (discussed below) and pathological cell death in other settings, although,

[2] The term necroptosis is admittedly unwieldy, but useful for searching the literature for this particular form of cell death.

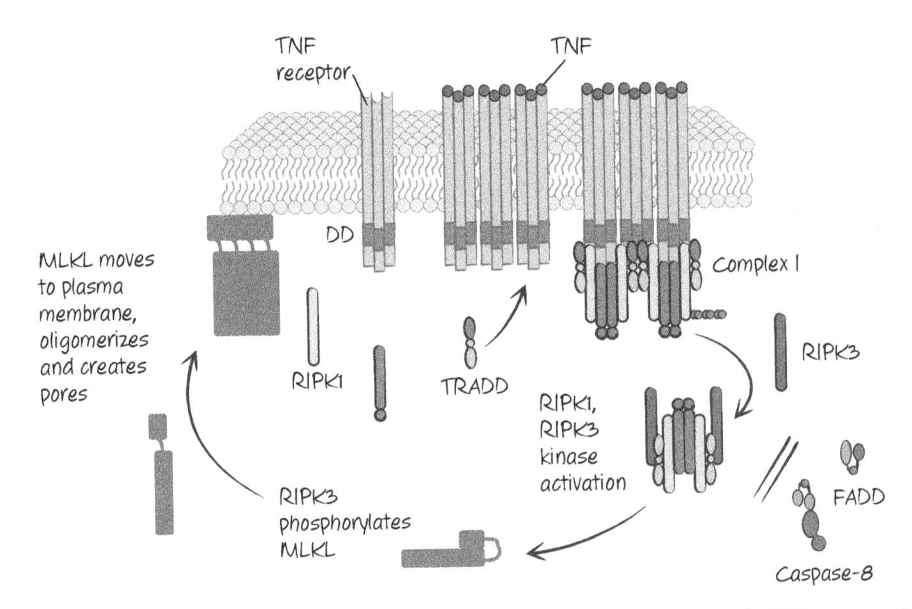

Figure 8.3. Tumor necrosis factor receptor (TNFR) signaling engages necroptosis if FADD–caspase-8–FLIP activity is blocked or overwhelmed.

in these cases, whether this is caused by TNF, another death ligand, or other signals that engage RIPK1 is not known.

RIPK1 recruits and activates another Ser/Thr-protein kinase, RIPK3, that is also required for TNF-induced necrosis. RIPK1 does not phosphorylate RIPK3, but instead the active RIPK1 binds to RIPK3, and this binding activates it. Cells lacking RIPK1 or RIPK3 do not die by TNF-induced necroptosis, and, if caspase-8-induced apoptosis is also inhibited, the cells survive TNF treatment.

OTHER STIMULI CAN INDUCE NECROPTOSIS

As it turns out, it is the activation of RIPK3 that is important for necroptosis, and there are ways in which RIPK3 can be activated independently of a requirement for RIPK1 (although, see below). Both RIPK1 and RIPK3 contain a small region called the RIP-homology interaction motif (RHIM) that mediates the interaction of RIPK3 with RIPK1 and other proteins. One of these is TIR-domain-containing adapter-inducing interferon (TRIF), an adapter molecule engaged by some Toll-like receptors (TLRs). Another is an intracellular sensor of viral nucleic acids called Z-DNA-binding protein 1 (ZBP1, also called DAI) (Fig. 8.4). By the binding of these proteins to RIPK3 through RHIM–RHIM interactions, RIPK3 oligomerizes and activates its kinase activity to promote necroptosis. Interferons can also induce the activation of RIPK3, although the mechanism is less clear.

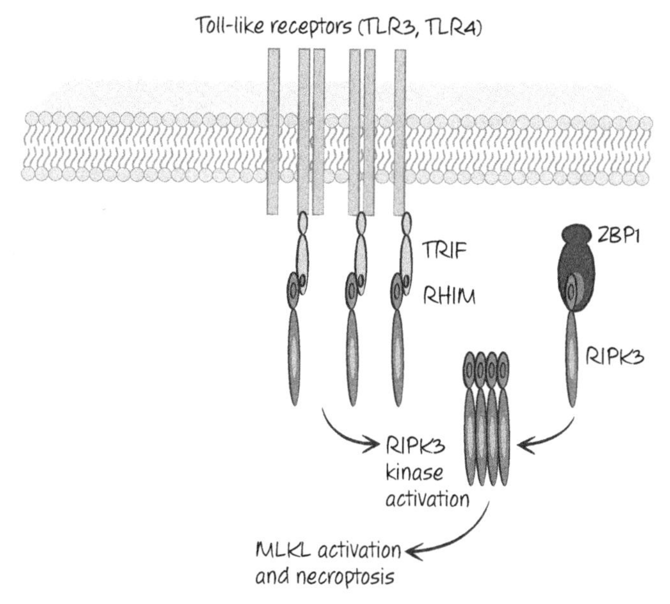

Figure 8.4. Other pathways can activate RIPK3 and necroptosis. Some Toll-like receptors (e.g., TLR3, TLR4) engage the adapter protein TRIF when they are activated by their ligands. TRIF contains a RIP-homology interaction motif (RHIM) that interacts with the RHIM of RIPK3 directly, activating the kinase. ZBP1 (DAI) is an intracellular sensor of viral nucleic acids, and it too can directly activate RIPK3 via RHIM–RHIM interactions. The figure shows activation of necroptosis in the absence of RIPK1. When RIPK1 is present, the interactions are more complex. The function of MLKL is discussed below.

MLKL IS THE "WEAPON" OF NECROPTOSIS

The crucial substrate for RIPK3 in necroptosis is a pseudokinase called MLKL (for "mixed-lineage kinase domain-like protein"). A pseudokinase structurally resembles kinases but lacks kinase activity. When MLKL is phosphorylated by RIPK3, it undergoes a conformational change (essentially identical to that of an activated kinase) that exposes a unique region of the protein, located near its amino terminus (Fig. 8.5). This region comprises a "bundle" and a "brace." The brace region can self-associate, and this functions to oligomerize the activated protein.

One way that the function of the amino terminus has been studied is by the use of artificial dimerization methods (further discussed in Chapter 12). If only the amino-terminal bundle of MLKL is fused to a so-called dimerization domain, a dimeric drug that binds this domain will force the bundle into dimers, and this is sufficient to kill the cell (Fig. 8.6).

When activated, MLKL oligomerizes, and the amino terminus of the protein interacts with the plasma membrane, disrupting it. As a result, the cell dies by necroptosis. But, along the way, something interesting happens. As in apoptosis, this cell

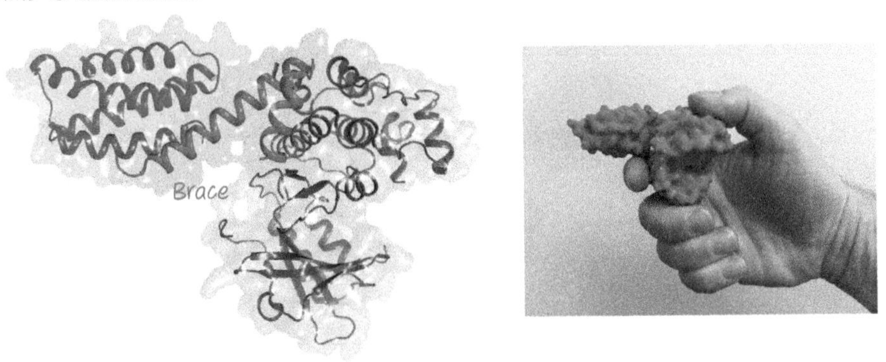

Figure 8.5. Structure of the mixed-lineage kinase domain-like protein MLKL. (*Left*) The amino-terminal bundle, brace, and pseudokinase domain are shown. (*Right*) MLKL is the weapon of necroptosis.

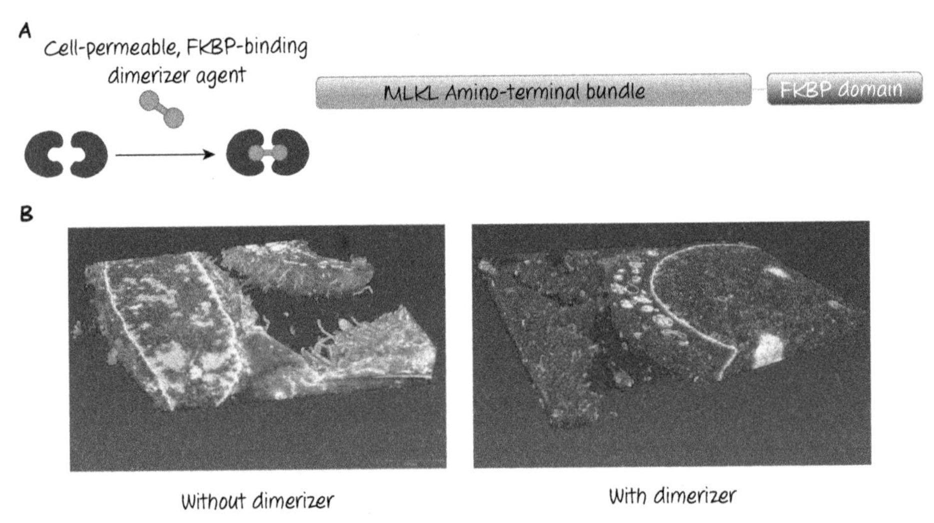

Figure 8.6. Dimerizing the amino-terminal bundle of the mixed-lineage kinase domain-like protein MLKL. (*A*) Scheme of a chimeric protein that can be expressed in cells. The amino-terminal bundle of MLKL is fused to a domain (FKBP) that can be bound by a cell-permeable dimeric drug that can connect two chimeric monomers. (*B*) A three-dimensional electron micrograph of cells expressing the construct in *A*. Without the dimerizer agent, the plasma membrane remains intact (*left*), but, upon addition of the dimerizer (*right*), the plasma membrane is rapidly destroyed, although intracellular membranes appear unaffected.

death process induces the exposure of phosphatidylserine on the outer leaflet of the plasma membrane, before the loss of plasma membrane integrity. Neither the lipid scramblase involved in apoptosis (Xkr8; see Chapter 2) nor the calcium-ion-responsive scramblase (TMEM16F) are involved in this effect, and it is possible that it is active MLKL, itself, that causes the externalization of phosphatidylserine.

It is likely, although not definitively known, that the activation of MLKL is the final step in necroptosis—that is, that MLKL is the "weapon" that destroys the plasma membrane (Fig. 8.5). However, MLKL might have one or more additional targets that actually do the killing. Some studies have suggested that MLKL interacts with and opens ion channels in the cell, and these are what ultimately kill the cell. It remains possible, however, that this is a side effect of the action of MLKL as attempts to prevent death by reducing ions in the extracellular milieu (thus limiting their influx) appear to be cell-type-specific, and at best only partially work. Furthermore, although there is an influx of calcium ions upon MLKL activation, this is dispensable for the externalization of phosphatidylserine that precedes the loss of plasma membrane integrity.

RIPK3 is the only kinase known to activate MLKL (the pathways are shown in Figs. 8.3 and 8.4). But RIPK3 (and RIPK1) have other effects in the cell, including the induction of inflammatory mediators. It is therefore difficult to define a role for necroptosis in a biological process (or disease) based on roles for RIPK1 or RIPK3. MLKL involvement is perhaps a more definitive measure, and the field has moved to define necroptosis as "cell death that is dependent on MLKL." Should we find conditions in which MLKL is activated independently of RIPK3, we will most likely consider the resulting cell death necroptosis.

Necroptosis appears to be a fairly recent evolutionary "invention," found only in some vertebrates. The key components, RIPK1, RIPK3, and MLKL, are found in bony-jawed fishes, amphibians, and reptiles, but RIPK3 is absent in birds, and MLKL is present in some fish but not others. Although most mammals have these components, marsupials lack RIPK3 and MLKL, and carnivores lack MLKL. Why some vertebrates have the necroptosis machinery and others lack it is a mystery.

Why do cells have this backup mechanism for cell death? One likely answer relates to viruses. As we mentioned, viruses deploy strategies to inhibit caspase-8 as a way to block apoptosis that is induced in host cells by cells of the immune system. When this occurs, the activation of MLKL and the destruction of the cell by necroptosis is a good alternative to combat the infection.

CASPASE-8 AND FLIP INHIBIT NECROPTOSIS

Inhibition of caspase-8 allows TNF to induce necroptosis rather than apoptosis, and therefore caspase-8 would appear to inhibit this pathway. Indeed, caspase-8 cleaves RIPK1, preventing its activity, and this appears to be how caspase-8 blocks necrosis. But, if activation of caspase-8 commits the cell to die by apoptosis anyway, can inhibition of necroptosis preserve the cell? When TNF binds to TNFR1, it induces the activation of nuclear factor-κB (NF-κB), and this transcription factor induces the expression of FLIP (FLICE-like inhibitory protein), which, as we have seen, is a protein that is related to the caspases but lacks a catalytic site. Like caspase-8, FLIP can be recruited to the death-inducing signaling complex (DISC) in the death receptor pathway. At the DISC, FLIP forms a complex with caspase-8 that is proteolytically active but unstable because caspase-8 is not cleaved; consequently, FLIP prevents apoptosis. (As we discussed in Chapter 6, FLIP can prevent apoptosis by not allowing the formation of oligomers of caspase-8.) However, this complex also prevents necroptosis. The caspase-8–FLIP dimer cleaves RIPK1 and RIPK3 in the complex and thereby ensures that the cell undergoes neither apoptosis nor necrosis and therefore survives (Fig. 8.7). Caspase-8–FLIP also cleaves CYLD, the deubiquitinase (DUB) that removes ubiquitin from RIPK1 (discussed in Chapter 6), and this cleavage might also function to restrict necroptosis.

Mice lacking caspase-8, FADD, or FLIP die during embryonic development because of extensive cell death in the endothelium (discussed in more detail in

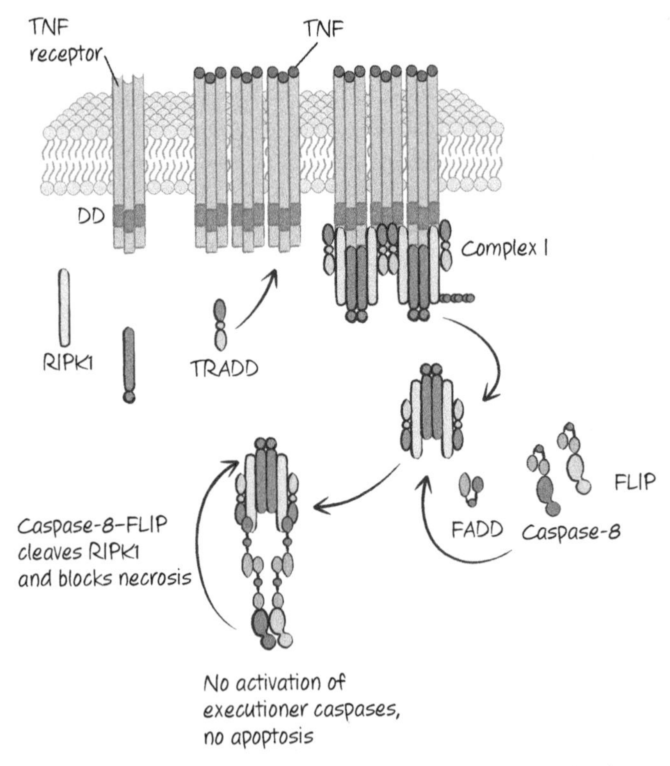

Figure 8.7. The FADD–caspase-8–FLIP complex inhibits necroptosis. RIPK1 binds to FADD, which in turn is bound to caspase-8–FLIP. The catalytically active caspase-8–FLIP cleaves RIPK1 to prevent necroptosis.

Chapter 10). Strikingly, if mice lacking FADD or caspase-8 also lack RIPK3 or MLKL, they survive. Furthermore, mice lacking FLIP can also survive if both FADD (promoting apoptosis) and either RIPK3 or MLKL are removed. Formally, these data allow the conclusion that the embryonic lethality caused by loss of caspase-8 or FADD is dependent on RIPK3 and MLKL. This makes sense in the context of the pathways we discussed. In the case of the embryonic lethality caused by loss of FLIP, we can say that it is dependent on RIPK3 and MLKL, and on FADD (and probably caspase-8).[3]

Again, this makes sense.

In each case, the mice develop and are born normally, but, as they age, they display the symptoms of acute lymphoproliferative syndrome (ALPS; see Chapter 6), with massively enlarged lymphoid organs. This disease is seen in mice and humans with defects in CD95 or its ligand CD95-L. We can conclude from these observations that the function of CD95 and CD95-L in controlling ALPS is likely to be dependent on cell death signaling by the receptor.

TWO FACES OF RIPK1

Mice that lack RIPK1 do not die during development but instead die shortly after birth. Deletion of RIPK3 or MLKL does not prevent this perinatal lethality, but animals lacking RIPK1, either FADD or caspase-8, and either RIPK3 or MLKL survive well into adulthood, until they succumb to ALPS.

These findings tell us several important things, all of which have been confirmed in experiments with cells: RIPK3 can be activated (and can engage MLKL) without a requirement for RIPK1 (see Fig. 8.8); RIPK1 can inhibit such RIPK3 activation; and RIPK1 also functions to inhibit apoptosis mediated by FADD and caspase-8. When RIPK1 is absent, cells become sensitive to both caspase-8-dependent apoptosis and some triggers of necroptosis.

It is likely that RIPK1 inhibits necroptosis in at least two ways. First, it recruits FADD through DD–DD interactions, which recruits caspase-8 and FLIP (as we have seen). When RIPK1 is absent, necroptosis induced by signals that directly engage RIPK3 (such as TLR, ZBP1, or interferon signaling) proceeds even if caspase activity is not prevented. Second, when RIPK1 is in its inactive conformation (e.g., upon addition of necrostatins), the RIPK1 appears to block RIPK3 activation. Again, when RIPK1 is absent, necrostatins have no effect on necroptosis induced by direct activation of RIPK3.

RIPK1 also prevents the activation of apoptosis, especially by the ligation of death receptors. Most likely this is because RIPK1 functions in the activation of NF-κB (as we have seen), which promotes expression of anti-apoptotic proteins, such as FLIP.

But RIPK1 can also *promote* both necroptosis and apoptosis in some settings. As described above, necroptosis induced by the ligation of death receptors requires

[3] This was not tested for the technical reason that FLIP and caspase-8 are very close to each other in the genome, and therefore the double-null allele cannot be generated by simple crosses.

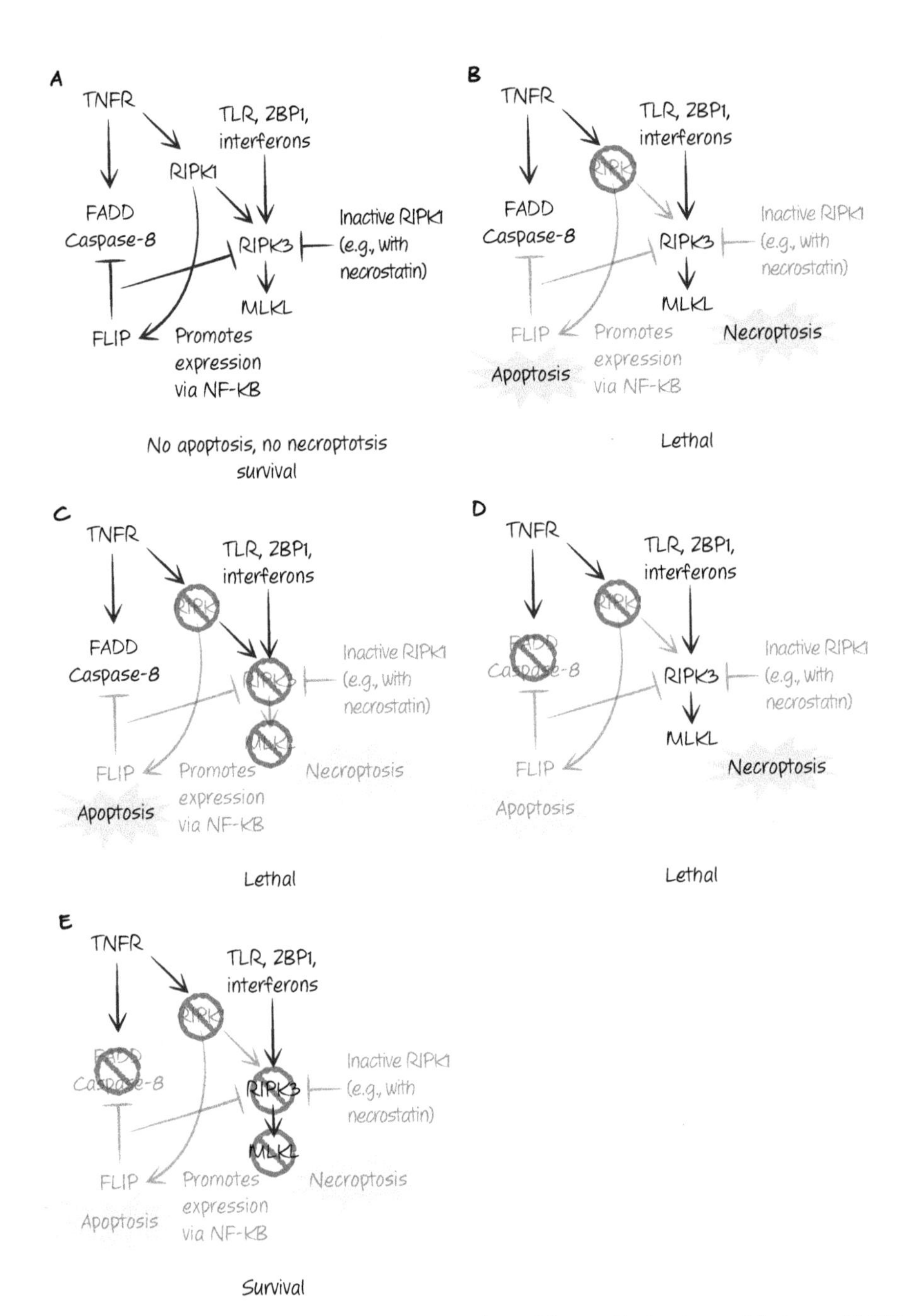

Figure 8.8. The two faces of RIPK1. (*A*) The complex inhibitory "balancing act" between FADD–caspase-8–FLIP and RIPK3–MLKL interactions depends on RIPK1 to prevent both apoptosis and necroptosis, resulting in normal development. (*B*) When RIPK1 is removed, both apoptosis and necroptosis occur, and signals promote lethality. (*C*) Removal of either RIPK3 or MLKL in RIPK1-deficient animals (or cells) does not restore survival in RIPK1-deficient animals (or cells), as apoptosis still occurs. (*D*) Removal of either FADD or caspase-8 in RIPK1-deficient animals does not restore survival, as necroptosis still occurs. (*E*) Removal of either RIPK3 or MLKL *plus* either caspase-8 or FADD allows RIPK1-deficient animals to survive.

RIPK1 to bind and activate RIPK3. When the kinase activity of RIPK1 is inhibited (e.g., by necrostatin), it prevents not only death-receptor-induced necroptosis, but also necroptosis induced by signals that directly engage RIPK3 (such as TLR, ZBP1, or interferon signaling). This inhibitory activity of RIPK1 has confused the interpretation of effects of RIPK1 inhibitors, leading to conclusions that RIPK1 is always required for necroptosis.[4] The studies with RIPK1-deficient animals and cells show that this is often not the case.

Because RIPK1 also binds to FADD (to recruit and activate caspase-8–FLIP heterodimers), RIPK1 can also promote apoptosis when FLIP is limiting. Interestingly, RIPK3 that is kinase inactive (owing to use of inhibitors or mutations) can activate this pro-apoptotic effect of caspase-8. Mice with a kinase-inactivating mutation in RIPK3 die during embryogenesis at the same time as do FADD- or caspase-8-deficient mice. However, removing caspase-8 protects these RIPK3 mutant animals, which are born and develop to adulthood normally (but, as with animals deficient in caspase-8 and RIPK3, these mice succumb to ALPS).

NECROPTOSIS AND DISEASE

There is compelling evidence that necroptosis plays roles in many diseases, but this is complicated. All of the proteins involved in necroptosis (RIPK1, RIPK3, and MLKL) have additional signaling features that can induce inflammation by the expression of cytokines and other mediators. Therefore, any interpretation that "necroptosis causes disease" must be treated with care.

Conditional deletion of caspase-8 in the intestinal epithelium or in the skin causes massive inflammation that is completely prevented by removing RIPK3, and therefore might be due to necroptosis in these tissues (with the caveat above). Similarly, conditional deletion of caspase-8 in endothelium causes embryonic lethality, which is again dependent on RIPK3 and MLKL. In contrast, several tissues and cell types are resistant to the effects of deletion of caspase-8, including heart, skeletal muscle, liver, and neurons.

Although the liver (in mice) appears to be resistant to necroptosis, this is not the case when animals are fed a high-fat diet. Animals with conditional deletion of caspase-8 in the liver become sensitive to liver damage and inflammation when fed on such a diet, and this is prevented by genetic removal of *RIPK3*. One effect of the diet is to induce expression of MLKL in the liver, thus sensitizing it for necroptosis. Another consequence of a high-fat diet is atherosclerosis, and animals that are genetically predisposed to this disease are protected by deletion of *RIPK3*.

[4] There is another problem with a commonly used RIPK1 inhibitor, necrostatin-1. It turns out that this inhibitor can have effects that are independent of RIPK1 (e.g., it can inhibit ferroptosis). This has led to gross misinterpretations in the literature. A more specific inhibitor of RIPK1 is necrostatin-1s. Thus, care should be taken in considering conclusions that are based only on the use of necrostatin-1.

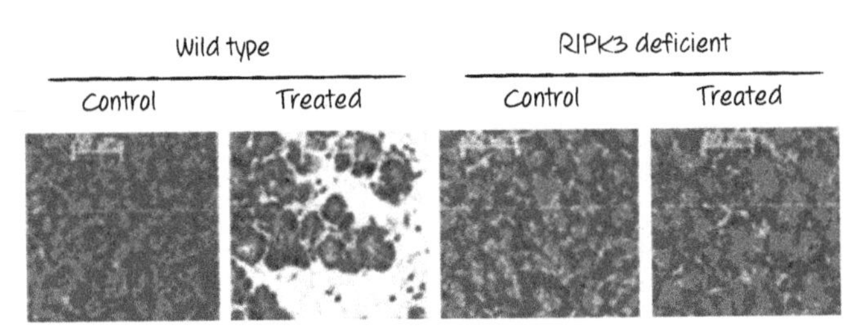

Figure 8.9. RIPK3 and pancreatic injury. Wild-type (*left*) or RIPK3-deficient (*right*) mice were treated with an agent that induces necrotic injury in the pancreas. Animals without RIPK3 were protected. This protection has also been described in animals lacking the mixed-lineage kinase-like protein MLKL.

Other liver diseases also appear to depend on RIPK3, including ethanol- and acetaminophen-induced liver damage (however, given the resistance of liver to necroptosis in conventionally fed animals, it is possible that this is not due to necroptosis).

Other models of disease also appear to be dependent (at least in part) on RIPK3. These include necrotizing pancreatitis (Fig. 8.9), retinal detachment, and acute peritonitis. Intriguingly, in two mouse models of amyotrophic lateral sclerosis (ALS), animals with a kinase-inactive RIPK1 (which inhibits necroptosis) were strikingly protected.

Necroptosis can also be protective. As we noted, most evidence suggests that necroptosis evolved as a mechanism of defense against viruses, and the death of an infected cell can limit disease. Animals lacking RIPK3 are sensitive to several types of viruses, including vaccinia, influenza, and West Nile virus. Interestingly, mice lacking MLKL do not show this sensitivity to influenza, but mice lacking both FADD and MLKL are as sensitive (or more so) as animals lacking RIPK3. At least in this case, the role for RIPK3 in influenza infection might be both in apoptosis and in necroptosis. In the case of West Nile virus, however, the protective role of RIPK3 ablation does not arise through the inhibition of cell death but, rather, because RIPK3 promotes harmful inflammation. Again, it is important that care is used when concluding that necroptosis is involved in a pathology based on effects of removing RIPK3.

Antibodies that detect the active, phosphorylated form of human MLKL are likely to be useful as indicators of where necroptosis occurs in diseased human tissues. These studies are under way.

PARP AND NECROSIS

The repair of damaged DNA is energetically costly, and, if the costs of repair exceed the resources available to the cell (ATP and NADH), necrosis can result. In particular, the DNA repair enzyme poly(ADP-ribose) polymerase 1 (PARP1) engages processes that consume a large amount of NAD^+, which the enzyme uses to generate the ADP-ribose needed for the repair process. But NAD^+ is also crucial for energy metabolism, and many metabolic reactions are dependent on a form of this molecule

(NAD^+, NADH, $NADP^+$, and NADPH). PARP can therefore promote necrotic cell death in some circumstances by using up NAD^+, especially if nutrients are limited.

DNA damage can also trigger apoptosis through signaling mechanisms, including the action of p53 (see Chapter 11). However, if DNA damage is very extensive, this can result in a form of necrosis[5] that can be blocked by pharmacological inhibitors of PARP or additional sources of NADH. For example, the addition of nicotinamide can reduce necrosis induced by high-dose radiation.[6]

Reactive oxygen species (ROS) can damage DNA, and high levels of these can also cause cell death that depends on PARP. Mice lacking PARP develop normally but are resistant to neuronal damage induced by hydrogen peroxide, one source of reactive oxygen. We will see below other ways in which ROS that lead to necrosis can be generated.

EXCITOTOXICITY

Neurons can undergo active necrosis in response to high levels of glutamate, which functions as a neurotransmitter in the brain. This can occur as a consequence of ischemic injury (i.e., stroke) in the brain (discussed in detail below). Glutamate-induced neuronal necrosis is often referred to as excitotoxicity or excitotoxic death.

When neurons are exposed to high levels of glutamate, this causes an influx of calcium ions into the cells. This, in turn, activates a complex enzyme, NADPH oxidase, that produces ROS. Originally, NADPH oxidase was said to occur only in neutrophils, which use it to destroy bacteria and other pathogens. But we now realize that many cells in the body express this or a related form of this enzyme, which is more simply referred to as Nox.

Nox produces ROS by moving an electron from NADPH to oxygen to make superoxide. But, unlike the electron-transport chain of mitochondria, this does not generate energy but instead produces protons. These must be transported out of the cell, which generates a high charge difference across the plasma membrane. This charge is seen as excitation in glutamate-exposed neurons. As a result of the superoxide, PARP is activated, and NAD^+ is depleted (Fig. 8.10). The loss of energy in the cells promotes their death by necrosis. Inhibitors of Nox prevent both PARP activation and death by excitotoxicity.

ISCHEMIA–REPERFUSION INJURY

Ischemic injury occurs when the blood supply to a tissue, such as the heart or brain, is disrupted. Not surprisingly, deprivation of oxygen and nutrients can cause necrosis.

[5] It can be confusing to suggest that apoptosis is superseded by necrosis under some circumstances, and, in real life, the effects can be mixed. However, as we noted, the mechanisms are the key, and blocking one mechanism (e.g., apoptosis) will not necessarily block the other (e.g., necrosis).

[6] Nicotinamide has multiple roles in cells, and it might be simplistic to conclude that its effect on cell death is only as a precursor for NADH and NADPH.

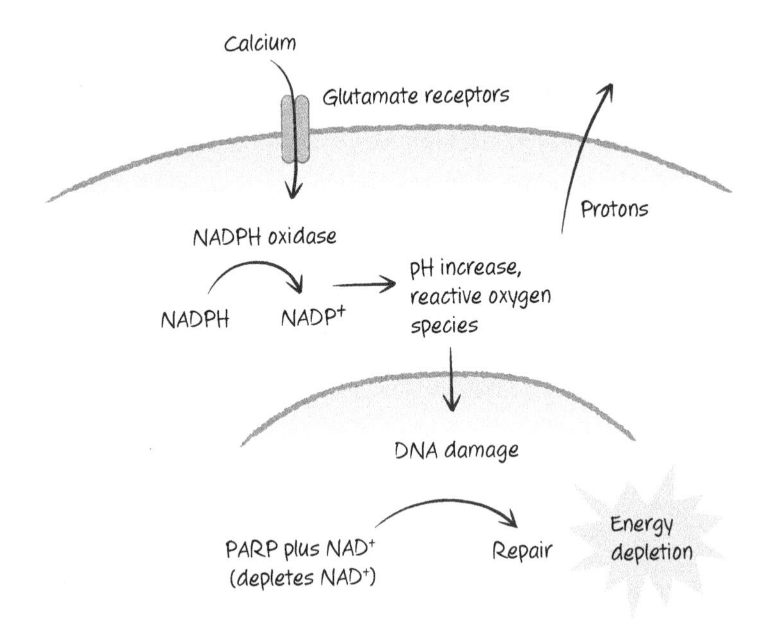

Figure 8.10. One view of excitotoxicity. Calcium influx triggers NADPH oxidase, which produces reactive oxygen species (damaging DNA and inducing the poly(ADP-ribose) polymerase PARP) and hydrogen ions (creating plasma membrane charge). Necrosis occurs as a consequence of energy depletion.

However, when the blood supply is restored ("reperfusion"), this often causes a wave of cell death that can have catastrophic consequences. In the heart, for example, the extent of cell death is directly linked to remodeling events that, over time, cause heart enlargement, accompanied by thinning of the muscle walls and ultimately heart failure. This is why heart attacks so often predispose individuals to subsequent heart attacks. Controlling the extent of damage from ischemia–reperfusion injury is a major therapeutic goal.

Why does ischemia–reperfusion cause so much damage? Part of the answer relates to changes in the levels of potassium ions in the cells. Following reperfusion, potassium channels open in the plasma membrane, causing a drop in intracellular potassium that triggers a range of changes, including an increase in intracellular calcium ions. As we have seen, this can activate NADPH oxidase.[7] In addition, the increase in calcium can also activate the protease calpain, which at high levels can cause necrosis (Fig. 8.11). Animals that are treated with inhibitors of calpain, or that lack this enzyme, show reduced damage as a result of ischemia–reperfusion injury.

[7] This is Nox1, a relative of the Nox that we have been discussing.

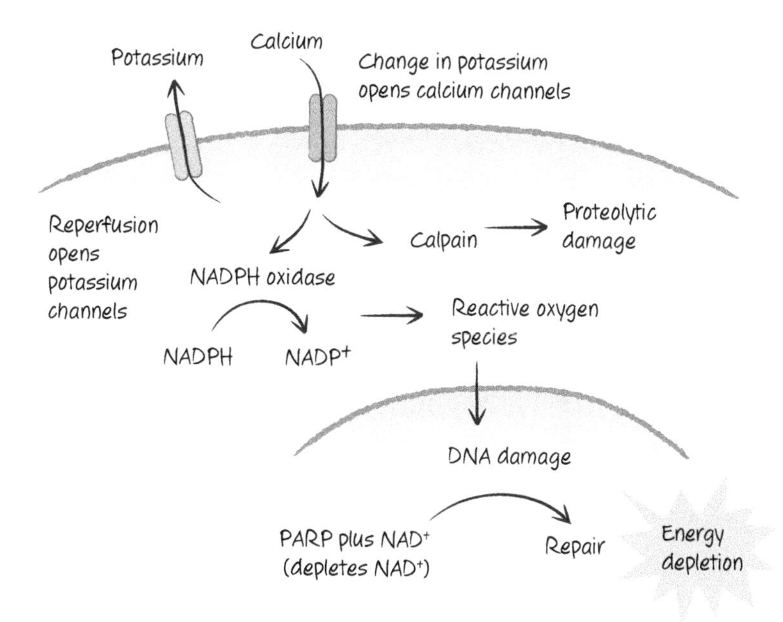

Figure 8.11. One view of ischemia–reperfusion injury. Reperfusion induces an opening in potassium channels, and, in turn, the change in potassium opens calcium channels. The influx of calcium triggers NADPH oxidase, and its consequences, and activates calpain, a protease. Death can be a consequence of energy depletion, calpain action, or both. Other forms of cell death contribute to this form of injury, including apoptosis and probably necroptosis and ferroptosis (see below).

Mitochondria are important in apoptosis (see Chapters 4 and 5), but they can also have an important role in necrosis. As we discussed in Chapter 4, mitochondria can respond to high levels of calcium, ROS, and other signals by opening a channel in the inner membrane, the permeability transition pore (PTP). This causes the mitochondrial matrix of the organelle to swell, eventually rupturing its membranes, resulting in the so-called mitochondrial permeability transition (MPT) (Fig. 8.12). One protein that we saw to be clearly involved is the peptidylproline isomerase cyclophilin D. Mice lacking cyclophilin D are developmentally normal, but mitochondria from these mice show defective PTPs.[8] These mice are strikingly resistant to ischemia–reperfusion injury of the heart and brain. The simplest explanation for this is that the MPT has a role in the damage. The MPT can contribute to necrosis by both destroying mitochondrial energy generation and ablating the ability of mitochondria to scavenge reactive oxygen (Fig. 8.12).

[8] Remember that cells from these animals display normal apoptosis, and we therefore suspect that this is not an important mechanism for MOMP in apoptosis (see Chapter 4).

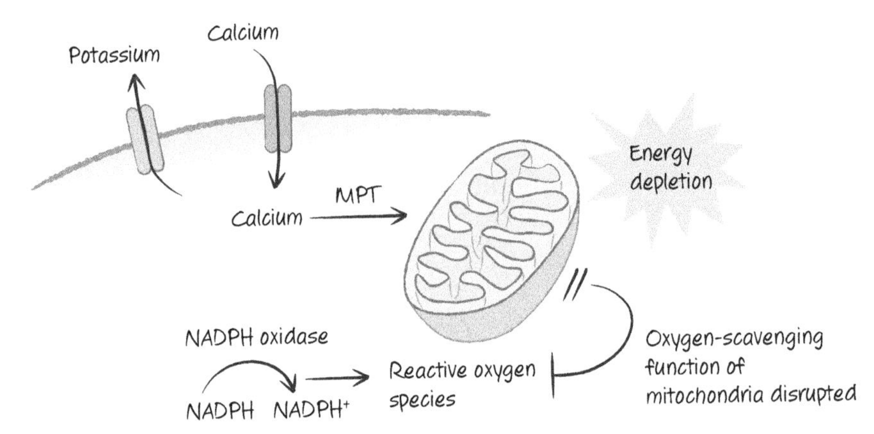

Figure 8.12. The mitochondrial permeability transition (MPT) in ischemia–reperfusion injury. Elevated calcium induces the MPT, which disrupts energy generation and curtails scavenging of reactive oxygen species (ROS). Mice lacking the MPT are resistant to ischemia–reperfusion injury.

Cyclosporin A, a drug that inhibits cyclophilins, including cyclophilin D, limits ischemia–reperfusion injury in rodents and can reduce heart damage in humans. It is tempting to believe that this is because the drug blocks the MPT. However, it might not be so simple. Cyclosporin A undergoes a conformational change when it binds to cyclophilins, and the complex of the drug and cyclophilins in the cytoplasm is a potent inhibitor of the enzyme calcineurin, and this effect might be more important.[9]

Ischemia–reperfusion injury might also involve necroptosis. Animals lacking RIPK3 are somewhat protected from such injury in the kidney and heart. Such injury involves inflammation, and it is possible that this is an effect of TNF and perhaps other death ligands on the endothelium and the kidney epithelium.

The damage in ischemia–reperfusion injury is not limited to necrosis; indeed, the area around the initial damage undergoes extensive apoptosis. This might be because of signals such as ROS released from the necrotic cells or the effects of inflammatory cells that are activated by the necrosis (discussed in more detail in Chapter 10). Inflammatory cells also produce ligands for death receptors (see Chapter 6). Mice lacking one of the death receptors, CD95, show reduced injury following ischemia–reperfusion. As we have seen, however, signaling from death receptors can also induce necroptosis. As mentioned above, necroptosis can also contribute to the damage observed in ischemia–reperfusion injury.

[9] Calcineurin is important for the activation of a set of transcription factors called NFAT (it is the inhibition of NFAT activation that accounts for the ability of cyclosporin A to block T-lymphocyte activation and tissue rejection, for which it is used). NFAT also has important roles in the development of blood vessels and other functions, and, at this point, we simply do not know how much these contribute to the beneficial effects of the drug in ischemia–reperfusion injury.

Another type of cell death that can be important in ischemia–reperfusion injury is ferroptosis. This form of regulated necrosis is considered next.

FERROPTOSIS

All eukaryotes and some bacteria synthesize glutathione, which in its reduced form (GSH) is an important antioxidant for scavenging ROS. The oxidized form (GSSG) can be enzymatically converted to the active, reduced form if NADPH is available. Therefore, glutathione and NADPH are essential to control ROS in our cells.

Glutathione is produced by an enzymatic pathway from three amino acids: glutamate, cysteine, and glycine. Cysteine is taken up in cells as cystine, in large part by the function of a cell-surface transporter, known as the cystine/glutamate antiporter or simply system Xc^-. If cells are deprived of cystine, or system Xc^- is impaired, glutathione levels decline and cells die.

It turns out that this cell death is not apoptotic, but is instead a regulated form of necrosis, called ferroptosis. Iron that is taken up in cells by transferrin and its receptor is required for ferroptosis (hence the name), and this metal reacts with hydrogen peroxide to generate ROS. These act on lipids, in particular polyunsaturated fatty acids in the cell, to generate lipid peroxides that are toxic (Fig. 8.13). This is because, if unchecked, lipid peroxides can act on other lipids to create more lipid peroxides, in a chain reaction that can damage the membranes of the cell, as well as producing other toxic species.

A glutathione peroxidase enzyme, GPX4, uses glutathione to reduce lipid peroxides and is the only known way that these can be reduced in cells. Inhibition of GPX4 also induces ferroptosis, without a concomitant loss of glutathione.

Ferroptosis appears to be important in many forms of ischemia–reperfusion injury. It also suggests an alternative mechanism for glutamate-induced excitotoxicity (discussed above) as excess extracellular glutamate can inhibit the function of system Xc^- to import cystine.

AUTOPHAGIC (TYPE II) CELL DEATH

Autophagic cell death is characterized by the presence of large vacuoles in the cytoplasm, as well as molecular markers of autophagy (see below). In general, the characteristic features of apoptosis, including plasma membrane blebbing, nuclear condensation, and chromatin fragmentation, are not seen, and caspases do not have a role.

Autophagic cell death might be a "victim" of its name; for the most part, neither autophagy (primarily a survival mechanism) nor molecular components of the autophagy pathway are responsible for autophagic cell death. There are exceptions and some tantalizing observations, as we will see. But most instances of autophagic cell death appear to be accompanied by autophagy, rather than being caused by it.

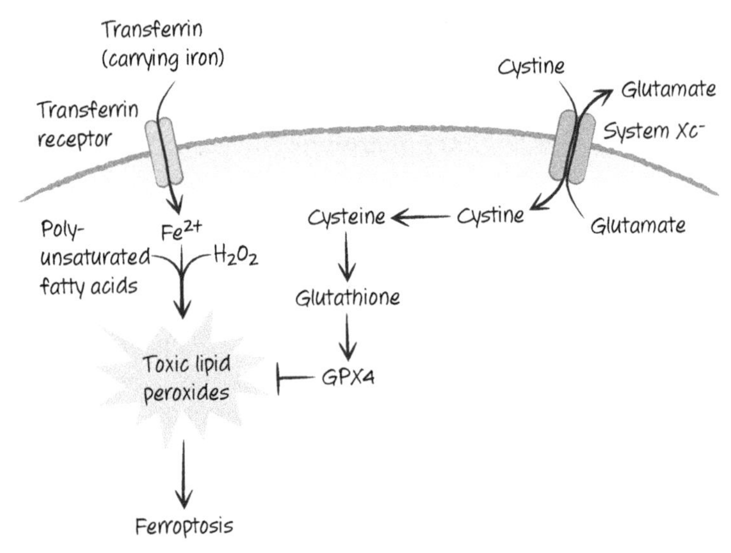

Figure 8.13. Ferroptosis. Iron (Fe^{2+}) is transported into cells by transferrin, where it can catalyze the oxidation of lipids by hydrogen peroxide (H_2O_2) in the cell to form toxic lipid peroxides. System Xc^- imports cystine (exporting glutamate), which is converted to cysteine and used in the generation of glutathione. The enzyme GPX4 reduces lipid peroxides but requires glutathione to regenerate its reducing potential. Disruption of System Xc^-, glutathione synthesis, or GPX4 function can therefore result in ferroptosis.

In fact, autophagy appears to antagonize cell death, and inhibition of autophagy can dramatically increase it. This is a controversial area, and there are sure to be important revelations in the coming years.

AUTOPHAGY AS A SURVIVAL MECHANISM

In general, autophagy is considered a survival mechanism for cells. Formally, there are three types, but here we are only concerned with macroautophagy.[10] This is usually referred to simply as "autophagy," and we use this nomenclature here.

Autophagy is found throughout the eukaryotes, and there is remarkable conservation of the components of the central pathway. It provides energy when external nutrients are depleted or the pathways for taking them up are disrupted, removes excess or damaged organelles and other cellular components, and is involved in the degradation of long-lived proteins and protein aggregates in cells. Autophagy also functions in cellular defense to isolate and destroy invading organisms.

[10] The other two types are called microautophagy and chaperone-mediated autophagy.

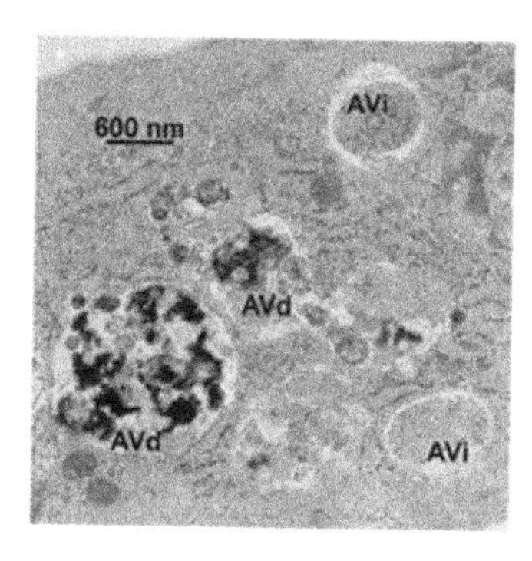

Figure 8.14. Autophagosomes. Shown are early autophagosomes (AVi) and autophagolysosomes (AVd).

Because autophagy is an important survival mechanism, it is useful to go into some detail about how it works. This will also help us to understand how it might be engaged to kill cells as well.

THE AUTOPHAGY PATHWAY

Autophagy involves the generation of membrane vesicles in the cell, called autophagosomes, that enclose cytoplasm, organelles, protein aggregates, or invading organisms and carry them to lysosomes, with which the autophagosomes fuse (forming autophagolysosomes). The degradative enzymes in the lysosomes break down the contents that are then reused as sources of energy and raw materials by the cell (Fig. 8.14).

Autophagy in mammals occurs by the hierarchical action of three complexes that form in a region of the endoplasmic reticulum (facing the cytosol). The first is a complex that contains a Ser/Thr-protein kinase (ULK1) together with proteins required for its kinase function. This activates a second complex that contains a lipid kinase, the phosphoinositide 3-kinase (PI3K) VPS34, together with additional proteins, including beclin-1. The action of this complex recruits a third complex, which functions as a ligase (akin to the E1–E2–E3 ligases of the ubiquitylation pathway). Its function is to place a small protein (LC3 or a related protein) onto a lipid (phosphatidylethanolamine) present in the endoplasmic reticulum membrane. It is likely that cytoskeletal components now "pull" the membrane as a sheet. Each complex recruits the next,[11] ultimately extruding a double-membrane structure called a "phagophore." The phagophore seals to engulf cytoplasm, the complexes dissociate, and the result is an autophagosome (Fig. 8.15).

ENGAGING THE AUTOPHAGY PATHWAY

Autophagy is inhibited in several ways, and how this inhibition is itself regulated provides some clues as to how autophagy is engaged in the cell. Although we consider

[11] This is only a model. However, the model is useful for understanding how autophagy might, in principle, work, and there is evidence to support it.

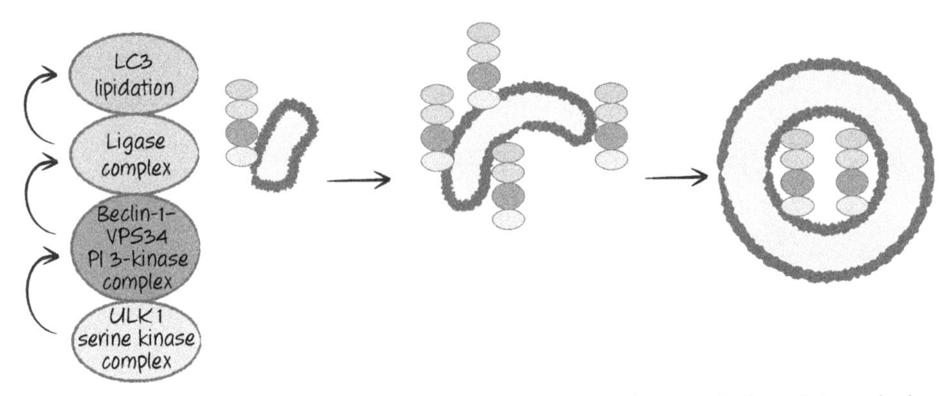

Figure 8.15. Simplified, hierarchical autophagy pathway. (*Left*) At the cytosolic face of the endoplasmic reticulum (ER), the serine kinase complex, activated, for example, by conditions of nutrient restriction, recruits and activates the phosphoinositide 3-kinase complex, which, in turn, promotes the assembly of the ligase complex. The latter directly conjugates the small molecules of the LC3 family to lipid in the ER membrane. (*Center*) The membrane is then extruded to form the phagophore. (*Right*) The phagophore seals to form a double-membrane structure, the autophagosome, trapping cytosolic material.

two mechanisms for induction of autophagy, it is worth noting that there are almost certainly other ways in which this pathway can be triggered.

One primary regulator of autophagy is also a key regulator of cell growth, metabolism, and protein synthesis, a kinase called mTOR. It is found in two different protein complexes, TORC1 and TORC2, but it appears to be TORC1 that controls autophagy by phosphorylating and inhibiting ATG13, a component of the ULK1 kinase complex. When nutrients are plentiful, mTOR is active—cells grow and autophagy is inhibited. However, when energy levels are low, adenosine monophosphate (AMP) accumulates, and this activates a Ser/Thr-protein kinase, AMPK, that inhibits mTOR, and

MAMMALIAN AUTOPHAGY IN DETAIL

Autophagy in mammals is engaged by the pre-initiation complex.[12] This contains the kinase mentioned above, ULK1, plus ATG13 and a protein called FIP200. When autophagy is activated by starvation or cellular damage, ATG13 dissociates and recruits the next complex in the pathway. This complex contains beclin-1 (see Chapter 5) and a class III PI3K called VPS34, together with other components. This generates the lipid phosphatidylinositol 3-phosphate (PtdIns3P, or PI3P) that recruits the next components of the autophagy pathway by a bridge molecule (WIPI2) that binds to PI3P.

[12] Also called the ULK1 complex because ULK1 is the catalytic subunit.

(Continued)

(Continued from previous page)

ATG7, an enzyme similar to a ubiquitin E1-ligase, binds to a small protein called ATG12 and passes it to an E2-like enzyme, ATG10, which then places it on another protein, ATG5. The ATG5–ATG12 complex recruits a further protein, ATG16, and this forms the E3-ligase, necessary for the next step in the process (Fig. 8.16). ATG16, carrying ATG5–12, is bound by WIPI2, bringing the E3-ligase to the membrane.

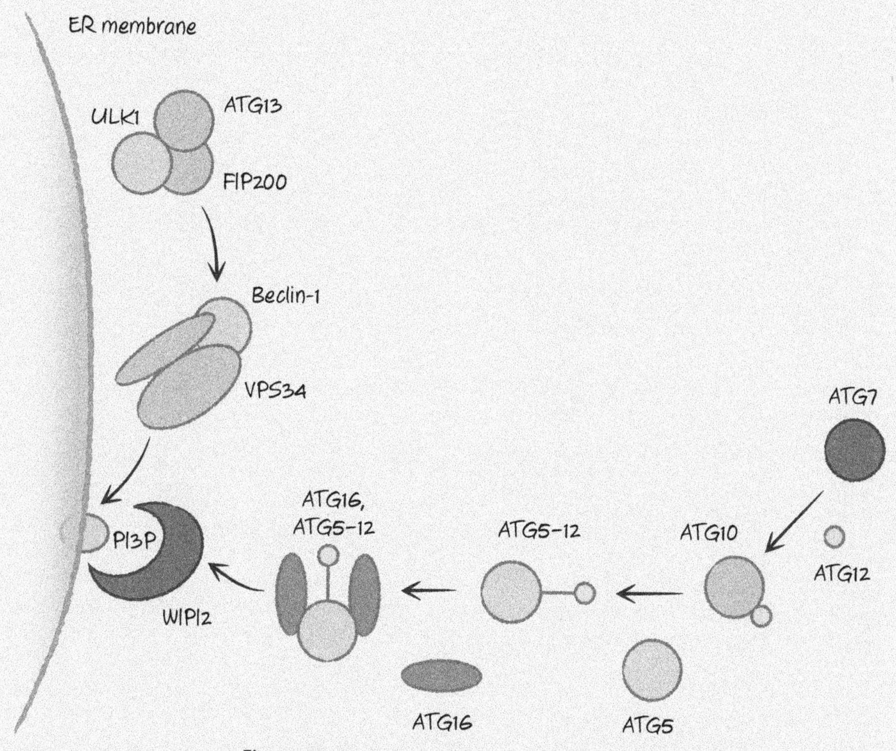

Figure 8.16. The initial steps in autophagy.

ATG7 again acts as a ligase, passing LC3 or related molecules[13] to ATG3, which then places LC3 onto a lipid, phosphatidylethanolamine (PE).[14] This LC3–PE complex is the building block of the phagophore, which begins to grow (Fig. 8.17). When it is complete, the other proteins peel off from the membrane, and it is now an autophagosome, ready to fuse with a lysosome.

[13] There are several LC3-like molecules that function in this regard. LC3 is actually MAP1LC3 and has several family members: LC3A, LC3B, and LC3C. Other LC3-like molecules that are similarly lapidated are GABARAP, GABARAPL1 and Gate16 (also called GABARAPL2), and GABARAPL3. It is possible that the LC3 and GABARAP subfamilies have different functions in the process, and both subfamilies are required for the biogenesis of autophagosomes

[14] Another view holds that the ATG5–12 pathway and the LC3–PE pathways are parallel, both contributing to the construction of the phagophore without the hierarchical organization proposed here.

(Continued)

(Continued from previous page)

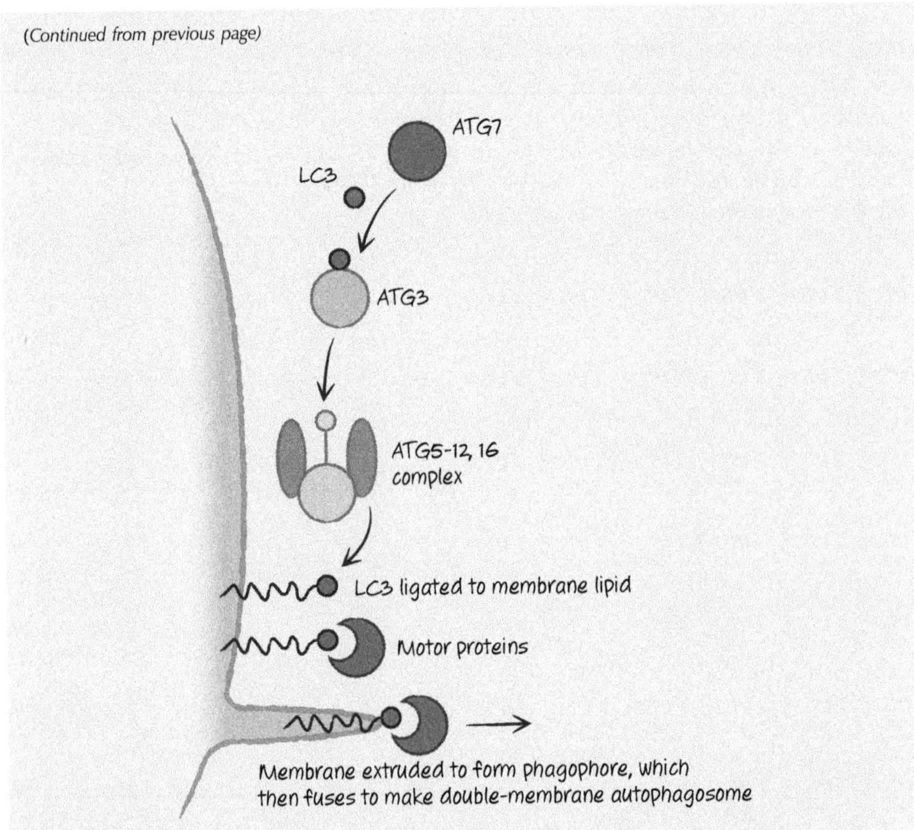

Figure 8.17. LC3 and the phagophore. The interaction of motor proteins with lipidated LC3 in the membrane to promote membrane extrusion is speculative. LC3 proteins also function at a later step in the fusion of the phagophore to form the autophagosome.

autophagy is engaged. AMPK also phosphorylates ATG13, but this is at a site different from that used by TORC1, and this actually activates ATG13. Similarly, growth-factor receptor signaling activates AKT (see Chapter 5), and this inhibits a complex, TSC1/2, that is an inhibitor of mTOR. When AKT is inactive, TORC1 is inhibited, and autophagy proceeds (Fig. 8.18). The drug rapamycin is a potent inhibitor of TORC1 (in fact, mTOR stands for "mammalian target of rapamycin"), and the addition of this drug to cells induces autophagy.

Another way in which autophagy is regulated involves BCL-2 (and possibly other anti-apoptotic BCL-2 family proteins), which can bind and sequester beclin-1 using the same BH groove used to sequester BH3-only proteins in apoptosis. Indeed, beclin-1 has a BH3 domain, although this binds weakly to BCL-2 compared with the binding strength of others that we have discussed in Chapter 5. It seems that if BH3-only proteins are

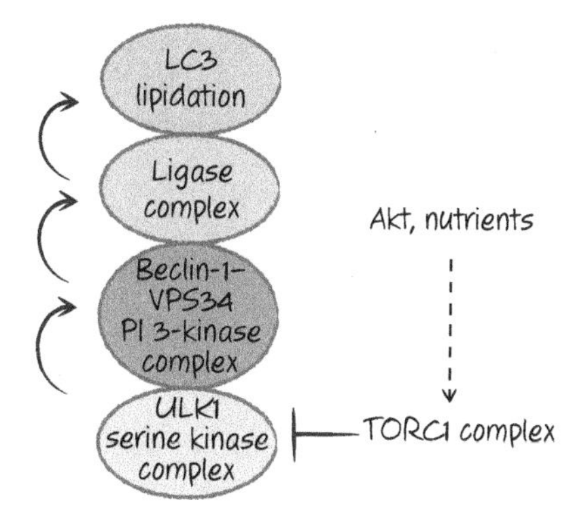

Figure 8.18. mTOR and autophagy.

expressed, they can neutralize the interaction between BCL-2 and beclin-1, and, if MOMP does not occur (i.e., BAX and BAK are not activated), autophagy can result.[15]

Autophagy can also be caused by DNA damage. This induces UVRAG, a protein that promotes the activity of VPS34 in the beclin–PI3K complex (see above). Autophagy in this setting might provide additional energy needed for DNA repair.

In each of these scenarios, autophagy protects cells. If it is blocked, apoptosis ensues. However, when apoptosis is blocked and autophagy is active (e.g., by removal of BAX and BAK), autophagy will keep the cells alive for extended periods, wearing them down to mere "skeletons" of their former selves (Fig. 8.19). If autophagy in stressed cells is also blocked, the cells die by necrosis.

MITOPHAGY

Mitophagy is a selective form of autophagy that specifically captures damaged mitochondria in autophagosomes and destroys them in the lysosomes. It is important for the quality control of mitochondria; defects in mitophagy lead to accumulation of

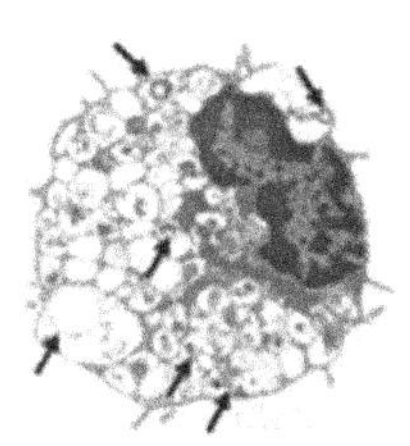

Figure 8.19. An autophagic survivor. Arrows indicate autophagosomes.

[15] In Chapter 5, we outlined one case in which this effect could be important. A BH3-only protein called NIX, which is not a potent potentiator of apoptosis, is important for the removal of mitochondria during the development of mammalian red blood cells. It is possible (although not formally proven) that NIX does this by disrupting the interaction between beclin-1 and BCL-2. Another BH3-only protein, BNIP3 (which is also a poor inducer of apoptosis), is related to NIX, and this protein might also be involved in the autophagic removal of mitochondria in some settings.

defective mitochondria that produce ROS and can inflict damage on the cell. As we saw in Chapter 5, mitophagy is also engaged under conditions in which the numbers of mitochondria must be reduced, such as in hypoxia and in the development of some cell types such as red cells.

There are several mechanisms by which mitophagy is engaged, one of which is through the functions of BNIP3 and NIX (Chapter 5). The targeting of defective mitochondria, however, operates through a different process.

Healthy mitochondria generate a charge across the inner mitochondrial membrane ($\Delta\Psi m$) that is produced as a consequence of the electron-transport chain or by reversal of the ATPase complex (if electron transport is not active). $\Delta\Psi m$ is essential for the import of proteins into the mitochondria and for other mitochondrial functions. One protein that is imported into mitochondria is a Ser/Thr-protein kinase, PINK1. When PINK1 is imported into the mitochondria, it is degraded by proteases; however, if the import machinery is inactive (owing to loss of $\Delta\Psi m$), PINK1 accumulates on the surface of the damaged mitochondrion. PINK1 then recruits and activates an E3-ligase, parkin, which ubiquitylates proteins on the mitochondria, including a protein involved in mitochondrial fusion, MFN2, which is also involved in this form of mitophagy (in this case, the fusion activity of MFN2 is not involved in the process). The autophagy machinery is then recruited to the damaged mitochondrion, which is then removed (Figs. 8.20 and 8.21).

Because mitochondria are important in several forms of cell death, the process of mitophagy and its role in sustaining mitochondrial quality is thereby linked to cell death pathways.

WHAT IS AUTOPHAGIC CELL DEATH?

The above considerations help to clarify why a stressed cell that can undergo apoptosis engages the autophagic pathway as a result of the stress. Therefore, autophagy and apoptosis can occur in the same cell, but in such cases this is apoptosis and not autophagic cell death. However, cells often die without displaying features of apoptosis, instead displaying features of autophagy. In such cases, this is called

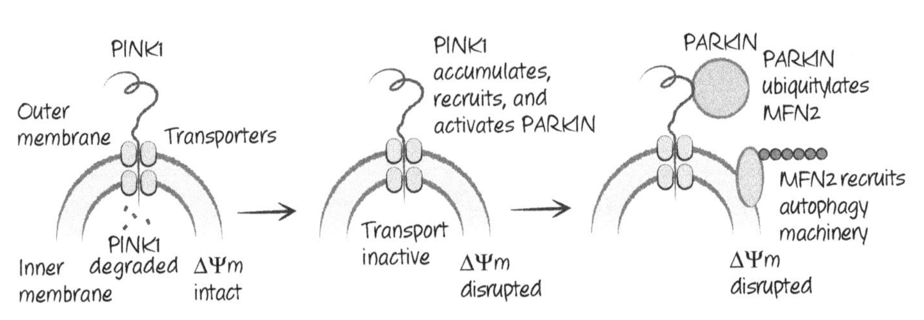

Figure 8.20. Mitophagy mediated by PINK1–PARKIN.

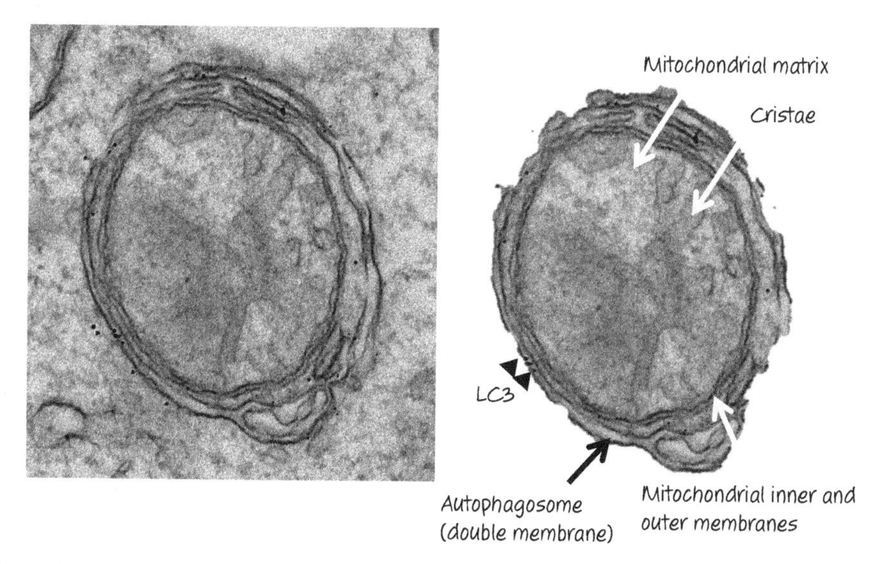

Figure 8.21. Mitophagy in action. Immunoelectron microscopy showing double-membrane autophagosome with associated LC3 (black dots), enclosing a damaged mitochondrion.

autophagic cell death (type II cell death), and it is seen in the response of tumor cells to several therapeutic agents. Usually, this is cell death that is associated with, but not caused by, autophagy, and if autophagy is blocked, the cells die by necrosis. Most autophagic cell death has this feature.[16]

WHEN AUTOPHAGY KILLS

Nevertheless, in some cases, it appears that autophagy can indeed promote (type II) cell death. The best-studied case is in the metamorphosis of the fruit-fly *Drosophila* (discussed in Chapter 11). Larvae have very large salivary glands that die during metamorphosis. Some of this death is by apoptosis, but most of it has the form of type II cell death. If autophagy genes are defective in the developing fly, much of the cell death does not occur and the development is prevented.

In mammals, there are several examples wherein type II cell death appears to be promoted by autophagy.[17] One way that this can occur is if the autophagic process is

---

[16] This might change as more systems are explored in vivo and new situations in which autophagy actually kills cells are identified.

[17] Often, this has been shown by knocking down levels of expression of key proteins with siRNA. This is not without problems because siRNA can have off-target effects, including interferon responses and other unexpected consequences that could affect cell death. Here, however, we will assume that there is something to the idea that autophagy can promote cell death in mammals.

activated but for some reason is not completed before the contents of the autophagosome are degraded in lysosomes. As a result, the cytoplasm is disrupted, but the cell does not gain nutrients from the process, leading to death.[18] Alternatively, it has been proposed that another way in which autophagy promotes cell death is by the removal of catalase, a long-lived protein that neutralizes some ROS. We can also imagine that, if autophagy removes other key survival proteins as well, cell death could result.

Although mitophagy, as discussed above, is important in sustaining healthy mitochondria in a cell, excessive mitophagy can kill a cell, either by apoptosis or necrosis. The antibiotic valinomycin is an ionophore that induces apoptosis that is dependent on PINK1 and parkin (see above), and causes widespread mitophagy. It is not known how this results in apoptosis, but it is likely that cytochrome *c* is released in the process, as the apoptosis depends on APAF1. In other settings, extensive mitophagy can compromise the metabolic fitness of a cell, resulting in a loss of ATP and necrosis.

There is another interesting possibility, albeit untested. As we have discussed, autophagy involves fusion of the autophagosome with lysosomes in the cell. It is conceivable that, if the process accelerates, so that incompletely formed autophagosomes engage lysosomes, the contents of the latter might be released into the cytosol. This might result in autophagic cell death as the destructive contents of the lysosome set to work. As mentioned above, large vacuoles are often associated with autophagic cell death, and it is possible that they are produced by such an interaction. Support for this notion comes from studies on a protein called DRAM. DRAM is associated with lysosomal membranes, and its expression is induced in some pathways leading to cell death. In some cases, inhibition of the expression of DRAM can protect cells from what appears to be type II cell death.

Perhaps the most interesting possibility relates to necroptosis. As we discussed, cells lacking caspase-8 can die following exposure to TNF in a manner that depends on RIPK1 kinase activity. In some studies, inhibition of autophagy prevents this cell death. Similarly, T lymphocytes that lack caspase-8 or FADD fail to proliferate and die following activation, and this cell death is blocked (and proliferation restored) by inhibition of RIPK1 kinase. Intriguingly, there is evidence that disruption of autophagy can also rescue these T cells. It appears that the ATG5–ATG12 complex in the autophagic pathway can recruit the adapter FADD. FADD, caspase-8, and FLIP associated with ATG5–ATG12, might cleave and inhibit RIPK1 kinase to prevent cell death when autophagy is engaged. If caspase-8 is not active, RIPK3-mediated necroptosis ensues. This intriguing, but untested, scenario links at least some forms of autophagic cell death to necroptosis (Fig. 8.22).

---

[18] This can be performed pharmacologically, by the addition of an autophagy inducer such as rapamycin and a lysosome inhibitor such as chloroquine. This strategy is being explored as a therapeutic regimen for cancer. Some viruses also inhibit lysosomes, and the metabolic stress caused by infection might result in defective autophagy that kills the cell in this manner.

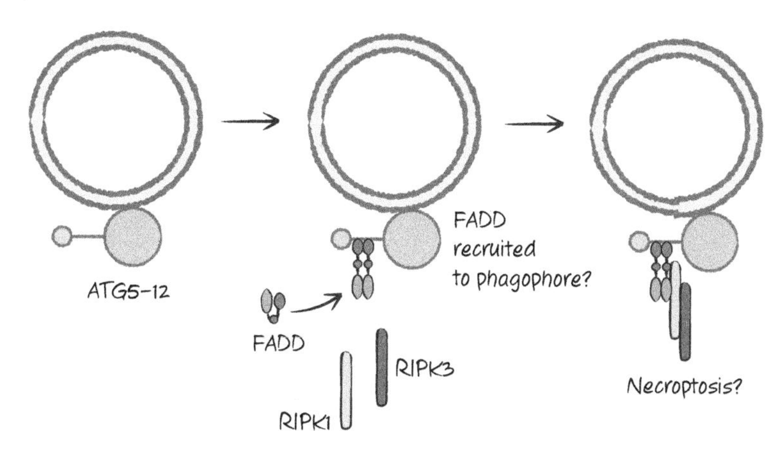

Figure 8.22. Hypothetical connection between autophagy and necrosis. Abundant ATG5–ATG12 on the phagophore can bind FADD. This might recruit RIPK1 and RIPK3 to signal for necroptosis.

Normally, during autophagy, ATG5–ATG12 dissociates from the membrane when the autophagosome forms. If the model shown in Figure 8.22 is correct, it might be that a determining factor in whether autophagy contributes to cell death is the continued presence of ATG5–ATG12 on the autophagosome membrane. This is only speculation, however.

MITOTIC CATASTROPHE

During mitosis, the nuclear membrane dissolves and chromosomes are segregated to the poles of the dividing cell. If this process is disrupted, cells die, and this is often called "mitotic catastrophe." This can occur if, for example, the DNA is extensively damaged, the cell cycle machinery is stalled, or microtubules are dysfunctional.

Often, mitotic catastrophe results in apoptosis. How the mitotic machinery is linked to the control of apoptosis is not clear, but one possible mechanism is intriguing. Cyclin-dependent kinase 1 (CDK1), the kinase that orchestrates mitosis, phosphorylates caspase-2 (see Chapter 7), preventing its activation. When mitosis is not completed on schedule, CDK1 activity declines and caspase-2 becomes active. Cells might therefore use caspase-2 to activate apoptosis when mitosis fails. There are likely other mechanisms as well.

If cells are faced with defective mitosis and apoptosis is blocked, the cells die nevertheless, showing features of autophagic cell death or necrosis. An example of this is seen in the intestines of mice lacking BAX and BAK following irradiation (Fig. 8.23). How this cell death occurs remains obscure.

In this chapter, we considered forms of cell death that are not apoptosis. Different paths to cell death are interesting (and can have fundamental repercussions for therapeutic intervention), but does it matter to the body how a cell dies? The consequences

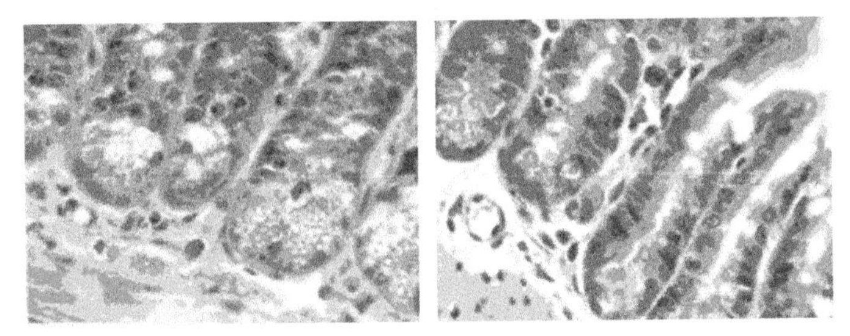

Figure 8.23. γ-Irradiation causes nonapoptotic cell death in intestines lacking BAX and BAK. Mice with (*left*) or without (*right*) intestinal BAX and BAK were irradiated. Although apoptosis (green stars) was decreased, cell death nevertheless occurred in the deficient cells. This was associated with abnormal mitosis.

of cell death go beyond the cell, and, indeed, the mode of cell death can influence these sequelae. We explore such consequences next.

CHAPTER 9

The Burial
Clearance and Consequences

DEATH MATTERS

Cell death has consequences beyond simply the loss of the cell itself. Dying cells trigger their removal by other cells, and this can be accompanied by an inflammatory response. The cells can also induce other cells to proliferate to compensate for the loss in cell number. And, in vertebrates, they can signal to the immune system to regulate adaptive immune responses to foreign or altered proteins associated with the dying cell. Dying cells are not simply debris to be discarded—they are resources to be recycled, and this process of waste management is tightly controlled.

In this chapter, we consider the ways in which dying cells are cleared from the body and the consequences for the organism after the cell is gone. As we will see, there are many possible outcomes, and these depend in part on the cell and how it dies.

DISPOSING OF THE CORPSE

In some cases, cells that die are simply sloughed off, and, in vertebrates, this happens to the many cells that die in the skin and the gut epithelium as a consequence of normal tissue homeostasis. But most cells that die in the body have no place to go; they must be cleared by other cells.

Regardless of the mode of cell death, the corpses are engulfed[1] by phagocytic cells, by which they are digested. Presumably, some breakdown products of the

[1] A quick note on the use of our term "engulfment" is warranted. In general, when a cell eats a particle of any kind, this is referred to as phagocytosis ("cellular eating"). There are two main types of phagocytosis: opsonization and macropinocytosis. In opsonization, the phagocyte binds to the particle and effectively "zips" the phagocyte's membrane around it, stripping away anything loosely associated with it and excluding surrounding fluids. In contrast, macropinocytosis involves "gulping" the particle together with surrounding fluids and molecules. In the case of removal of cell corpses, both processes have been described (and have different consequences), leading to some confusion in the literature. For this reason, many who study the process prefer the term "engulfment" as a way to sidestep the issue. As we will see, there are several ways in which dying cells are removed by phagocytes. A general term for engulfment of dying cells is "efferocytosis."

dead cell are recycled, but others represent a problem for the cell doing the clearing and might require a form of waste management. The phagocytes ("eating cells") are often "professionals," such as macrophages (as well as dendritic cells and neutrophils in vertebrates). However, they can also be "amateurs," such as epithelial cells. Some of the consequences of cell death depend on which cell does the eating.

Although any dead cell can be removed by phagocytosis, apoptosis provides a means to clear the dying cell before plasma membrane integrity is lost, thereby avoiding "secondary necrosis" (see Chapter 8). As we have seen, necrosis can promote inflammation through the release of intracellular contents. However, apoptosis involves effectively preparing the corpse for degradation by preprocessing some of its components while preserving the plasma membrane. The consequences of caspase activation—which leads to DNA fragmentation, compaction, and the "blebbing off" of pieces of the dying cell—can all be seen as preparation for clearance.

The signals produced by apoptotic cells echo Pete Townsend's "see me, feel me, touch me, heal me" from the classic rock opera *Tommy*.[2] The steps here, however, are "find me, bind me, eat me, clear me."

"FIND-ME" SIGNALS

Often, cells that die are simply engulfed and cleared by neighboring cells, and no recruitment of phagocytes is necessary. However, in the case of apoptosis, the facilitation of clearance can involve the production of find-me signals by the dying cell that recruit macrophages and other phagocytes.

One such find-me signal is the phospholipid derivative lysophosphatidylcholine (LPC), which attracts macrophages, among its many other activities. Many cells that undergo apoptosis produce LPC in a caspase-dependent manner. Caspases cleave and activate calcium-independent phospholipase A_2 by removing an inhibitory domain on the enzyme, and this catalyzes the production of LPC (Fig. 9.1). Although the generation of this find-me signal is by an elegant mechanism, high levels of LPC are required to attract macrophages, and there are already high levels of LPC in the circulation. Other find-me signals are likely to be important. Another lipid that attracts macrophages is sphingosine-1-phosphate, produced by apoptotic cells, although how apoptosis induces its production is not known.

Dying cells can also release damage-associated molecular patterns (DAMPs; see Chapter 7). Some examples that act (probably indirectly) as find-me signals are ATP and another nucleotide, UTP, that is converted to UDP in the extracellular environment. Both of these bind to receptors on phagocytic cells. Another DAMP, released from necrotic cells, is uric acid, and this can also generate find-me signals. It is likely

[2] Readers unfamiliar with the reference are urged to listen to The Who's classic rock opera *Tommy*, while reading this chapter. Others who have not heard it for a long while might wish to do the same.

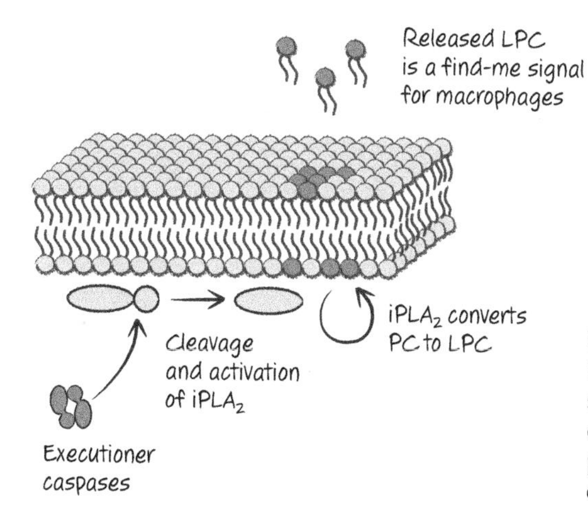

Figure 9.1. Executioner caspases induce production of the "find-me" signal LPC. LPC, lysophosphatidylcholine; iPLA$_2$, calcium-independent phospholipase A$_2$; PC, phosphatidylcholine.

that, in these cases, cells that are near the dying cell are activated by the DAMPs to produce molecules such as cytokines and chemokines, which are the actual signals that recruit more phagocytic cells. In general, the DAMPs that do this are released from necrotic cells, and whether they have roles in generating find-me signals for other forms of cell death has not been established.

"BIND-ME" SIGNALS ON DYING CELLS

As cells die, changes in the plasma membrane act as signals for other cells to engulf the corpse. The most important of these signals is a lipid—phosphatidylserine. In healthy cells, many plasma membrane lipids are arrayed asymmetrically, residing predominantly in either the inner or the outer leaflet of the membrane. Phosphatidylserine is kept to the inner leaflet by an ATP-dependent aminophospholipid translocase that flips phosphatidylserine and other aminophospholipids from the outer to the inner membrane. If energy in the form of ATP is unavailable, or if the plasma membrane is disrupted, phosphatidylserine accumulates by diffusion in the outer leaflet. In cells undergoing apoptosis, however, externalization of phosphatidylserine occurs in a rapid, caspase-dependent manner, before plasma membrane integrity is lost. The "scramblase" responsible for this phospholipid scrambling is the XK-related protein Xkr8, as we saw in Chapter 2. Interestingly, Xkr8 is homologous to a protein in the nematode *Caenorhabditis elegans*, CED-8, which was found to be important in dying cells for their removal. Caspases also destroy the phospholipid translocase (ATPC11) to prevent movement of phosphatidylserine from the outer to the inner leaflet of the plasma membrane. It turns out that the appearance of phosphatidylserine on the cell surface, before a loss in plasma membrane integrity, also occurs in cells undergoing necroptosis as a consequence of the active pseudokinase MLKL (see Chapter 8).

DETECTING PHOSPHATIDYLSERINE BY USE OF ANNEXIN V

The appearance of phosphatidylserine on apoptotic cells can be detected with a probe, annexin V, that binds to phosphatidylserine and can be coupled to fluorescent dyes for detection (Fig. 9.2). It is important to note, however, that annexin V is not how apoptotic cells are recognized by the body. Apoptotic cells can be stained with annexin V before they become permeable to other dyes, that is, before plasma membrane integrity is lost. This is also seen in necroptosis. In contrast, cells undergoing other forms of necrosis expose phosphatidylserine at the same time that they lose the plasma membrane barrier.

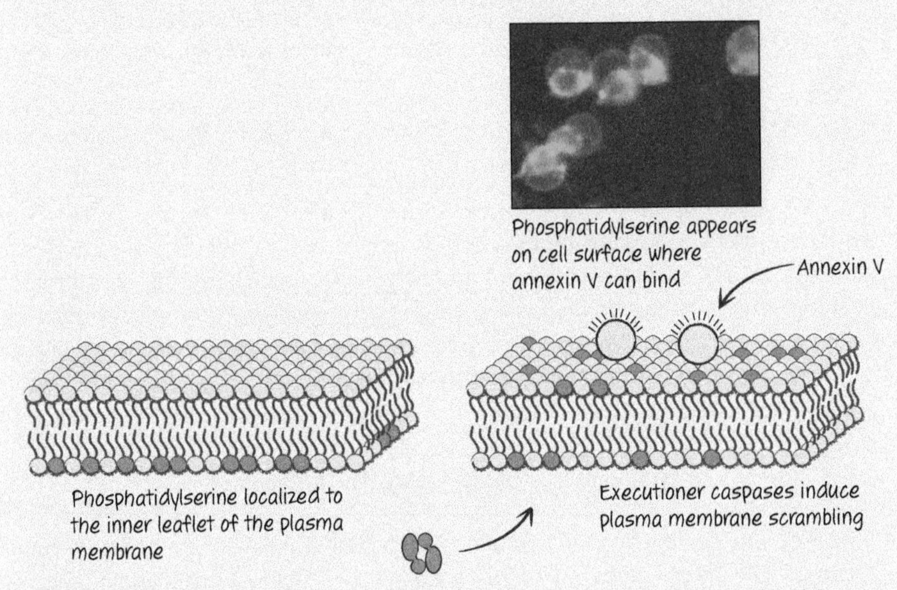

Figure 9.2. Plasma membrane scrambling and its detection in dying cells. Phosphatidylserine "scrambled" from the inner to the outer membrane can be detected by probes, such as fluorescently labeled annexin V.

Annexin V is not the only way to detect phosphatidylserine on the surface of dying cells. The bridging molecule MGF-E8 (discussed below) can also be used as such a probe.

Because membrane disruption also causes the exposure of phosphatidylserine on the surface, a second dye that is membrane impermeable is generally used to monitor the integrity of the plasma membrane. Some examples are propridium iodide, 7-AAD, and Sytox Green (although several such dyes can be used) (Fig. 9.3).

During apoptosis, phosphatidylserine is exposed before any loss of plasma membrane integrity, and this condition is often used as a measure of apoptosis. As the dying cell continues to deteriorate, the plasma membrane loses its coherence, and the cell stains with both dyes.

However, in necroptosis (see Chapter 8), phosphatidylserine is also exposed on the cell surface before a loss of plasma membrane integrity. Therefore, to assess apoptosis in this way,

(Continued)

(Continued from previous page)

the effects of caspase inhibitors on the staining should be evaluated to determine whether the effect is caspase dependent (Fig. 9.3)

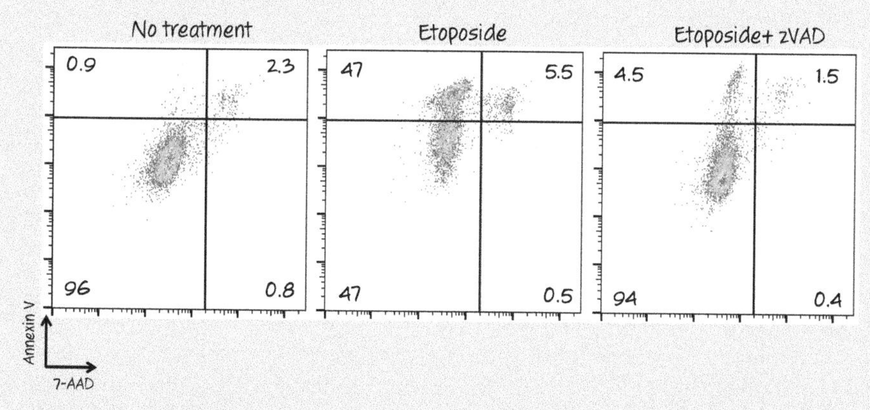

Figure 9.3. Detection of phosphatidylserine exposure by fluorescence-activated cell sorting (FACS). Cells were treated with the chemotherapy agent etoposide to induce apoptosis (± the caspase inhibitor zVAD-fmk) and then stained with annexin V (coupled with a fluorescent dye) and the vital dye 7-AAD. The fluorescence intensity of each cell is represented by a dot. Notice that the cell population becomes positive for annexin V before membrane integrity (measured by 7-AAD) is lost. This annexin V staining, and by implication phosphatidylserine exposure, is dependent on caspases, as treatment with the inhibitor zVAD largely prevented it. 7-AAD, 7-amino actinomycin D; zVAD, zVAD-fmk.

In other forms of necrosis, phosphatidylserine is exposed passively, as the plasma membrane is disrupted. As a result, all dying cells display phosphatidylserine.

Although phosphatidylserine is an important bind-me signal (probably the most important one for engulfment), other bind-me signals also appear on dying cells. Oxidation of lipids and changes in carbohydrates also occur and can contribute to recognition. Calreticulin is a protein within the endoplasmic reticulum, and cells that die in different ways often expose this protein on the cell surface. This can also act as a bind-me signal for phagocytes. Calreticulin also appears to be important later on for subsequent responses to the corpse, as we will see.

"DON'T-EAT-ME" SIGNALS ON CELLS

If phosphatidylserine exposure were the only signal for removal of dying cells, then we would have a problem—plasma membrane scrambling and externalization of phosphatidylserine occurs in perfectly healthy cells as well and might have roles in cell signaling. One way this occurs is through the activation of a phospholipid scramblase—TMEM16F (also called anoctamin-6)—by calcium ions (as discussed in Chapter 2, this is not involved in phosphatidylserine exposure during apoptosis, and, as it turns out, it is also not involved in such exposure during necroptosis). Activation of lymphocytes, for example, induces a calcium flux that causes a transient exposure of phosphatidylserine.

But healthy cells (at least in vertebrates) appear to produce "don't-eat-me" signals that are rapidly lost from dying cells (Fig. 9.4). One don't-eat-me signal is the leukocyte surface antigen CD47, a cell-surface molecule expressed on all healthy cells. CD47 is recognized by a receptor, SIRPα, on phagocytic cells, and this generates an inhibitory signal in the form of active tyrosine phosphatases. These signals appear to block phagocytosis as well as a range of other signaling pathways, and the CD47–SIRPα interaction is also involved in other processes. During cell death, CD47 is rapidly lost from the cell surface or partitioned away from externalized phosphatidylserine.

Mice lacking CD47 do not have a problem with spontaneous uptake of cells, other than a tendency of macrophages to remove living red cells from the circulation. However, neutralization of CD47 or SIRPα can cause living tumor cells to be engulfed by macrophages and destroyed, and this has been suggested as a novel route to cancer therapy. It seems that other signals (other than exposure of phosphatidylserine) cooperate with a loss of the don't-eat-me signal to cause this engulfment. One such signal on cancer cells is calreticulin (see above), perhaps exposed as a consequence of endoplasmic reticulum stress.

Another don't-eat-me signal is the cell-surface molecule CD31, which, when it interacts with CD31 on the phagocytic cell, generates an uncharacterized repulsion signal. The CD31 homotypic interaction is apparently disabled in dying cells.

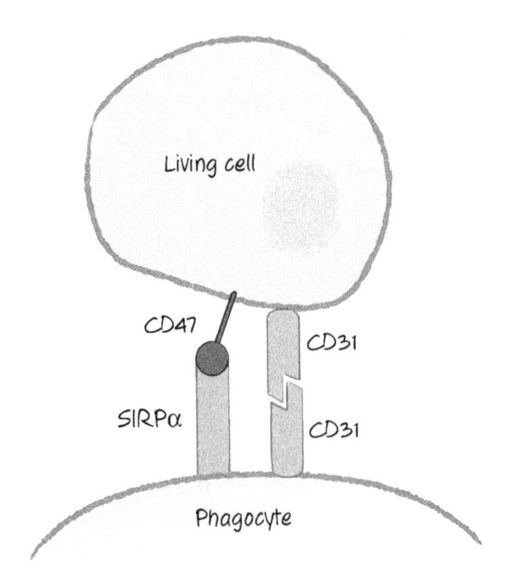

Figure 9.4. "Don't-eat-me" signals. These include the leukocyte surface antigen CD47, recognized by phagocyte cell-surface receptor SIRPα, and the homotypic interaction between CD31 molecules. SIRPα, signal-regulatory protein alpha.

There also appear to be "stay-away" signals that apoptotic cells produce. All of the find-me signals mentioned above attract neutrophils as well as macrophages, but neutrophils are not recruited to sites of apoptotic cell death. One stay-away signal is lactoferrin, which can selectively prevent neutrophil recruitment. How this is produced by apoptotic cells and how it acts on neutrophils are currently unknown.

BRIDGING MOLECULES RECOGNIZE "BIND-ME" SIGNALS

A number of soluble bridging molecules bind to dying cells because of the changes in the plasma membrane. Several of these bind to externalized phosphatidylserine on dying cells, and the most important of these is milk fat globulin-E8 (lactadherin/ MFG-E8). Mice lacking MFGE8, or animals injected with a dominant-negative mutant protein, accumulate apoptotic corpses. This is especially evident in one region of the lymphoid tissues, the germinal centers, where B lymphocytes proliferate during immune responses. In normal animals, germinal centers contain what are known as "tingible body macrophages" that turn out to be macrophages with engulfed apoptotic bodies that they are digesting (Fig. 9.5). Mice lacking MFG-E8 do not have these cells.

Several other bridging molecules bind to phosphatidylserine on dying cells. These include thrombospondin-1, Gas6, β1-GPI, and protein-S. Figure 9.6 illustrates the array of bridging molecules that can bind to phosphatidylserine on dying cells.[3] Altered carbohydrates on the dying cells are bound by mannose-binding lectin

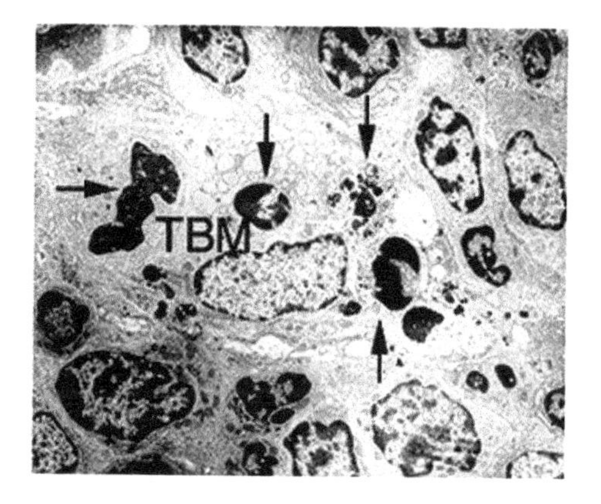

Figure 9.5. A single tingible body macrophage (TBM) in a germinal center engulfing four apoptotic cells (arrows).

[3] The bewildering variety of molecules and mechanisms for binding to dying cells can be taken as an indication of the importance of the process. It is also the consequence of an emerging field of research, in which the importance of particular interactions is not fully understood. If the reader prefers, it is enough for now to let the message be "there are lots of bridging molecules" that can decorate dying cells.

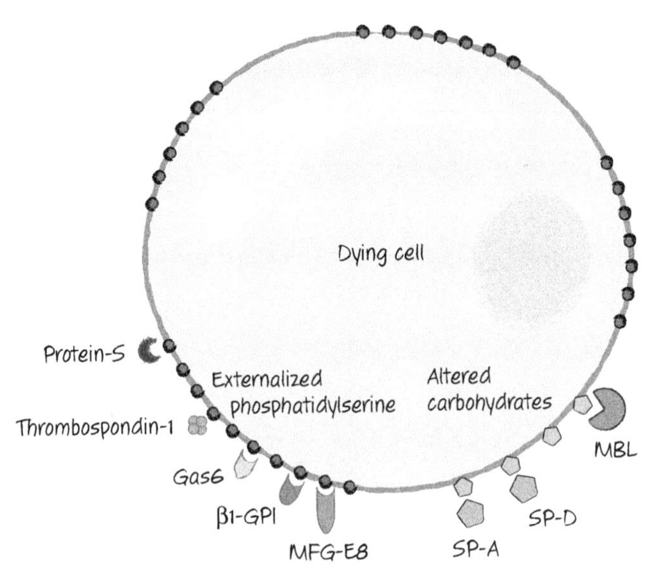

Figure 9.6. Bridge molecules are soluble molecules that bind to dying cells because of the changes in the plasma membrane. MBL, mannose-binding lectin; MFG-E8, lactadherin.

(MBL) and the lung surfactant proteins A and D (SP-A and SP-D) (Fig. 9.6). It is likely that not all dying cells are bound by all of these bridging molecules, but clearly there is great redundancy in the system.

Two other bridging molecules might also be involved in the recognition of some dying cells. These are proteins in the complement pathway—a set of circulating proteins that function in immune responses to bacteria and other foreign invaders. C1q, which binds to antibody clusters, appears to be involved, as is C3b, which normally binds directly to bacteria to trigger their removal (Fig. 9.7). In the case of dying cells, C3b appears to be bound and activated by MBL.

Of course, bridging molecules that decorate dying cells must themselves be recognized by receptors on phagocytic cells if they are to aid engulfment. As we will see, there are also receptors that directly recognize the surface changes in the dying cells.

THE "TETHER AND TICKLE" MODEL

Why are there so many signals for uptake of dying cells? Relatively few of these actually trigger engulfment. A useful idea is that removal of cell corpses proceeds by a process of "tether and tickle." In this model, many of the receptors on phagocytes that recognize bridging molecules and the dying cells themselves simply tether the corpse. Other receptors, then, actually generate the responses leading to engulfment. These "tickle"—that is, they produce intracellular signals that cause the phagocyte to eat the dying cell.

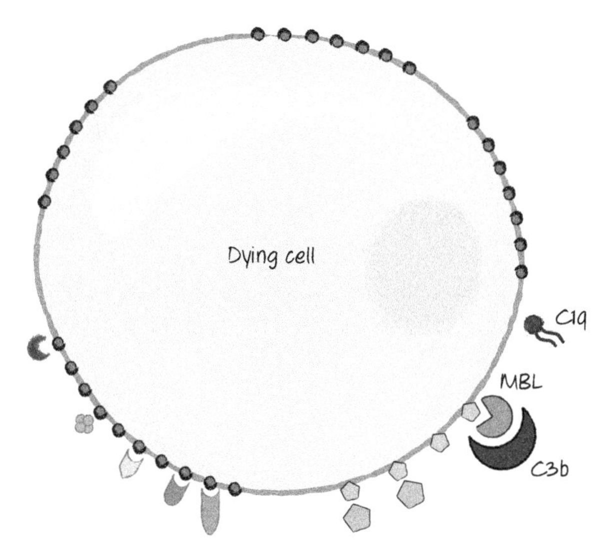

Figure 9.7. Complement components C1q and C3b can also bind to dying cells to act as bridge molecules.

This will prove a useful idea as we continue our survey to include the receptors on the engulfing cells.

PHAGOCYTES HAVE RECEPTORS FOR "BIND-ME" SIGNALS AND BRIDGING MOLECULES

Many different receptors on the phagocyte bind to bridging molecules; others appear to bind directly to phosphatidylserine and signals on the dying cells. No one phagocytic cell has all of the relevant receptors, and therefore the rather complex picture that emerges might be a bit misleading. It is worthwhile to keep this in mind in the discussion that follows.

At least three receptors on phagocytes bind to phosphatidylserine and participate in the uptake of dying cells. These are TIM4 (T-cell immunoglobulin and mucin-domain-containing molecule 4—confusingly, residing on phagocytes), BAI1 (brain-specific angiogenesis inhibitor 1—confusingly, not brain specific), and stabilin-2 (Fig. 9.8). TIM4 appears to be particularly important; mice lacking TIM4 accumulate dying cells, and macrophages derived from bone-marrow precursors take up apoptotic cells less efficiently if they lack TIM4.

The phosphatidylserine-binding bridging molecule MFGE8 binds to two integrins on the phagocyte surface: $\alpha_v\beta_3$ and $\alpha_v\beta5$. Gas6 binds to a different receptor on the phagocyte, a receptor tyrosine-protein kinase called Mer. Both the integrins and the Mer generate signals, and therefore these probably "tickle" the phagocytic cell to induce the subsequent engulfment. Macrophages from mice that lack these

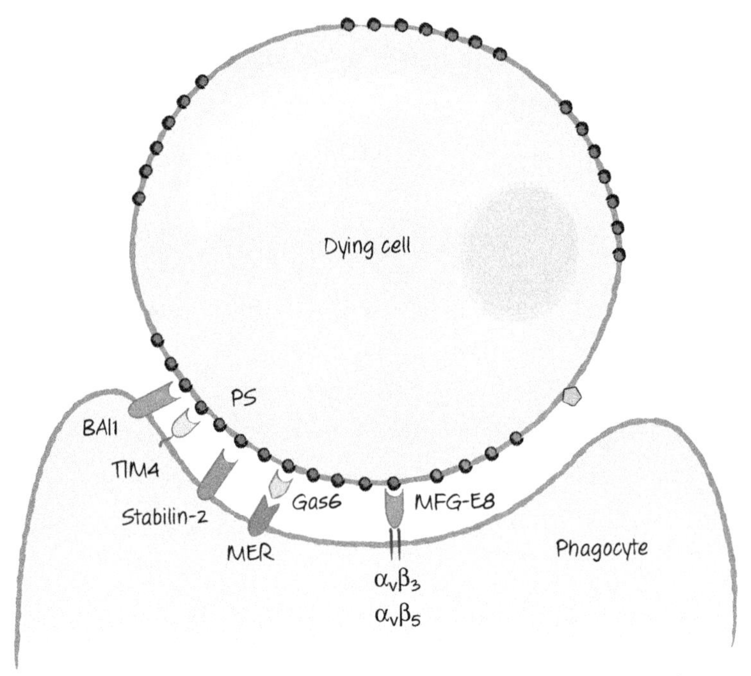

Figure 9.8. Recognition of phosphatidylserine (PS), exposed on dying cells, by phagocytes occurs either directly (e.g., for BAI1, stabilin 2, TIM4) or indirectly through bridge molecules (e.g., for MER, integrins).

integrins or MER display at least partially defective binding to dying cells. TIM4 does not have a signaling domain in its intracellular region, and therefore if it "tickles" the phagocyte for uptake of corpses, how this happens is not clear.[4]

Some of the other phagocyte receptors that are probably involved in tethering (although they might also have "tickle" functions about which we do not know) are the scavenger receptors, which include scavenger receptor-A, LOX1, CD68, and CD36. These normally bind to oxidized low-density lipids (oxLDLs) and clear them from the circulation. OxLDL-like sites on dying cells are also bound by these receptors. It is possible that these sites are actually lysophosphatidylserine (a find-me signal) that was not released from the dying cell (Fig. 9.9).

The complement molecule C3b is bound by complement receptors on the phagocyte. Because these clearly function in the engulfment of bacteria, it is likely that they similarly "tickle" phagocytes to engulf dying cells. In addition, C1q is also bound by a receptor comprising LRP1 (CD91) and the protein calreticulin (this time on the

[4] Another receptor for phosphatidylserine, PSR1, has also been described. However, subsequent studies have since shown that this is a nuclear factor and not directly involved in recognition of phosphatidylserine. It is mentioned here to help readers avoid unnecessary confusion when perusing the literature in this area.

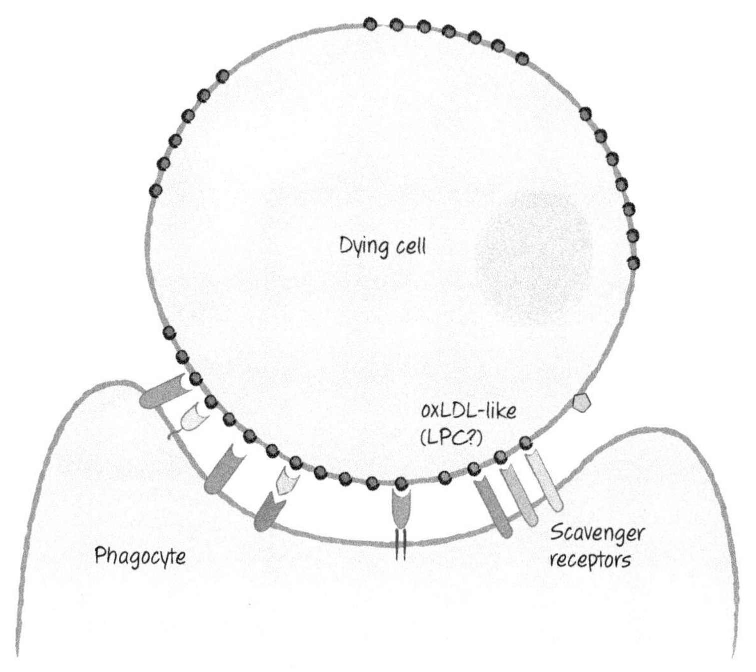

Figure 9.9. Scavenger receptors recognize dying cells expressing, for example, altered carbohydrates and exposed phosphatidylserine.

phagocyte). Calreticulin is also exposed on some dying cells, and it is likely that its interaction with LRP1 directly signals for engulfment. Figure 9.10 shows the array of receptors on phagocytes for dying cells and bridge molecules.

"BIND-ME" SIGNALS AND RECEPTORS IN OTHER ANIMALS

So far, we have focused entirely on the recognition of dying cells in vertebrate animals. However, much of what we know of the steps that follow binding has been gleaned from genetic studies of *C. elegans*. These have uncovered two complementary pathways for binding and engulfment of dying cells. Two of the genes identified (one for each pathway) encode the actual recognition receptors, CED1 and CED7. CED1 is the nematode homolog of LRP1, the potential receptor for cell-surface calreticulin, but we do not know whether calreticulin is the signal recognized by CED1. Alternatively, a bridging molecule that binds to phosphatidylserine, called transthyretin-like protein 52 (TTR-52), is bound by CED1 in nematodes. TTR-52 is a protein secreted by the endoderm; it forms clusters around dying cells and is required for their clearance.

CED7 is a cell-surface molecule in the ATP-binding cassette (ABC) family. It is important in both the dying and the engulfing cell and can engage in homotypic

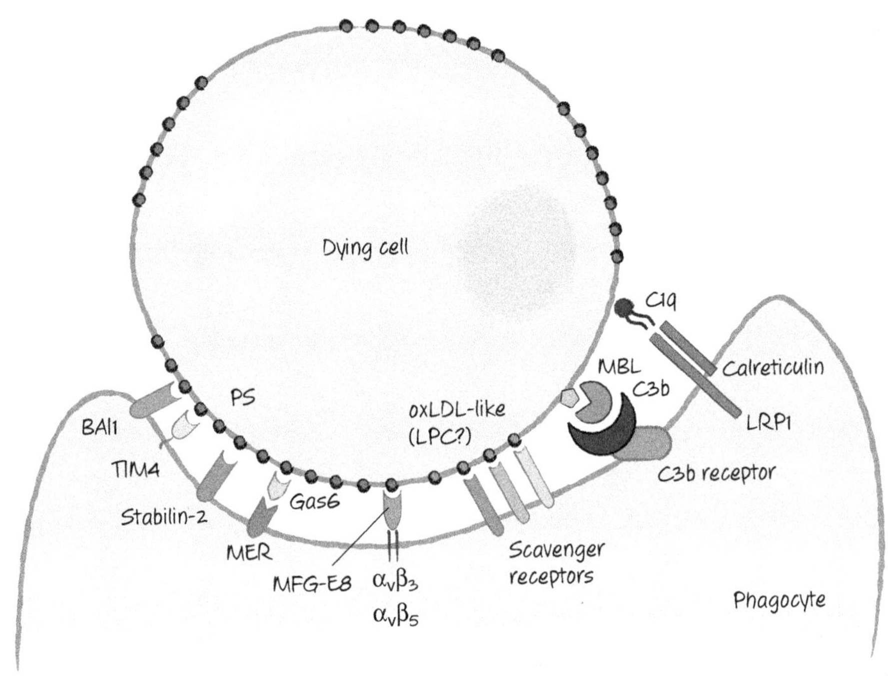

Figure 9.10. The array of "bind-me" receptors that can directly, or indirectly through bridge molecules, bind to dying cells.

interactions (that is, CED7 binding to CED7; there are, of course, more-complex possibilities). Some studies suggest that, in mammals, two related proteins, ABC1 and ABC7, function similarly in engulfment of dying cells and contribute to externalization of phosphatidylserine.[5] The recognition receptors of *C. elegans* are illustrated in Figure 9.11.

In *Drosophila*, an evocatively named protein, Croquemort ("undertaker"), can confer the ability to clear dying cells on a mammalian cell line on which it is otherwise lacking. Flies with mutations in the gene show defects in engulfment of apoptotic cells during development (Fig. 9.12). Croquemort is a homolog of the scavenger receptor CD36, which, as we have seen, is implicated in vertebrate recognition of dying cells.

"EAT-ME" SIGNALS PROMOTE ENGULFMENT

Engulfment requires extensive reorganization of the actin cytoskeleton of the engulfing cell to extend the phagocyte membrane around the corpse. The mechanisms that control this step are conserved in animals and involve small GTPases of the Rho family,

[5] How ABC1 and ABC7 might function in either phosphatidylserine externalization or engulfment is not known, and therefore these were not included in the mammalian bind-me scheme discussed above.

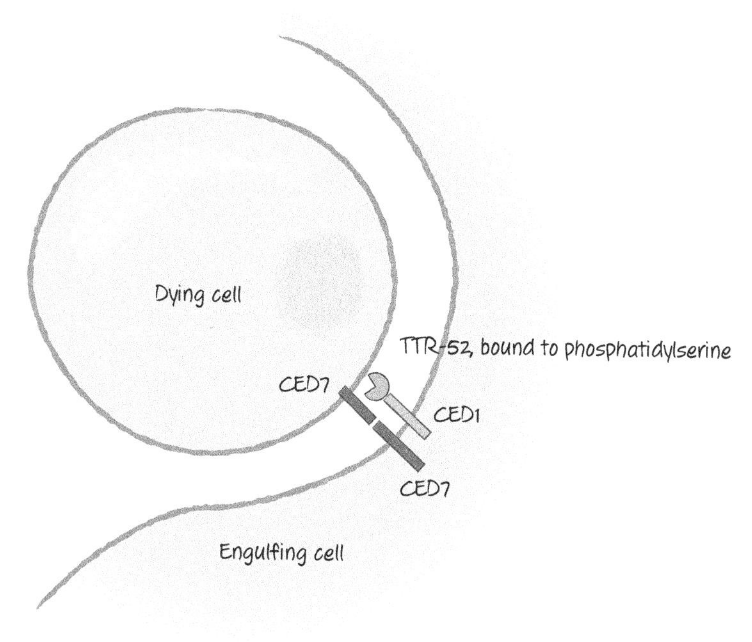

Figure 9.11. Recognition receptors CED1 and CED7 in nematodes. CED7 is important in both the dying and the engulfing cells, whereas CED1 is important in the engulfing cell. CED1 can bind to a bridging molecule, TTR-52, which recognizes phosphatidylserine and is required for clearance of dying cells. There is at least one additional receptor (not illustrated) involved in recognition and engulfment of dying cells.

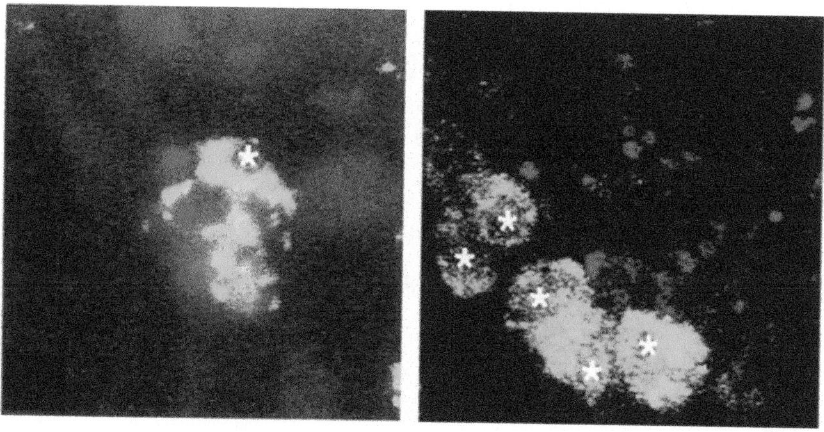

Figure 9.12. The *Drosophila* Croquemort mutant has defects in engulfment of dying cells. Here, phagocytic cells (green, with white stars) take up small dying cells (red) in wild-type flies (*left*) but not in mutants unable to express Croquemort (*right*).

especially RAC1 (CED10 in nematodes). This protein performs many functions in the cell, but we focus on its role in engulfment. In *C. elegans*, the two complementary pathways for engulfment converge on RAC1/CED10 (Fig. 9.13).

One of the pathways is linked to CED1 (LRP1 in vertebrates), which signals to RAC1/CED10 through the adapter molecule CED6, called GULP (get it?) in other organisms. Although we do not know whether LRP1 signals through GULP, another phosphatidylserine receptor, stabilin-2, appears to require GULP to induce engulfment (Fig. 9.14). How GULP–CED6 activates RAC1–CED10 is not known. In *Drosophila*, the CED1 homolog is called Draper, and this interacts with GULP–CED6 as well. But engulfment induced by Draper is also dependent on a tyrosine kinase,[6] and this might in time provide a clue as to how this signaling complex functions.

The other pathway that converges on RAC1–CED10 is linked to an unknown receptor in nematodes, but to BAI1 in mammals. This signals through a complex of three nematode proteins: CED2, CED5, and CED12. CED2 is CRKII in other organisms. It has two Src-homology 2 (SH2, phosphotyrosine binding) domains and a Src-homology 3 (SH3, proline-rich binding) domain and performs numerous functions in the regulation of actin. It interacts with CED5, called DOCK180 in other organisms, and CED12, called ELMO (for "engulfment and cell mobility") (Fig. 9.15).

Here is how it all appears to work.[7] DOCK180 is an unusual type of guanine nucleotide exchange factor (GEF) thought to load GTP onto RAC1 by forming a scaffold that binds several components, including RAC1. DOC180 is normally inhibited by the binding of its SH3 domain to its RAC1-binding region. When an appropriate receptor, such as the phosphatidylserine-binding receptor BAI1, is activated, a protein called TRIO (another GEF) loads GTP onto the small GTPase RHOG. RHOG–GTP then recruits ELMO (Fig. 9.16). ELMO then binds to the SH3 domain of DOCK180, allowing the latter to bind RAC1 as well as CRKII (Fig. 9.17). This complex, now localized to the contact site where the phagocyte binds to the dying cell, activates RAC1. This, in turn, reorganizes the actin cytoskeleton.

The phagocyte membrane envelops the corpse to form a vesicle called the phagosome. The actin is reorganized, and the phagosome now carries its cargo off to lysosomes for disposal (Fig. 9.18). We return to this part of the story later in this chapter.

ENGULFMENT CAN PROMOTE CELL DEATH

In some situations, cells that engage apoptotic (and perhaps other) pathways that promote their uptake might not be "dead" until the engulfment and degradation

[6] Intriguingly, this kinase is a homolog of the human kinase Syk, which is involved in the process of phagocytosis triggered by Fc receptors (the receptors on macrophages and other cells that bind antibodies). It is not known whether Syk participates in the uptake of dead cells in mammals.

[7] This model is not quite complete. Our discussion assumes some understanding of GTPases and how they work; if this is obscure to the reader, it will not greatly matter for general understanding.

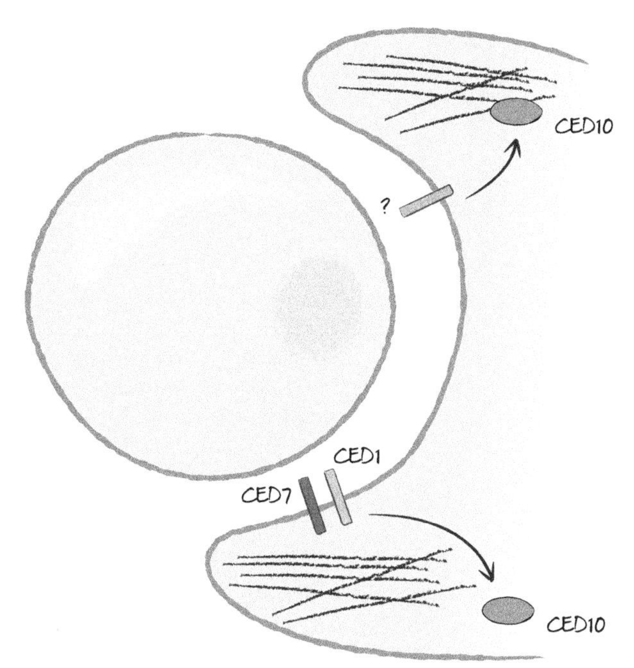

Figure 9.13. Two pathways converge on the Ras-related protein CED10 (Rac1) to induce actin reorganization and phagocytosis. Although one pathway is linked to CED1 and CED7, the other pathway is initiated in the nematode *Caenorhabditis elegans* by an unknown receptor.

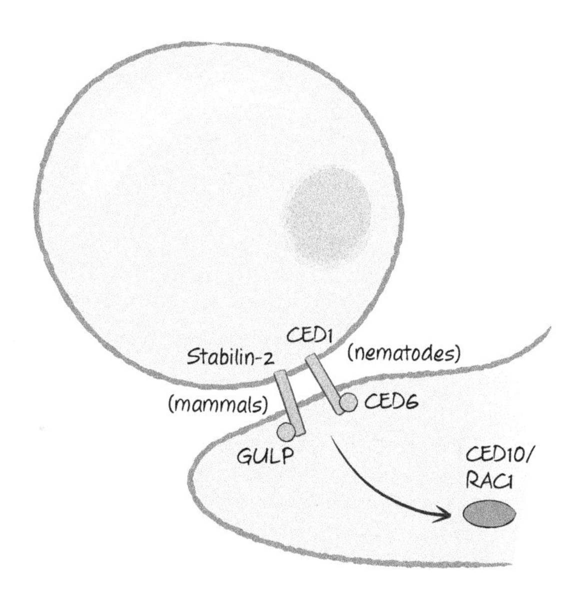

Figure 9.14. CED1 (nematodes) and stabilin-2 (mammals) engage an adapter molecule, CED6 or GULP, respectively, following recognition of dying cells. CED6/GULP activates CED10/RAC1 by an unknown mechanism.

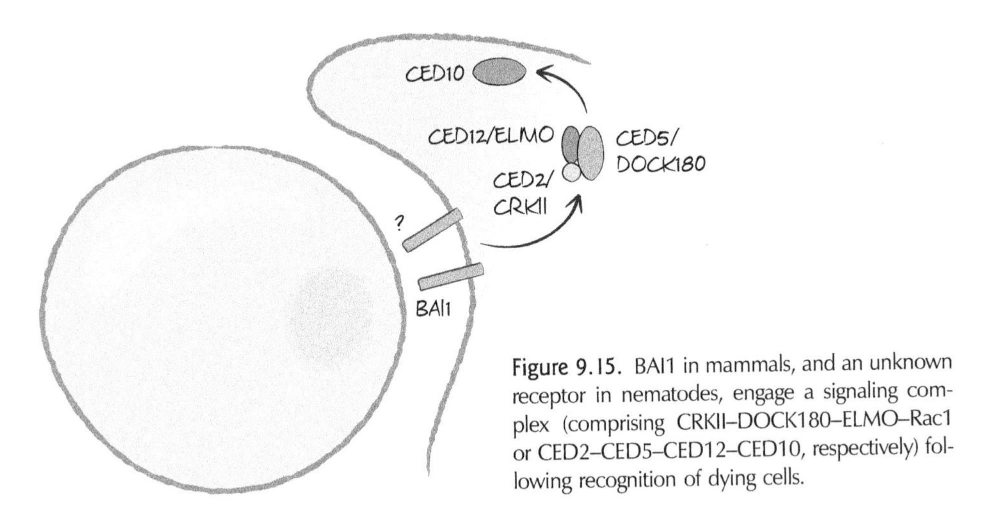

Figure 9.15. BAI1 in mammals, and an unknown receptor in nematodes, engage a signaling complex (comprising CRKII–DOCK180–ELMO–Rac1 or CED2–CED5–CED12–CED10, respectively) following recognition of dying cells.

machinery acts on them. This has been most compellingly seen in the case of apoptotic cell death in nematodes, especially when the activity of the executioner caspase (CED3) is limited. In such cases, defects in engulfment can result in the appearance of living cells that would not otherwise be there. It appears that engulfment is necessary to enforce cell death in such cases.

In *Drosophila*, some cell death is mediated by autophagy (discussed in Chapter 8). One of the molecules involved in engulfment of dying cells, Draper, is somehow necessary for this phenomenon, but in this case it must be expressed in the cell that is destined to die. At this point, we do not understand how this molecule functions in the cell death.

As we discussed above, disruption of "don't-eat-me" signaling (e.g., CD47, SIRPα) can also promote the engulfment of stressed cells, such as tumor cells, and

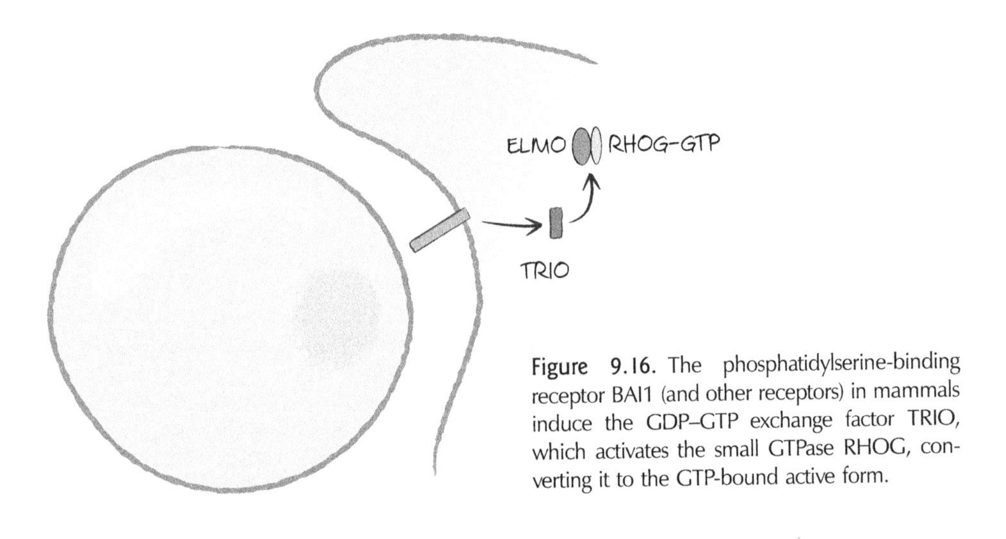

Figure 9.16. The phosphatidylserine-binding receptor BAI1 (and other receptors) in mammals induce the GDP–GTP exchange factor TRIO, which activates the small GTPase RHOG, converting it to the GTP-bound active form.

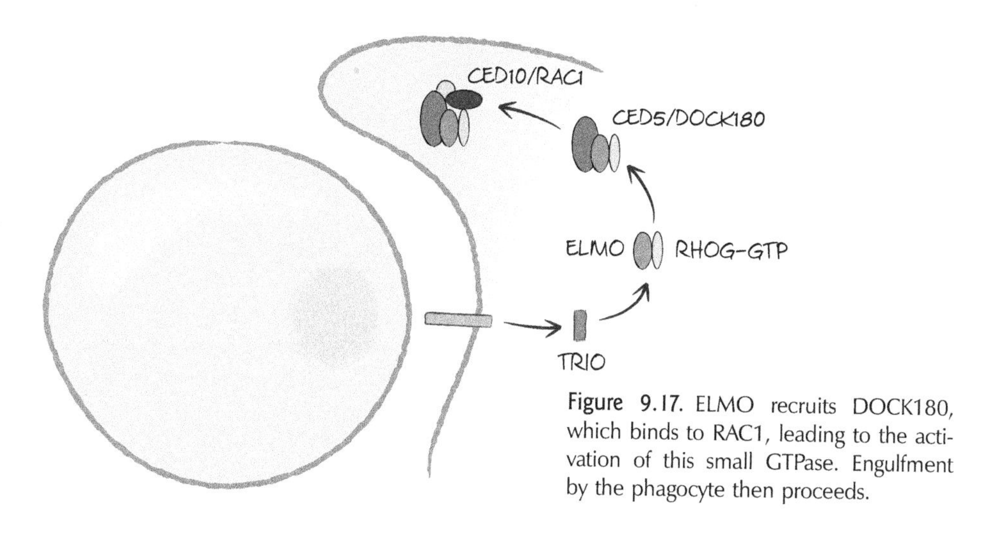

Figure 9.17. ELMO recruits DOCK180, which binds to RAC1, leading to the activation of this small GTPase. Engulfment by the phagocyte then proceeds.

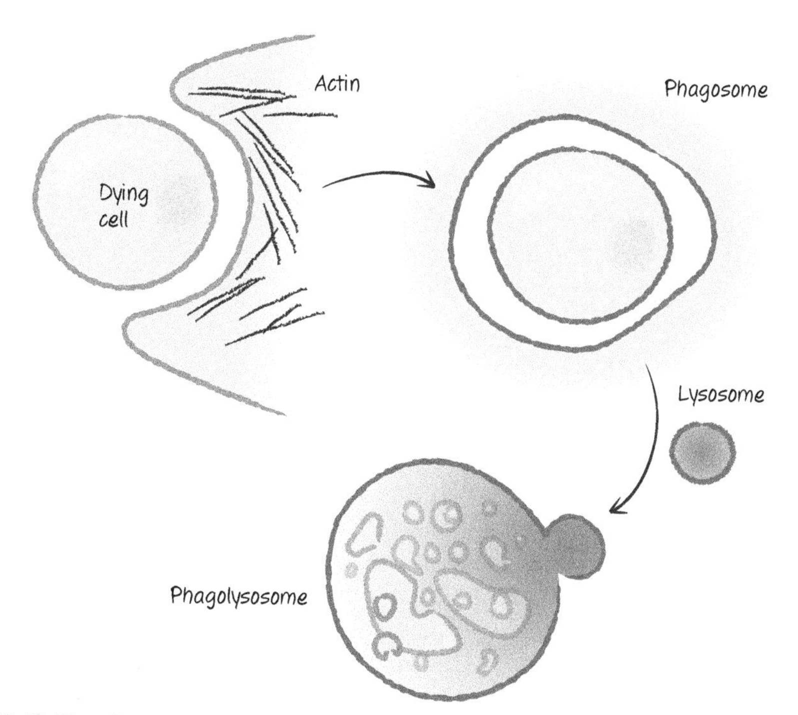

Figure 9.18. The phagosome and degradation. The phagocyte envelops the dying cell, forming an internal phagosome that, on rearrangement of the actin cytoskeleton and fusion with a lysosome, forms a phagolysosome, where disposal of the dying cell takes place.

their death. In mammals, we do not know how much this process might function in physiological situations.

There is another phenomenon that promotes death of cells by engulfment, although, in this case, the process is not dependent on the mechanisms by which phagocytes clear dying cells. This process (entosis) is discussed in more detail below.

THE "CLEAR-ME" PROCESS AND WASTE MANAGEMENT

Once the dying cell is engulfed in a phagosome, the engulfing cell must deal with the debris. This begins in the phagosome with the action of enzymes that degrade proteins and lipids, but the real work is performed when the phagosome fuses with lysosomes to form a phagolysosome.

Lysosomes contain an array of degradative enzymes including a nuclease, DNAse II. Apoptotic cells chop up their DNA by the caspase-dependent activation of the nuclease CAD (see Chapter 2). Cells lacking CAD or its inhibitor-chaperone iCAD do not effectively fragment their DNA. However, when such cells are engulfed by phagocytes, the DNA is nevertheless degraded. In contrast, mice lacking both CAD function and DNAse II show no DNA fragmentation of dying cells.

The iCAD–CAD complex is highly conserved in animals but is not present in nematodes. In the worm, a DNase in the phagosome digests the DNA. Digestion of DNA is an important waste-management problem in the clearance of dying cells.

Another waste-management issue is cholesterol, the levels of which effectively double when a phagocyte takes up a dying cell. Recognition of phosphatidylserine on the corpse generates signals that activate PPARγ (peroxisome proliferator-activated receptor γ), and LXR (liver X receptor), which are important regulators of cholesterol and other lipids in the cell. PPARγ shuts down a number of processes, including generation of cholesterol; and LXR activates expression of a transporter to remove cholesterol from the cell (Fig. 9.19).

THE IMPACT OF DYING CELLS ON THE INNATE IMMUNE SYSTEM

Massive numbers of cells die in vertebrates all the time, and were we to generate immune responses to these, we would face the severe consequences of autoimmunity.[8] But organisms must also be able to tell whether the cell death is a consequence of infection, so that an appropriate response can be rallied.

The innate immune response induced by dying cells involves Toll-like receptors (TLRs) and NOD-like receptors (NLRs), the pattern-recognition receptors (PRRs)

[8] To fully appreciate the impact of dying cells on the immune response, the reader will require a much more complete understanding of the immune system than can be provided here. Therefore, our overview is necessarily cursory.

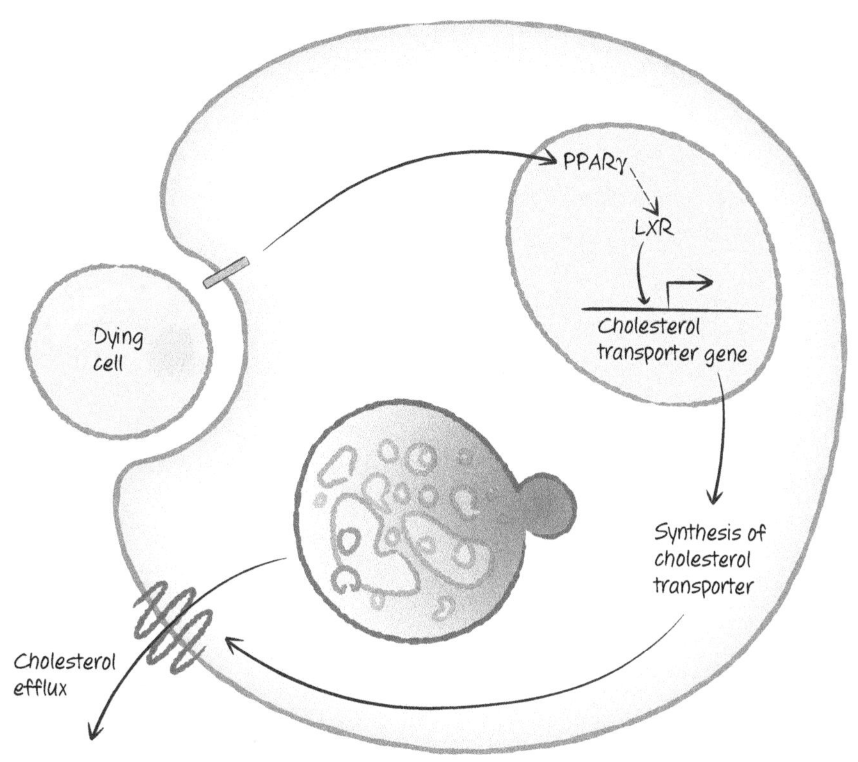

Figure 9.19. Uptake of dying cells induces cholesterol management. Activation of the peroxisome proliferator-activated receptor γ (PPARγ) induces the expression of the nuclear oxysterols receptor LXR-alpha (NR1H3), which in turn induces the expression of a cholesterol transporter operating at the plasma membrane.

discussed in Chapter 7. When these are engaged by pathogen-associated molecular patterns (PAMPs) or DAMPs, they generate signals that include inflammasome activation (see Chapter 7) as well as a variety of other effects. The latter include the production of cytokines and chemokines, molecules that control the differentiation and functions of immune and inflammatory cells.

Cells that die by necrosis release a number of DAMPs that trigger inflammatory responses. These DAMPs include uric acid, ATP, and another DAMP that induces the production of cytokines but not inflammasome activation, high-mobility group protein B1 (HMGB1). HMGB1 interacts with RAGE ("receptor for advanced glycation end products") and, perhaps, one of the Toll-like receptors (TLRs), TLR4, on phagocytes to trigger such responses.[9]

[9] TLR4 is activated by bacterial lipopolysaccharides. Because these frequently contaminate proteins, there is always a concern that ligands that are identified for TLR4 are actually artifacts because of such contamination by the lipopolysaccharides. Therefore, the ability of HMGB1 to directly stimulate TLR4 is controversial.

In contrast, apoptotic cells seem to have a calming effect on these responses. Not only do they not have a chance to release their DAMPs before engulfment, in some settings they also actively suppress inflammatory responses. There seem to be different ways in which they do this. Some apoptotic cells release the cytokine transforming growth factor β (TGF-β), which inhibits many immune responses. Alternatively, some phagocytic cells (macrophages and dendritic cells) produce TGF-β following interaction with apoptotic cells.

Another way apoptotic cells inhibit inflammatory responses may be through the activation of PPARγ (discussed above). PPARγ inhibits the production of several inflammatory mediators, including leukotrienes and prostaglandins. Yet a third mechanism involves Mer, the receptor for Gas6. Mer also interacts with the type I interferon receptor, and this activates transcription of an intracellular inhibitory protein, suppressor of cytokine signaling 1 (SOCS1), which can interfere with signaling by TLRs. Figure 9.20 shows the different ways in which apoptotic cells can inhibit inflammation.

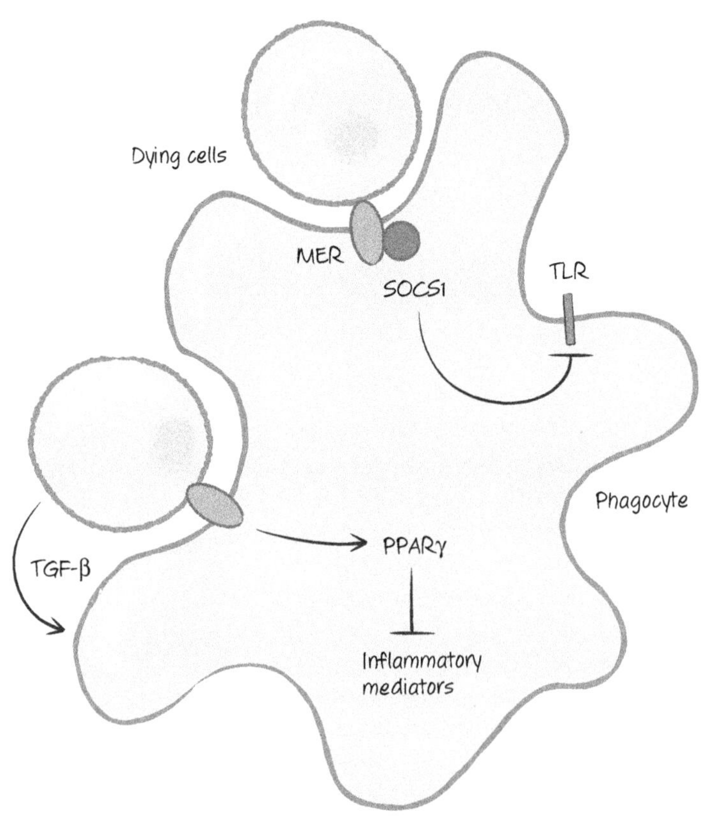

Figure 9.20. Apoptotic cells can inhibit inflammation. Mer, tyrosine-protein kinase Mer; PPAR γ, peroxisome proliferator-activated receptor γ; SOCS1, suppressor of cytokine signaling 1; TGF-β, transforming growth factor beta; TLR, Toll-like receptor.

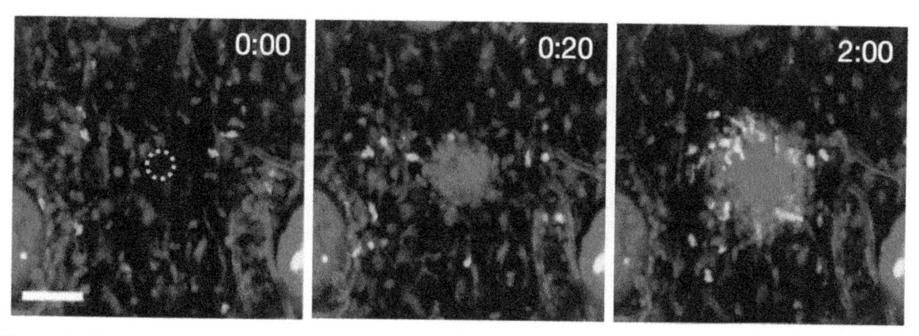

Figure 9.21. Neutrophils swarm to sites of cell damage. A single cell was damaged by laser light (dashed circle). Within seconds, the neutrophils (red) swarm to the site, followed by macrophages (green). The elapsed time is shown (minutes:seconds).

By such means, the extensive apoptosis that occurs during tissue turnover does not evoke inflammatory responses. In contrast, what does do so is necrotic cell death, which occurs in response to damage or pathological conditions, and the inflammatory response is rallied to deal with any infectious agent that might as a consequence gain access to the body.

What happens with other forms of cell death? As we know, activation of the inflammasome and caspase-1 can lead to death in the cells that are stimulated, and this death is accompanied by the production of cytokines (interleukin-1 and interleukin-18) that can stimulate inflammatory responses. It is generally assumed that such cell death does not inhibit innate immune responses.

A dramatic demonstration of the effect of cell death on the innate immune system can be seen when a cell is damaged by infection or injury and dies. Very quickly, neutrophils swarm to the site of the damaged cell (Fig. 9.21), and this depends on the production of a lipid signaling molecule, leukotriene-B4, produced as the cell dies (how this occurs is not clear). These cells function to seal the wound, and in time macrophages enter the area to clear the debris.

Cells undergoing necroptosis not only release DAMPs to induce inflammation, but also produce inflammatory cytokines (as a consequence of RIPK1 signaling). Finally, cells that undergo autophagic cell death are cleared from the body, but how these affect innate immune responses is not known.

CONSEQUENCES OF DYING CELLS FOR THE ADAPTIVE IMMUNE RESPONSE

When a phagocytic cell such as a macrophage or dendritic cell engulfs a particle, proteins associated with the particle are chopped into peptides and "presented" by the major histocompatibility complex (MHC) molecules on the cell surface. If a patrolling T lymphocyte recognizes the peptide, it can proliferate and generate an immune response, become refractory to stimulation, or actively inhibit immune responses (a state called immune tolerance). How it responds depends on other molecules on the presenting cell as well as the cytokines in the immediate environment.

PAMPs and DAMPs serve as maturation signals to dendritic cells, the "professionals" that present peptides to T cells and determine the T-cell response. Therefore, when cells that die by necrosis are eaten by dendritic cells, their proteins are processed to give peptides that are presented to T cells. Any peptides that are novel—for example, from an infectious organism that might have caused the necrosis—will be recognized, the T cells will proliferate, and an immune response will ensue.

Cells undergoing necroptosis are especially good at priming T-cell responses. This turns out to be dependent on the activation of NF-κB by RIPK1 during necroptosis. NF-κB induces a variety of cytokines in the dying cell that help to promote the T-cell response.

Apoptotic cells can similarly generate immune responses, but they can also actively inhibit them. The differences between immunogenic (the former) and tolerogenic (the latter) forms of apoptosis are not entirely clear. As with necrotic cells, the proteins associated with apoptotic cells are processed to give peptides that are presented on the surface of dendritic cells. Whether an immune response to any novel peptide occurs depends on several factors. One such factor is the presence of calreticulin on the dying cell. Some agents that induce apoptosis promote the exposure of calreticulin, whereas others do not. The reasons for this are not known, but the exposure of calreticulin appears to correspond with immunogenic cell death.

Another factor is the DAMP high-mobility group protein HMGB1. HMGB1 is expressed in all cells, and its release promotes immunity. However, this effect of HMGB1 can be blocked by oxidation (and perhaps other modifications) of the protein (Fig. 9.22). During apoptosis, a caspase-dependent burst of ROS can prevent the immunogenic effects, so that a tolerogenic response dominates. When they die, cells lacking HMGB1 tend to promote tolerance rather than an immune response, whether by apoptosis or necrosis (Fig. 9.22). In contrast, apoptotic cells in which the oxidation of HMGB1 is disrupted (e.g., by mutation of the caspase substrate NDUFS1 [see Chapter 2], which is responsible for the reactive oxygen burst) tend to promote immune responses rather than tolerance.

Cytokines also influence adaptive immune responses. As mentioned above, TGF-β is an inhibitory cytokine produced by some dying cells and by cells that engulf them. TGF-β induces T cells that encounter appropriate peptide antigens to differentiate into regulatory T cells (T_{reg}), which actively inhibit immune responses. However, other cytokines that can be induced by PAMPs can skew T-cell differentiation toward a type of T cell that promotes, rather than inhibits, inflammatory responses (Fig. 9.23).

LC3-ASSOCIATED PHAGOCYTOSIS AND CLEARANCE

As we discussed, autophagy is a process that forms double-membrane structures (autophagosomes) that trap cytoplasm and deliver it to the lysosomes for digestion

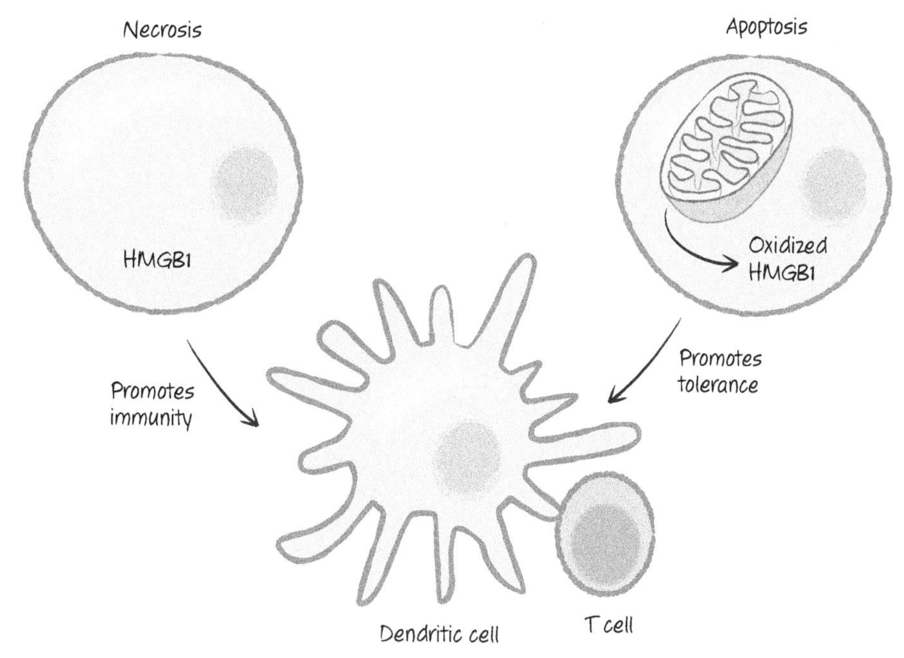

Figure 9.22. The high-mobility group protein B1 (HMGB1) in immunity versus tolerance. Apoptotic cells, unlike necrotic cells, end up oxidizing HMGB1, preventing it from promoting an immune response.

(see Chapter 8). When a dying cell is engulfed, some (but not all) components of the autophagy machinery are recruited to the membrane of the corpse-containing phagosome, where they decorate it with LC3 (on the single membrane of the phagosome). This process, called LC3-associated phagocytosis (LAP), is distinct from canonical autophagy,[10] as some molecules required for autophagy are not required for LAP, and vice versa. For example, LAP proceeds without a requirement for the ULK1 serine kinase complex, and the elements of the VPS34 phosphoinositide 3-kinase complex are distinct from those of autophagy (Figs. 9.24 and 9.25). These differences are useful in discriminating roles for LAP versus canonical autophagy in the clearance of dying cells.

Many of the events we have described so far in the clearance of apoptotic cells appear to depend on LAP. Apoptotic cells engulfed by macrophages that are defective for LAP are not rapidly degraded, do not induce the expression of the cholesterol efflux machinery, and do not induce the production of inhibitory cytokines. Instead,

[10] LAP is induced not only by the uptake of dying cells, but also by engulfed particles engaging some TLRs or, if antibody is present, by the binding of receptors for the antibody (FcRs). LAP plays a variety of roles beyond those of clearance of dead cells, but it is only the latter that concerns us here.

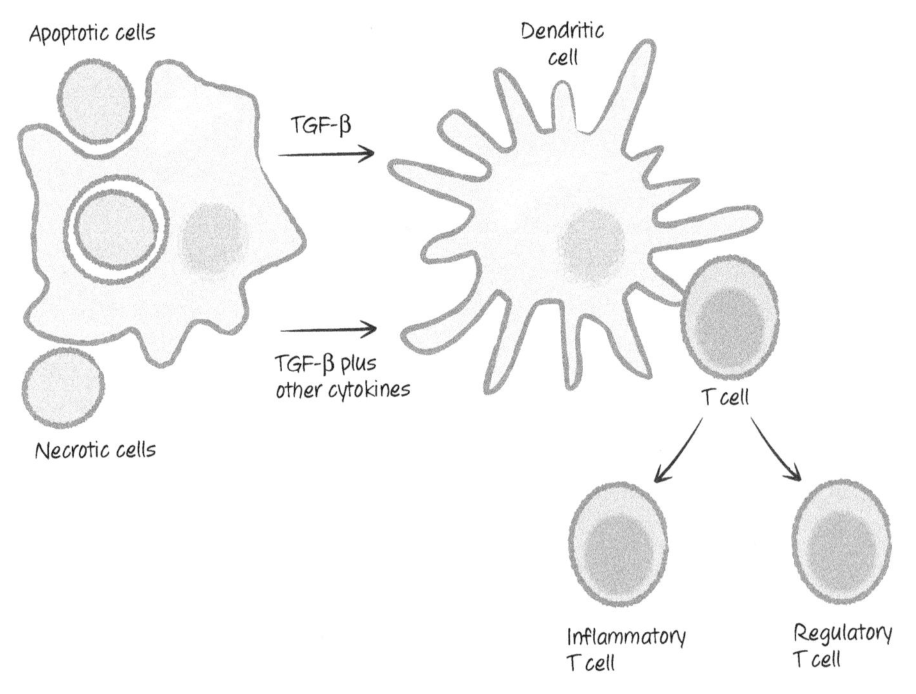

Figure 9.23. Dying cells influence T-cell functional differentiation. TGF-β, transforming growth factor beta.

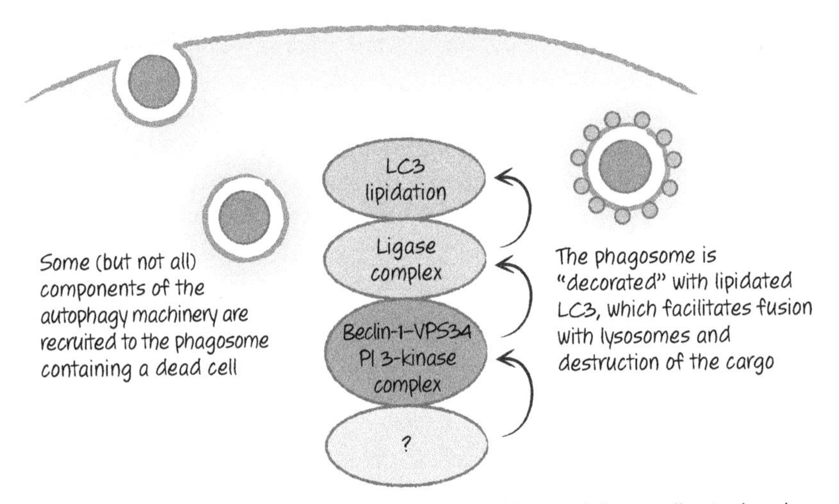

Figure 9.24. LC3-associated phagocytosis occurs on engulfment of dying cells. As the phagosome forms, an unknown signal is generated that recruits a complex containing beclin-1 and the phosphoinositide 3-kinase VPS34, and other molecules distinct from those involved in autophagy. As a result, the ligase complex is recruited and LC3 is conjugated to lipids in the phagosome membrane. This facilitates fusion with lysosomes and the degradation of the corpse.

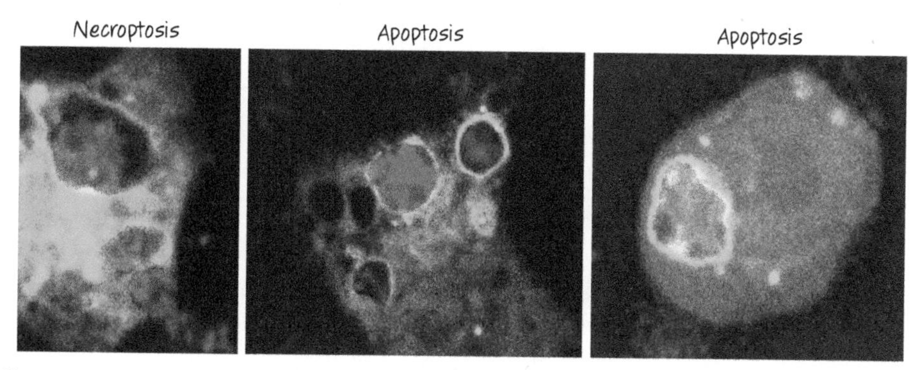

Figure 9.25. LC3-associated phagocytosis occurs on engulfment of dying cells. Macrophages expressing an LC3 fused to green-fluorescent protein (GFP) were cocultured with cells (stained red) that were dying by means of different mechanisms. Note the ring of green LC3 around the corpse-containing phagosomes.

such engulfment induces the expression of a variety of inflammatory cytokines, such as those that are produced on engulfment of necrotic cells. How, exactly, the process of LAP promotes these responses to apoptotic cells is not known.

CLEARANCE AND DISEASE

Do all these complex decisions really matter? The short answer is yes. Mice with defects in apoptotic cell clearance, such as those lacking MER, MFG-E8, or C1q, not only accumulate extra cell corpses but also show a form of autoimmunity resembling human systemic lupus erythematosus (SLE), a disease in which the immune system produces antibodies against DNA and other nuclear components. This effect is also seen in animals lacking LAP. In people with SLE, effective clearance of apoptotic cells frequently appears to be compromised. A diagnostic feature of SLE is the presence of so-called LE bodies in the circulation (see Fig. 9.26). These turn out to be free apoptotic corpses or corpses engulfed by macrophages (normally, clearance is sufficiently robust that such features are not seen in the circulation of healthy individuals). A feature of defective LAP is the persistence of the corpse in the phagosome. Intriguingly, polymorphisms in a gene involved in both canonical autophagy and LAP (*ATG5*) is associated with SLE in several human populations. This raises the interesting possibility that defects in LAP might be responsible for human SLE in some cases.

The inhibition of immune responses by apoptotic cells could have therapeutic value. A routine procedure in tissue transplantation is to infuse cells from the donor individual into the recipient. It is likely that these cells undergo apoptosis and they might dampen immune rejection of the graft.

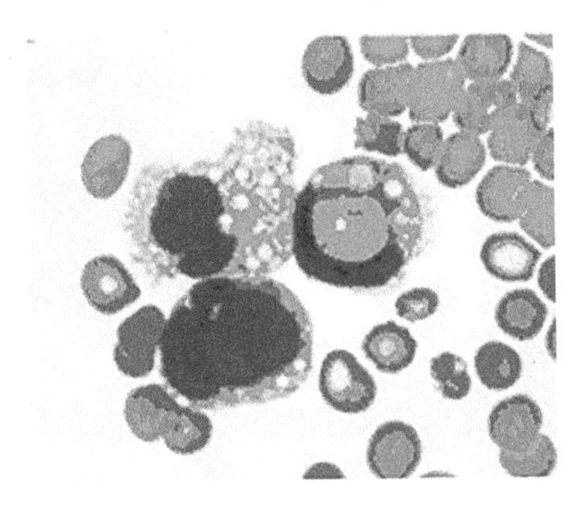

Figure 9.26. LE bodies. These are apoptotic cells in the circulation and are diagnostic of systemic lupus erythematosus (SLE).

Another therapeutic benefit involves the use of apoptotic cells to treat autoimmune diseases. In animal models, apoptotic cells coupled to proteins that are targets of the immune cells causing such diseases have proven effective in preventing or ameliorating autoimmunity in animals. Alternatively, it is possible to mimic the signals generated by apoptotic cells; drugs that activate a nuclear receptor downstream from PPARγ (see above) cure SLE in mice that have defective apoptosis. We will have to see whether these approaches work in humans.

Finally, when we treat cancers, we often use therapies designed to trigger apoptosis in the tumor cells. It is possible that, when such apoptosis is immunogenic, it contributes to eradication of the tumor.

ENTOSIS: EATEN TO DEATH[11]

Epithelial cells that lose contact with the basement membrane can undergo a process whereby they engulf each other, with the engulfed cell dying. In some developmental events and some cancers (discussed in Chapters 10 and 11, respectively), entosis is observed as the appearance of "cell-in-cell" structures, sometimes to such an extent that the engulfing cell is itself eaten, leading to a "Russian doll" situation—cell-in-cell-in-cell.

Unlike the engulfment of dying cells, entosis does not involve the usual machinery of eat-me signals and their receptors. Instead, the detached cells stick to each other by the interaction of adhesion molecules (cadherins), and the "losing" cell actively moves into the "winning" cell by actin rearrangement driven by ROCK1 (mentioned in Chapter 2 in another context) (Fig. 9.27).

[11] As entosis is an alternative form of cell death in some cases, it might have been considered in Chapter 8. However, given the role of LAP in the process, as well as apoptosis, we have chosen to discuss it here. Nevertheless, it is certainly a pathway of cell death.

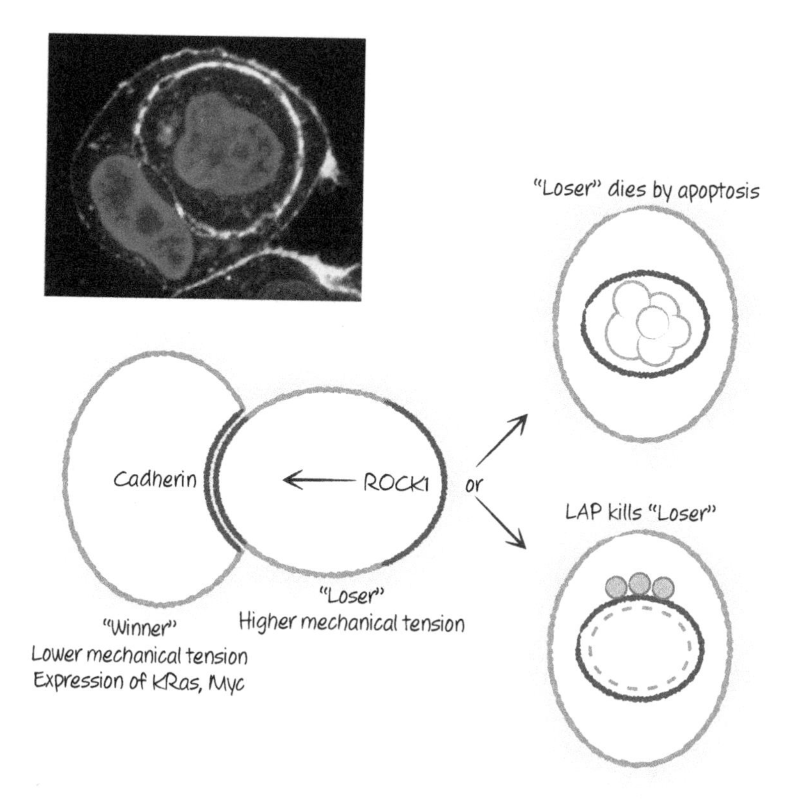

Figure 9.27. Pathways and molecules of cell death by entosis. LAP, LC3-associated phagocytosis.

Once eaten, the "loser" cell dies by a combination of processes. Because it is deprived of its normal growth factors, it can undergo apoptosis. Meanwhile, the "winning" cell engages LAP (see above) to promote the fusion of the cell-containing phagosome with lysosomes, ensuring the destruction of the "loser." If cells are engineered to both prevent apoptosis (e.g., by expression of anti-apoptotic BCL-2 proteins) and to avoid LAP (e.g., by deletion of molecules required for LAP), the engulfed cell survives and can escape, to either be engulfed again or do the engulfing. Cells that are normally dependent on attachment can grow in an anchorage-independent manner if both apoptosis and LAP are prevented or if entosis is prevented (e.g., by a ROCK1 inhibitor).

Several factors determine which cell "wins" or "loses" during entosis. Cells that have a lower mechanical tension (i.e., they are more "deformable") tend to engulf cells that have higher mechanical tension. Oncogenes, such as the GTPase K-Ras and Myc also promote a "winner" status, and indeed cells expressing these can engulf and kill normal cells that they recognize as "losers." These observations have potentially important repercussions in situations of cellular competition, including cancer (Chapter 11).

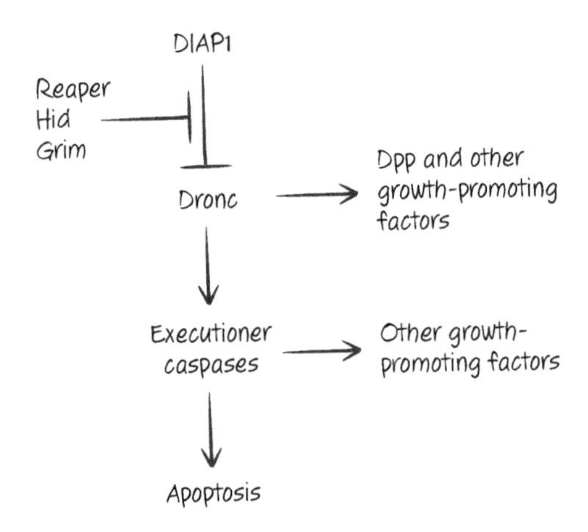

Figure 9.28. Caspase activation promotes compensatory proliferation in the fruit fly *Drosophila*. DIAP1, death-associated inhibitor of apoptosis 1; Dpp, Decapentaplegic.

COMPENSATORY PROLIFERATION

When cells die, we might expect this to generate a signal to make more of that cell type. Such compensatory proliferation has been verified in insects. Cells that undergo apoptosis in *Drosophila* produce factors that induce proliferation of surrounding cells. Interestingly, one of these factors, Decapentaplegic (Dpp), is a member of the TGF-β family (Fig. 9.28). This and the other factors[12] responsible for compensatory proliferation are produced in a manner that depends on the activation of caspases, but the precise mechanisms of Dpp release are not known.

A different mechanism occurs during wound healing, but also involves cell death. When a wound occurs in *Drosophila*, macrophages are attracted to the wound by hydrogen peroxide, produced at the site, and these promote healing. It turns out that this process requires that the macrophage had at one time engulfed apoptotic cells, leaving a "memory" of the event that primes the macrophage for this subsequent function. Macrophages from animals that are defective for apoptosis do not migrate to wounds, and this failure can be "corrected" by exposing them to apoptotic cells.

In vertebrates, apoptosis induces compensatory proliferation of adult stem cells in several tissues. Although TGF-β is sometimes produced by dying mammalian cells, it does not appear to be important for this effect. Instead, caspase-mediated activation of phospholipase A_2 (see Fig. 9.1) leads to the generation of prostaglandin E_2 (PGE_2), which is responsible for the effect. Although caspases are required for production of this mediator from dying cells, necrotic cells can induce PGE_2 production by

[12] The other pathways signal proliferation through Wingless (Wg; mammalian WNT is the homolog) and Hedgehog (Hh). Curiously, Dpp and Wg are produced in a manner that depends on the initiator caspase Dronc, but apparently not executioner caspases, whereas Hh signaling is dependent on the latter.

macrophages that engulf them,[13] and therefore it is likely that necrotic cells, too, can induce compensatory proliferation of stem cells.

Compensatory proliferation in response to cell death can also occur less directly. Living cells rely on growth and survival factors, as we saw in Chapter 5. It transpires that our cells do not only rely on these but, instead, compete for them; cells often take up and actively degrade such factors. When a cell dies, there are more of such factors available for the living cells, and these can then promote proliferation until the normal number of cells is achieved.

This process can be readily shown by removing cells from some tissues, such as surgical resection of part of the liver, or by introducing small numbers of normal lymphocytes into an animal in which they are lacking. In either case, rapid proliferation of the healthy cells occurs until normal tissue levels are attained.

Because dying cells are not directly responsible for the proliferation of healthy cells, this effect is not generally called "compensatory" but is, instead, often referred to as "homeostatic proliferation." As we will see in Chapter 11, even this indirect effect of cell death can have consequences for disease.

[13] However, as we noted, uptake of apoptotic cells seems to prevent production of prostaglandins. It is possible that, in addition to limiting inflammation, apoptosis in mammals tends to limit compensatory proliferation, whereas necrosis promotes it.

CHAPTER 10

Cell Death in Development

DEATH CREATES FORM

Michelangelo once described his art like this: "For sculpture, I mean that made by removing; that by adding is similar to painting."[1] Although animals are not fashioned from solid blocks by the removal of unneeded parts, such a process does contribute to the generation of form and function in the embryo. Of course, development is an immensely complex process involving cell proliferation, differentiation, migration, and adhesion as well as cell–cell interactions and cell death. But here, we are concerned primarily with the latter.

The goal here is not to catalog all the ways that cell death contributes to animal development but, instead, to explore how specific cues can activate the core cell death machinery, to program cell death in development. That is, we examine how a cell can be specified for death.

As noted in Chapter 1, the term "programmed cell death" refers to cell death that occurs at a genetically prescribed point during development (and has since come to mean apoptosis in many investigators' lexicons) and removes extra cells that can be present for a variety of reasons. For example, they might have inductive or scaffolding functions in developing tissues or be involved in selection steps. We discuss examples of each. Alternatively, animals can undergo metamorphosis during their life cycles, and cell death is involved in the repatterning that occurs. The specification of cells for death by developmental cues is best understood in invertebrate systems, and we begin with some examples from these.

CELL DEATH IN NEMATODE DEVELOPMENT

During the development of nematodes, cells follow a strict program in which every cell produced, beginning with the fertilized egg, has a predetermined fate. There

[1] In a letter to Benedetto Varchi, he wrote, "Io intendo scultura quella che si fa per forza di levare: quella che si fa per via di porre è simile alla pittura." Literally: "I mean, for sculpture, the one made by force of removing; the one through adding is similar to painting." (Thanks to Gerry Melino for the translation.)

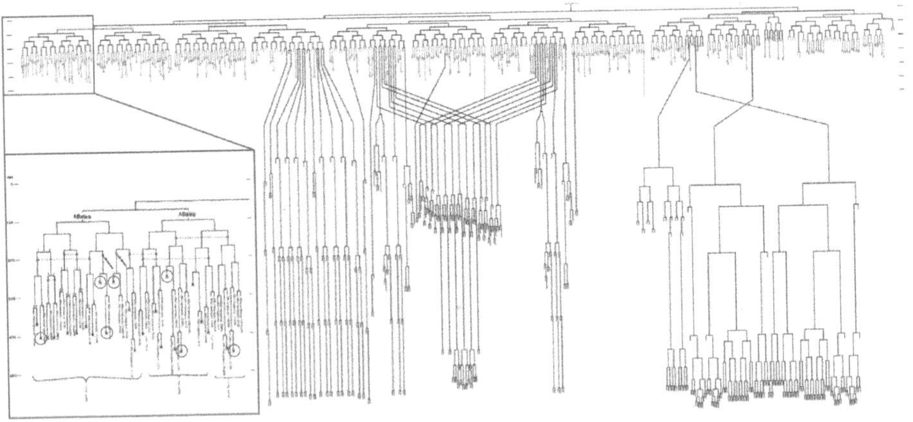

Figure 10.1. Developmental plan of the nematode *Caenorhabditis elegans*. During development, each cell that arises by cell division undergoes a set process of further division, differentiation, or cell death. (*Inset*) Part of the developmental plan, with seven cell deaths circled.

are 1090 cells produced, and 113 of these die during embryogenesis. After this, another 18 cells die on the way to adulthood. Some of the cell deaths that occur during embryogenesis are indicated in the inset in Figure 10.1. In the adult hermaphrodite, about half of the developing oocytes die, but there are no other cell deaths. In males (the other sex), no cell death occurs in adults.

All of the cell deaths that occur are by apoptosis, except for one—the death of the so-called linker cell. This cell is involved in the development of the male gonad and, based on morphology, appears to die by necrosis (Fig. 10.2). All of the other cell deaths depend on CED4 (the APAF1 homolog) and CED3 (the caspase).

Most of this cell death is in the neuronal lineages. The signals controlling the core apoptotic pathway are known for two of these: the neurosecretory motor neuron (NMN) sister cells and the hermaphrodite-specific neurons (HSNs). The latter only die in developing male worms.

In NMN sister cells and the HSNs destined to die, the helix–loop–helix transcription factors HLH2 and HLH3 induce EGL1, the nematode BH3-only protein. Another factor, CES1, inhibits expression of EGL1 in the NMN cells that do not die. Yet another

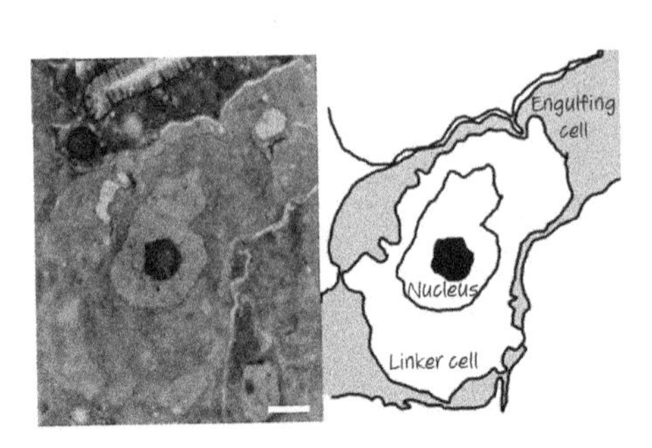

Figure 10.2. Necrotic-type linker cell death. In the micrograph (*left*), note that the dying linker cell has no features of apoptosis.

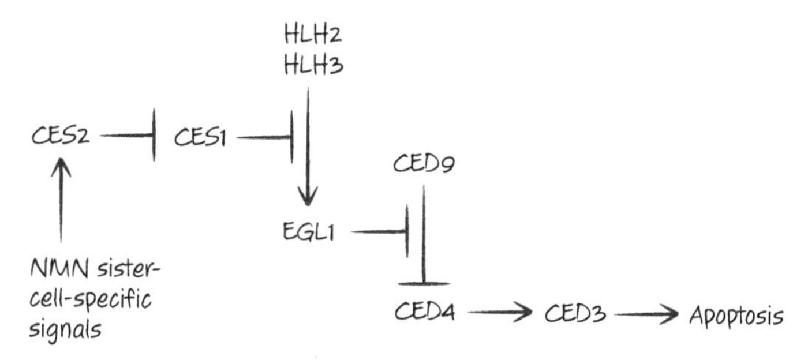

Figure 10.3. Neurosecretory motor neuron (NMN) sister cell death.

protein, CES2, is expressed in the NMN sister cells specified for death, and CES2 transcriptionally represses CES1. As a result, the NMN sister cells die (see Fig. 10.3).

In males, the pathway controlling HSN death is similar. Again, HLH2 and HLH3 induce expression of EGL1, leading to cell death. In hermaphrodites, expression of EGL1 in the HSN cells is blocked by expression of another factor, TRA1, that is not expressed in males (Fig. 10.4).

Additional pathways exist in other cell lineages. These probably also act through regulators of the core apoptosis pathway, but how this occurs is less clear. In a cell called the tail-spike cell, the timing of cell death appears to be controlled by transcriptional regulation of the caspase CED3 (Fig. 10.5). Although CED4 is required for the developmental death of this cell, EGL1 and CED9 (which are upstream) appear to have only minor regulatory effects. The control mechanism is a transcription factor, the homeobox protein PAL-1, that regulates CED3 and therefore apoptosis in this cell. EGL1 and CED9 therefore influence death but, in this case, are not targets of the specification.

The death of the tail-spike cell illustrates a concept that is useful for understanding the timing of cell death in development. If one or more components of the cell death pathway are limiting in a cell, factors that influence the expression and/or functions of these components can be crucial for specification of the cell for death. Although only one cell in *Caenorhabditis elegans* appears to be specified by the regulation of caspase

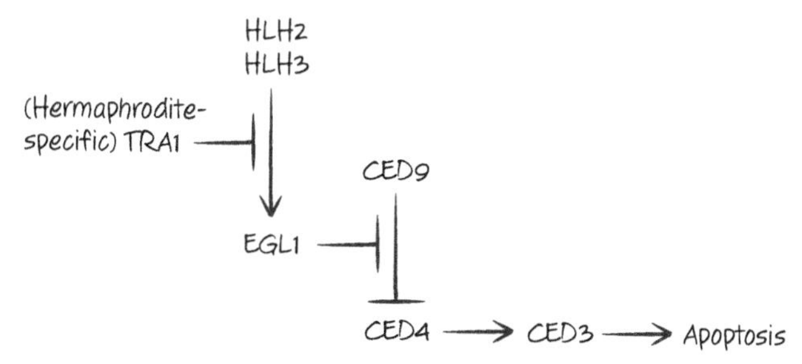

Figure 10.4. Hermaphrodite-specific neuron (HSN) cell death in males.

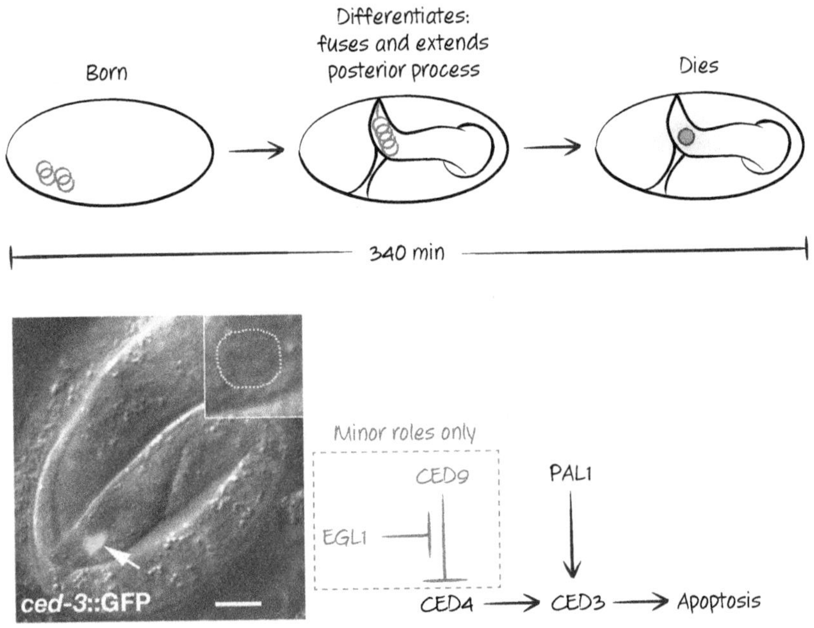

Figure 10.5. Cell death in the tail-spike cell. (*Top*) Timing of the appearance ("born"), differentiation, and death of the tail-spike cell during embryogenesis. (*Lower left*) Image shows expression of CED3 (arrow) in the tail-spike cell in an embryo in which the *ced3* gene is replaced with the gene for green fluorescent protein (and is therefore under the same regulation as *ced3*). (*Inset*) Death of the tail-spike cell in a wild-type embryo at the same stage. (*Lower right*) Schematic of the genetic pathway. Death in this cell is controlled mainly by the transcription factor PAL-1, which controls the expression of CED3, with little contribution from CED9 or EGL1.

availability, this general mechanism for the developmental regulation of cell death is probably involved in developmental cell death in other organisms.

CELL DEATH DURING METAMORPHOSIS IN *DROSOPHILA*

In insects, there is a dramatic change in morphology when the larva pupates and undergoes metamorphosis to form an adult. In the fruitfly *Drosophila*, this is triggered by two waves of the steroid hormone ecdysone, whose receptor is a ligand-dependent transcription factor. The second wave of ecdysone induces the expression of the transcription factors BR-C, E74, and E93. These up-regulate the expression of several genes encoding proteins involved in cell death, including ARK, Dronc, and Drice (the fly APAF1 homolog, caspase-9 homolog, and executioner caspase, respectively). These are probably not sufficient to cause cell death in metamorphosis because the caspases are inhibited by DIAP (see Chapter 3). However, the transcription factors also induce the expression of two DIAP inhibitors, Reaper and Hid, and as a consequence, apoptosis is engaged (Fig. 10.6).

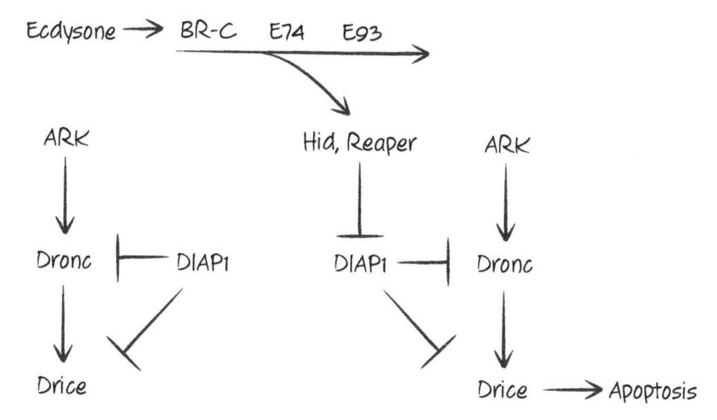

Figure 10.6. Activation of the apoptosis pathway by the hormone ecdysone. By inhibiting DIAP1, the proteins Hid and Reaper promote the cell death pathway leading to activation of the executioner caspase Drice.

Cell death by apoptosis proceeds in the anterior muscles, larval midgut, and larval salivary glands. Apoptosis is not responsible for all (or even most) of these cell deaths. Much of the cell death that occurs during insect metamorphosis displays characteristics of the second form of cell death, autophagic cell death (see Chapter 8). Figure 10.7 shows autophagic cell death in the metamorphosing salivary gland.

It turns out that BR-C, E74, and E93 also induce the expression of several proteins in the autophagy pathway, including ATG5 and ATG7. As we discussed in Chapter 8, autophagic (type II) cell death is accompanied by autophagy but is often not dependent on this process. In this case, however, the autophagy pathway is required for the cell death that occurs, and flies with defects in autophagy display fewer cell deaths in the salivary glands during metamorphosis (Fig. 10.8). We do not know precisely how the autophagy pathway kills these cells, but several tissues show this effect.

In addition to expression of proteins that promote death by apoptosis and autophagy, ecdysone triggers expression of Croquemort, the *Drosophila* CD36 homolog

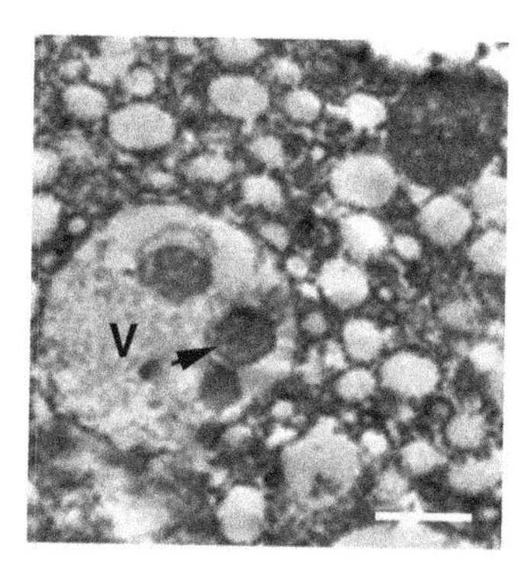

Figure 10.7. Type II death in the *Drosophila* salivary gland. Autophagosomes in dying cells (arrow) are abundant. V, vacuole-containing autophagosomes.

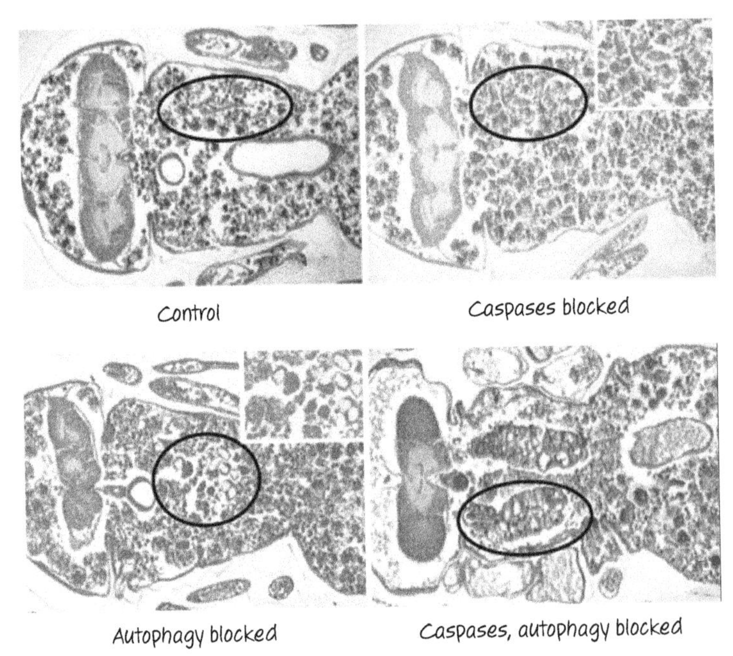

Figure 10.8. Cell death in the metamorphosing salivary gland of *Drosophila*. Inhibition of either caspases or autophagy in the animal promotes cell survival (circled areas, magnified in *insets*); inhibition of both produces additive effects.

involved in phagocytosis and clearance of dying cells (see Chapter 9). Presumably, Croquemort is expressed in cells that do not die, so that they can participate in corpse removal during metamorphosis. There is another possibility, however; it might be that autophagy and phagocytosis in the dying cells cooperate in their death. Flies with defects in phagocytosis accumulate living cells in the metamorphosing salivary gland, and intriguingly the CED1 homolog Draper (involved in engulfment of dying cells) is required for autophagic cell death in this tissue.

CELL DEATH IN VERTEBRATE DEVELOPMENT

Cell death occurs in many cell types throughout vertebrate development, and, in the majority of cases, this is by apoptosis. In invertebrates, cell death has roles in formation of structures, removal of structures, and control of cell number. In vertebrates, it can also have fundamental roles in selecting cells for emergent functions of a cell population (discussed in the next section).

A classic example of cell death in the formation of a structure is the generation of digits in the vertebrate limb. During development, a limb paddle (or autopod) forms in which tissue fills the areas between the skeletal elements. Bone morphogenetic

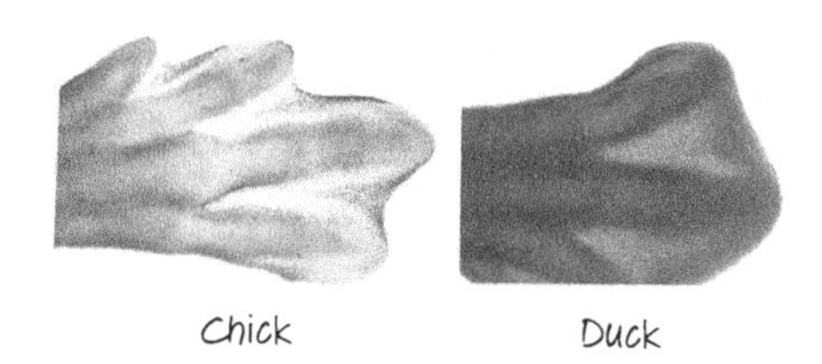

Figure 10.9. Cell death in developing chick and duck autopods. Dying cells in the autopods of both birds stain with red dye at the web tips. An artifact creates the more general staining in the duck interdigital webs.

proteins (BMPs), members of the transforming growth factor beta (TGF-β) family, trigger an apoptotic pathway in the interdigital webs, but how this specifies cells for death is not known. Figure 10.9 shows examples from chick and duck development in which the extent of apoptosis creates characteristically different outcomes.

Although the precise mechanisms have not been delineated, this interdigital cell death is almost certainly mediated by the mitochondrial pathway of apoptosis. Mice lacking the pro-apoptotic BCL-2 family effectors BAX and BAK usually die as embryos, but rare mice that survive retain persistently webbed feet (Fig. 10.10). Remarkably, this is also seen in mice that lack two of the BH3-only proteins, BIM and BMF. As discussed in Chapter 5, there are several BH3-only proteins, but, in this case, it appears that the developmental cell death is dependent only on two of them, which function redundantly. Mice lacking APAF1 or caspase-9 show a delay in digit formation during embryogenesis, but the cells nevertheless die by caspase-independent cell death (see Chapters 4 and 8).

Another example of cell death creating form is cavitation, the process by which some areas hollow out during development. Examples include the formation of the pro-amniotic cavity just after implantation in mammals and the hollowing of ducts in the mammalian breast. An unknown pro-death signal is generated, but cells that are in contact with the basement membrane receive survival signals that block the death. Only the cells in the center undergo apoptosis (Fig. 10.11).

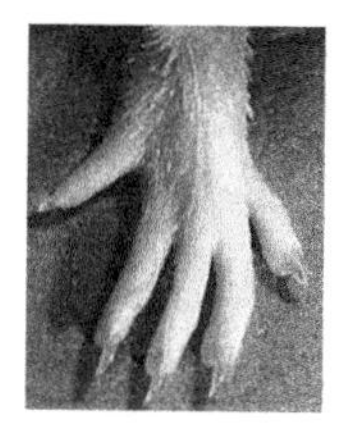

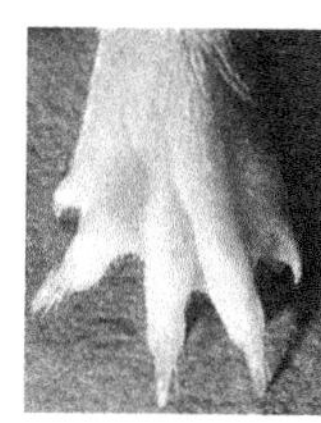

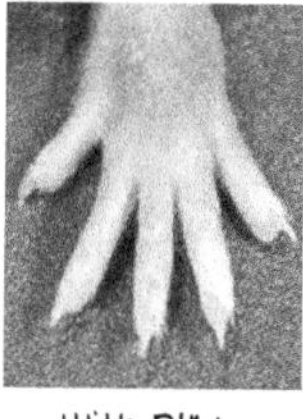

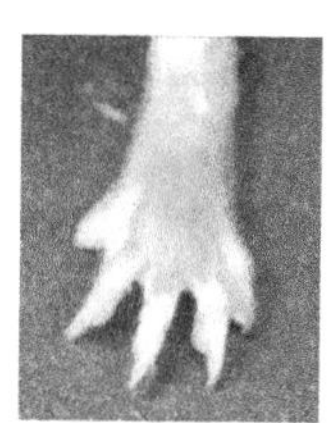

Figure 10.10. Webbed feet in mice lacking BAX or BAK, and mice lacking BIM or BMF, compared with mice expressing either (or both) of the genes.

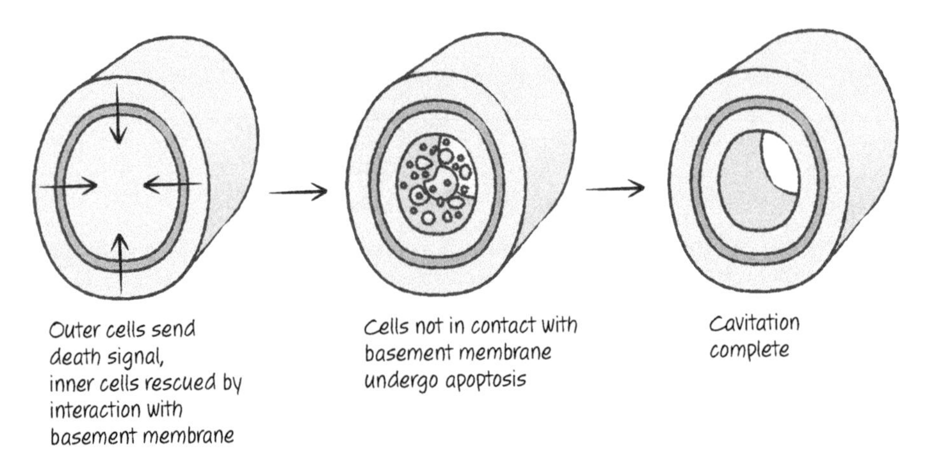

Figure 10.11. A basic plan depicting the regulatory signaling events leading to hollowing/cavitation.

In the case of the mammalian breast, the cell death signal appears to be provided by the BH3-only proteins BIM and BMF. Anti-apoptotic effects of MCL-1 can prevent the death of cells in contact with the basement membrane.

In the acini (the milk-producing structures) of the breast, cells that move away from the basement membrane and its extracellular matrix can undergo anoikis (see Chapter 5) because they no longer receive protective survival signals. The concept is a useful one and probably applies to many other developmental events. The survival signals provided by growth factors and the extracellular matrix help to define the boundaries to which a tissue can extend.

Other interesting examples of developmental cell death in vertebrates are the loss of the tail and restructuring of the intestine during amphibian metamorphosis (Fig. 10.12). Here, the signal for metamorphosis and cell death is thyroid hormone. This triggers a gene-expression program in which several caspase genes are induced, and there are changes in the expression of BCL-2 family proteins as well as a variety of tissue proteases. How cells are specified for death is not well understood, nor is the initiation of apoptosis. It is likely that the mitochondrial pathway of apoptosis is important because removing the amphibian caspase-9 homolog severely perturbs metamorphosis.

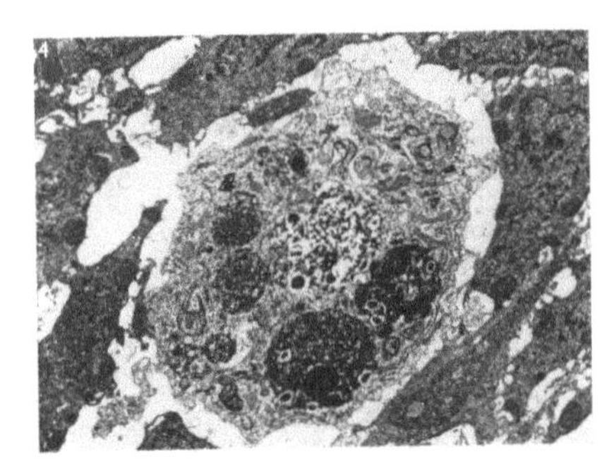

Figure 10.12. Frog metamorphosis and cell death. An apoptotic cell in the tail of a metamorphosing tadpole.

Not all developmental cell death in mammals is apoptotic. When the embryo implants, cells of the uterine epithelium must die for implantation to occur. However, no caspase-3 cleavage (indicative of apoptosis) occurs in these cells. Instead, it appears that embryonic cells (trophectoderm) engulf the epithelial cells (and it is possible that this is entosis [see Chapter 9]), thus killing the cells of the epithelium.

APOPTOSIS AND SELECTION IN DEVELOPMENT

Collections of cells can have emergent properties that far exceed the sum of their parts. By overproducing cells and then selecting for functions, the body can take advantage of such emergent properties. Selection can be positive, negative, or both, and most (if not all) selection involves cell death. Two examples of such selection in vertebrate development produce connectivity among neurons and self–nonself discrimination in the immune system. The latter probably represents the best example available of specification for cell death in a vertebrate system.

SELECTING NEURONS

In the developing nervous system, the axons of neurons extend to form connections with the dendrites of other neurons, and proper connectivity is essential for higher-order functions. One important way this can come about is proposed in the neurotrophism model. In this model, more neurons than will be ultimately used are produced, and these extend their axons toward another "target" population of neurons. Those that hit their targets then receive signals necessary to sustain their survival, whereas those that miss their targets die (Fig. 10.13).

Neurotrophic survival factors, such as nerve growth factor, released by the target cells, bind to receptors (e.g., TrkA) on the recipient axon and are transported to the cell body. There, the signals generated by the receptor are easily transmitted to the mitochondria and the nucleus. We can mimic this effect by culturing neurons and then depriving them of survival signals. When developing sympathetic neurons are deprived of nerve growth factor, they undergo apoptosis. This apoptosis depends, in part, on the BH3-only proteins BIM, BAD, and HRK and the pro-apoptotic effectors BAX and BAK (see Chapter 5). Mitochondrial outer membrane permeabilization (MOMP) then occurs, and caspases are activated through the mitochondrial pathway. As discussed in Chapter 5, MOMP can result in caspase-independent cell death in many cell types. However, neurons deprived of nerve growth factor can survive post-MOMP if the mitochondrial pathway is blocked or disrupted. This might be why mice lacking APAF1, caspase-9, or caspase-3 sometimes have brain abnormalities and excess neurons (see Chapter 4). However, an alternative explanation for these extra neurons is discussed in more detail below.

There is a problem with the view that the mitochondrial pathway of apoptosis is crucial for neuronal development. Mice lacking BAX and BAK cannot engage the

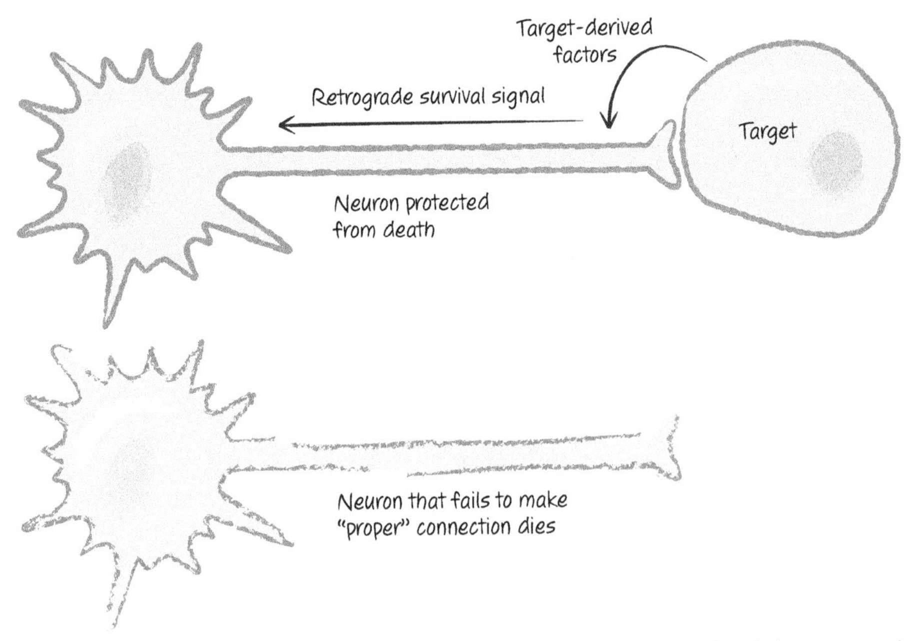

Figure 10.13. Cell death as a mechanism of neuronal selection. Neurons that fail to contact the appropriate target cell do not receive factors that promote their survival.

mitochondrial pathway, but the animals sometimes survive, and, although these animals display increased numbers of neurons, their behavior is not obviously disturbed. We have no good explanation for this. Perhaps cell death is not essential for neurotrophic selection and other mechanisms are involved.

APOPTOSIS IN LYMPHOCYTE SELECTION

The immune system has a remarkable ability to discriminate between self-tissues and foreign invaders, and this self–nonself recognition is essential for survival in an opportunistic world.[2] How this comes about involves a number of mechanisms, but we will focus on a central process that specifies developing lymphocytes to undergo apoptosis if they have the potential to recognize self-tissues that would obviously endanger the organism.

[2] The concept of self–nonself discrimination by the immune system was proposed at the beginning of the 20th century and was so persuasive an idea that almost 50 years passed before we recognized that some diseases are actually caused by a failure of the mechanisms involved, leading to autoimmunity. It was around this time that we first realized that it is only an approximation of self–nonself discrimination that is learned by the immune system. It is a very complex process, and we concern ourselves here with only one of the steps in the learning mechanism and how cells are specified for apoptosis. Interested readers would do well to consult an immunology textbook for a detailed treatment.

Vertebrate T lymphocytes express specialized T-cell receptors capable of recognizing small peptides presented to them on the surfaces of other cells (see Chapter 9). These receptors are generated in developing T cells by a random process involving rearrangements of gene segments that provide each T cell and its progeny with copies of a unique T-cell receptor of indeterminate specificity. The immature T-cell progenitors arise throughout life from hematopoietic stem cells that move to the thymus and differentiate into T cells.

At one point in the process, after the T-cell receptor has been generated and expressed on the cell surface, the cell undergoes a test in the form of the ligand for its receptor—a peptide bound to a major histocompatibility complex (MHC) molecule on another cell (see Chapter 9). If the T cell recognizes this ligand, it undergoes apoptosis and is removed from the population. This process is called negative selection. In effect, the ligand for the T-cell receptor on a given cell can specify the cell for death at this stage of lymphocyte development. Later, after the cell matures, recognition of ligand by the T-cell receptor instead produces an immune response (Fig. 10.14).

Negative selection helps self–nonself discrimination emerge in T cells. Peptides derived from proteins from the organism itself are always present, and therefore negative selection ensures that no T cell matures that can make immune responses to these self-peptides (at least, those that are present). Novel peptides that appear (e.g., during infection) are detected by T cells that developed when these were not present, and hence survived the test. Therefore, superficially, the discrimination is between peptides that are always present (self) and those that are only sometimes present (nonself).

How T-cell receptor signaling results in apoptosis is at least partially understood. When stimulated, the receptor complex generates signals that cause a rapid elevation of calcium ions in the cytosol. Calcium stabilizes the BH3-only protein BIM, allowing it to accumulate.[3] In addition, signaling also activates the mitogen-activated kinase JNK, which further stabilizes BIM and also phosphorylates BCL-2, and this decreases the anti-apoptotic activity of the latter (BIM is probably also transcriptionally induced in these cells, but the mechanisms have not been delineated). BIM then triggers the mitochondrial pathway of apoptosis. Mice that lack JNK or BIM do not display apoptosis associated with negative selection. However, the surviving cells do not mature to produce autoimmune disease; so, clearly, there are additional checkpoints that control this important process if apoptosis fails.[4] T-cell negative selection in the thymus nevertheless represents a clear example of specification for cell death mediated by a well-defined selection signal (Fig. 10.15).

[3] In Chapter 5, we described some ways in which BIM is regulated. Another is through calcium, which appears to activate a phosphatase to remove a phosphate group from BIM that otherwise promotes its degradation.

[4] In some experimental systems, defects in the recognition system that initiates negative selection permit such cells to mature, and the consequences include devastating multiorgan autoimmunity. It is likely that, in the absence of BIM-mediated apoptosis, stimulated cells differentiate into immunosuppressive regulatory T cells that inhibit any autoreactivity (see Chapter 9).

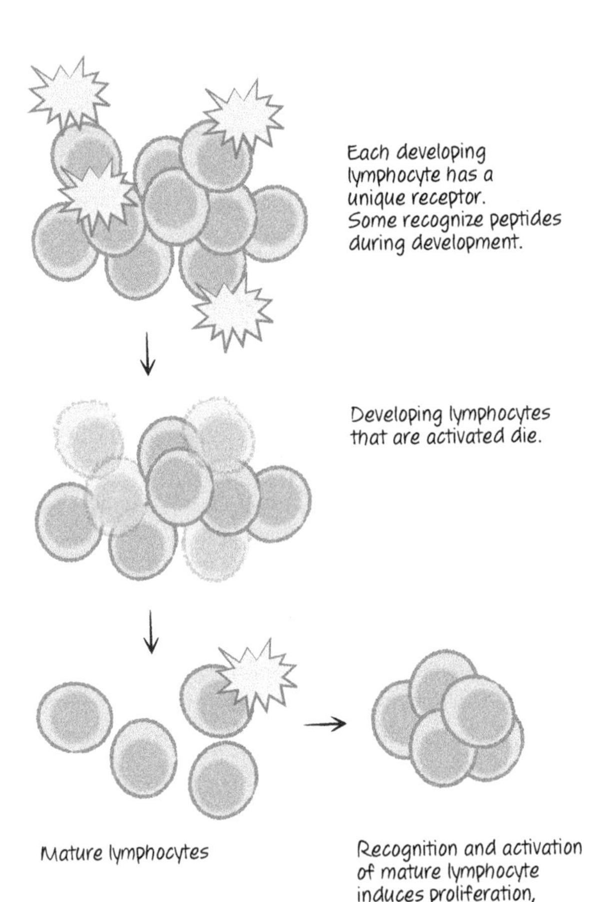

Figure 10.14. The concept of negative selection. Cell death of activated lymphocytes during their development produces emergent self–nonself discrimination.

Once the T cell matures, BIM remains in place, but now elevated levels of the anti-apoptotic protein BCL-2 appear to keep the cell alive. Mice that lack BCL-2 develop T cells, but these die in the first weeks after birth. If these animals also lack even one allele of *Bim*, however, the T cells mature normally and survive to function in immune responses.[5] In normal individuals, the recognition of a ligand by a T cell causes it to proliferate rapidly, but this is followed a few days later by a population-contraction phase because of apoptosis of the clone, and this limits the immune response. This contraction phase is also dependent on BIM, and animals lacking BIM do not show this effect (Fig. 10.16).

[5] The story is more interesting than related here, with additional insights into apoptosis and development. Mice that lack BCL-2 develop fairly normally to birth, but turn gray within 3 weeks and die of kidney failure a few weeks later. If the animals lack one allele of *Bim*, they survive (but have gray hair). If they lack both alleles of *Bim*, their coat color is normal.

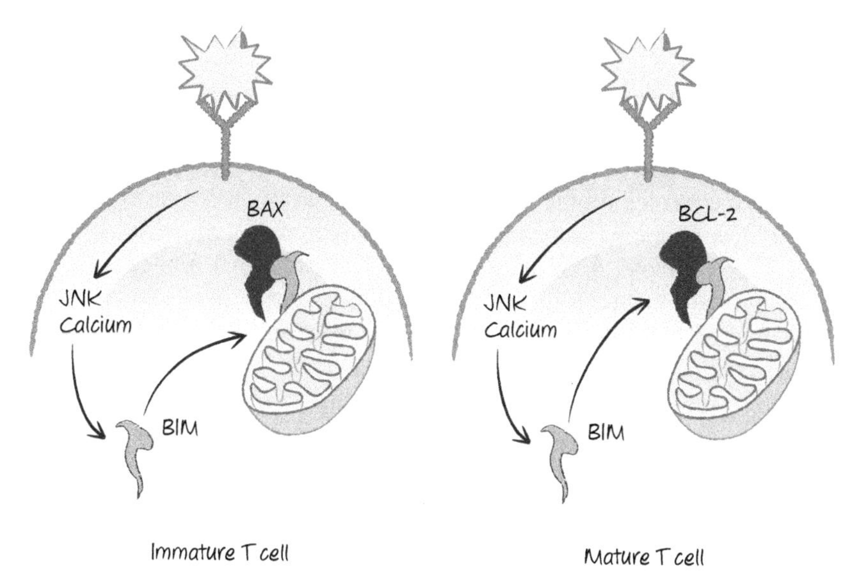

Figure 10.15. BIM and negative selection in T cells. T-cell activation induces JNK and elevated calcium, both of which stabilize BIM. In immature T cells, BIM induces apoptosis, but in mature T cells, this is blocked by BCL-2.

Those cells that persist mediate immune memory to guard against the appearance of the ligand again. We do not know what factors determine which cells die after the initial expansion versus those that persist to preserve immune memory.

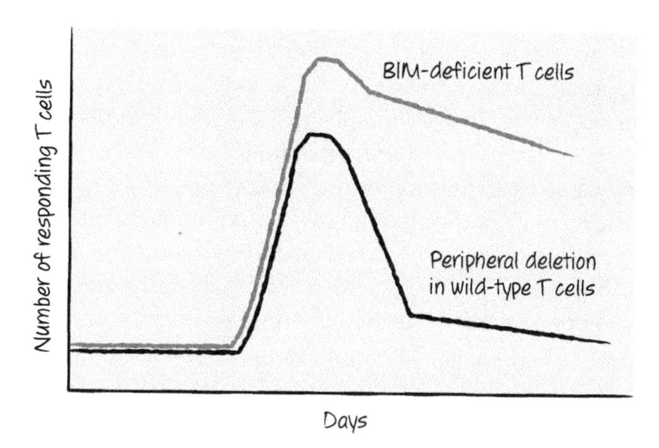

Figure 10.16. Peripheral deletion and BIM. The absence of the BH3-only protein BIM inhibits the peripheral deletion of T cells that would otherwise reduce the numbers of responding cells.

WHEN IS APOPTOSIS IMPORTANT IN MAMMALIAN DEVELOPMENT?

It is evident that cell death occurs during development in embryos and homeostasis in adults, but how crucial is apoptosis to these processes? We have discussed a number of striking developmental defects associated with removal of key components of the apoptotic machinery, some of which are necessary for survival. It is useful to revisit these to see how apoptosis participates in the developmental processes.

As we saw in Chapter 4, mice lacking APAF1 or caspase-9 or mice with a mutation in cytochrome *c* that prevents APAF1 activation all display developmental defects that result in forebrain outgrowth, a catastrophic developmental abnormality. However, in some strain backgrounds, the same gene ablations have much smaller effects, and many mice survive into adulthood. As it turns out, this forebrain outgrowth is not due to a failure of the neurons to die through the mitochondrial pathway of apoptosis. Instead, the defect is in the rate and efficiency of closure of the neural tube during development. Upon closure, the neural tube emits signals that halt the proliferation of developing neurons, and when this closure does not occur on schedule, the neurons continue to proliferate. Efficient apoptosis in the neural tube cells guides the closure of the neural tube and ensures that it occurs at the right developmental time. A delay in death can therefore result in the observed abnormalities.

The same is true for animals that lack BAX and BAK. Most of these animals (~85%) die during development or shortly after birth. Triple deletion of BAX, BAK, and BOK is even more devastating; only ~5% of these animals mature.[6]

Many of the animals that die around birth do so as a result of a severe cleft palate, resulting in them being unable to feed. As with the example above, apoptosis in the soft palate appears to be important in "guiding" the bones of the palate into proper position.

Nevertheless, some animals lacking BAX, BAK, and BOK survive into adulthood. These animals accumulate excessive myeloid and lymphoid cells (as do BAX, BAK double-deficient mice that survive), and therefore the mitochondrial pathway of apoptosis is important for homeostasis of these cells. However, most other tissues in these animals appear normal. Although the mitochondrial pathway of apoptosis is clearly important for normal development in most cases, the fact that some animals (and most of their tissues) can develop normally raises questions of how important this form of cell death is for most developmental and homeostatic processes.

Of course, there are other forms of cell death besides the mitochondrial pathway of apoptosis. At this point, we do not know whether such forms of cell death are important in development. We discussed the role of necroptosis in the embryonic lethality of animals that are deficient in caspase-8, FAS-associated death domain protein (FADD), or FLICE-like inhibitory protein (FLIP) (and this applies to a number of

[6] As we mentioned in Chapter 5, this finding indicates that there are signals (yet to be discovered) that engage BOK during development.

other effects of deletions of genes that regulate necroptosis), and there is no evidence that necroptosis, per se, is important in normal development. Mice that are deficient in RIPK3 or mixed lineage kinase domain-like protein (MLKL) cannot engage necroptosis in their cells, but develop normally.

Our foray into cell death in development has touched on how apoptosis and other forms of cell death are specified and how they function in some normal organismal processes. But cell death and survival also figure prominently in disease. We have seen how cell death is involved in ischemic injury and other forms of damage (see Chapter 8). We now turn to a discussion of how cell death and its regulation contribute to another important problem of defective cell and tissue homeostasis—cancer.

CHAPTER 11

Cell Death and Cancer

WHY IS CANCER RARE?

Most of us have been touched by cancer, whether directly or in people close to us; so at first this question seems bizarre. But humans are composed of approximately 10^{14} cells, most of them turning over continuously, and given that mutations arise at a frequency of about one in a million (and note that we produce about a million cells every second), why are we not all riddled with cancers every day?

The answer, of course, is that multiple checkpoints are in place to prevent cancers from occurring. In fact, if we try to produce cancer experimentally—for example, by introducing a known cancer-causing gene (an oncogene) into a tissue—often cancers arise only after a delayed period and from a single cell that has acquired additional mutations. So, these checkpoints are efficient.

Cancer is one of two diseases in which a single, dysregulated cell can, in theory, lead to lethal disease (the other is autoimmune disease, which can, in theory, be caused by one lymphocyte and its progeny).[1] It is better, in either case, to sacrifice the cell rather than try to manage the consequences. The induction of apoptosis is therefore a key strategy for preventing the emergence of cancer.

APOPTOSIS AND PROLIFERATION

It is a frequent misconception that, if the cell cycle is suddenly halted, this directly activates apoptosis. However, any links between the machinery of the cell cycle and those of the core apoptotic pathways are at best obscure (one such link is mitotic catastrophe; see Chapter 8). The cell cycle, per se, does not directly regulate apoptosis. That said, there are connections between apoptosis and the signals that engage the cell cycle, and these are fundamentally important for our discussion of cancer.

Myc is a nuclear factor required for entry into, and probably passage through, the cell cycle in animal cells. Mammals have three forms of Myc (Myc, L-Myc, and

[1] Although this is not how autoimmunity occurs, in general.

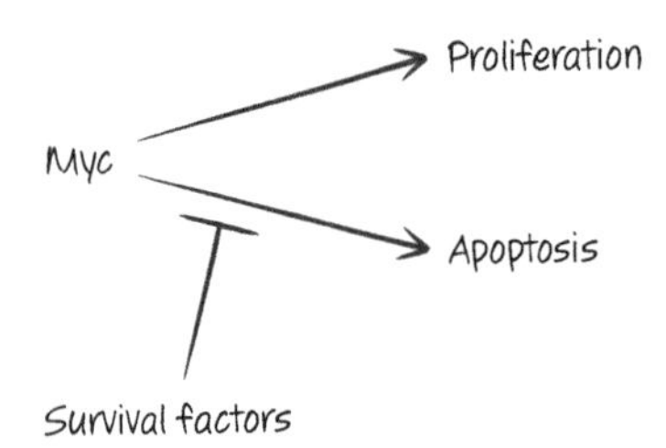

Figure 11.1. The principle of antagonistic pleiotropy.

N-Myc), and all function as transcription factors that cause cells to proliferate. It is therefore logical that, if Myc is expressed, cells will accumulate. But things are not that simple; the expression of Myc can also induce cells to undergo apoptosis.

Relatively low levels of Myc are sufficient to engage the cell cycle, but the apoptosis that would also be triggered by Myc in these cells is blocked by survival signals. Any of a variety of extracellular molecules that engage cell-surface receptors can produce anti-apoptotic signals (Fig. 11.1). For example, activation of the serine/threonine-protein kinase AKT by survival factors[2] induces the expression and/or stabilization of anti-apoptotic BCL-2 proteins and the repression and/or destabilization of pro-apoptotic BCL-2 proteins (see Chapter 5).

The underlying idea is termed "antagonistic pleiotropy."[3] If a cell activates the cell cycle (e.g., by engaging Myc), it will die unless other cells provide signals instructing it to survive and, thereby, allow it to generate increasing numbers of cells as it proliferates. Therefore, the "society of cells" collectively determines how much expansion is allowed, based on the availability of survival factors (Fig. 11.1).

Is it simply apoptosis that restricts cell accumulation? One approach to answering this is to replace survival factors (that have other effects as well, including metabolic effects) with an anti-apoptotic BCL-2 protein. This can also promote cell accumulation in this setting by blocking cell death (Fig. 11.2).

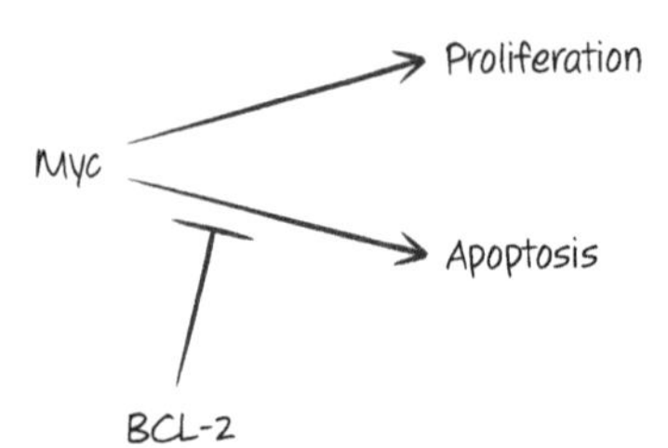

Figure 11.2. Antagonistic pleiotropy is regulated by BCL-2 proteins.

[2] Survival factors include cytokines (such as interleukin-2), growth factors (such as epidermal growth factor), and hormones (such as insulin) that bind to receptors on the cell surface and induce signals that ensure cell survival.

[3] There is another form of antagonistic pleiotropy that is an important model for the evolution of aging, and the two might be related. This other form does not concern us in this discussion, but it is worth looking up!

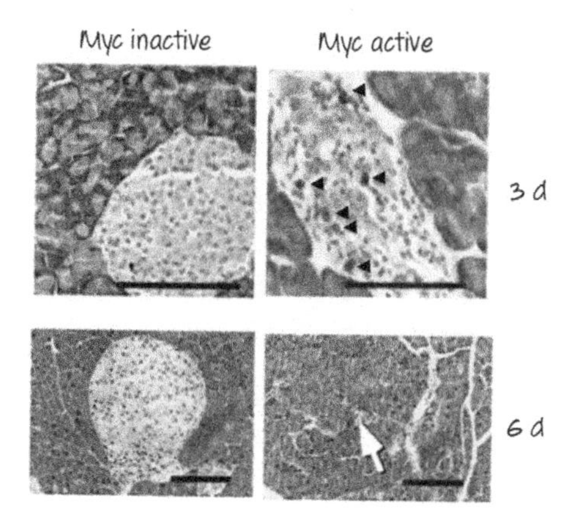

Figure 11.3. Myc in pancreatic islets causes apoptosis. Myc was activated in pancreatic islets and apoptosis was assessed at 3 days and 6 days postactivation. (*Top right*) Apoptotic cells marked with black arrowheads. After 6 days, the islets had involuted (*lower right*, white arrow).

The apoptosis-inducing effects have been dramatically shown in mice in which an inducible form of Myc was expressed in the β-islet cells of the pancreas (the cells that make insulin). Following activation of Myc, the cells died by apoptosis and the animals became diabetic (Fig. 11.3). However, if the cells were also provided with an anti-apoptotic signal such as BCL-2, the effect of activating Myc was very different; in this case, every islet expanded dramatically and displayed all of the features of a tumor (Fig. 11.4).

This extremely informative model tells us that two signals (one for proliferation and one to block apoptosis) might be all that is needed to promote cancer. Tumors, however, are not generally this simple, and many other changes contribute to the aggressiveness of a tumor and, ultimately, the harm it causes. But the underlying concept is a useful one. Thus, blocking apoptosis can have profound effects that promote oncogenesis.

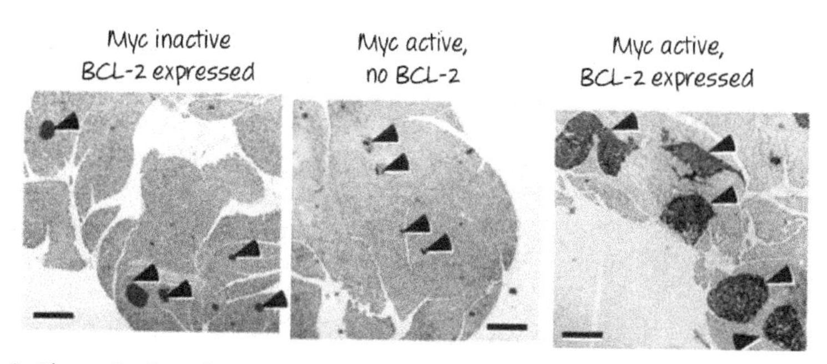

Figure 11.4. The activation of Myc in pancreatic islets expressing BCL-2 results in tumor-like expansion. (*Left*) Islets (black arrowheads) expressing BCL-2 appear to be normal. (*Middle*) Activation of Myc without BCL-2 (as for Fig. 11.3) for 7 days causes islet involution. (*Right*) Activation of Myc together with BCL-2 results in pronounced tumor-like expansion.

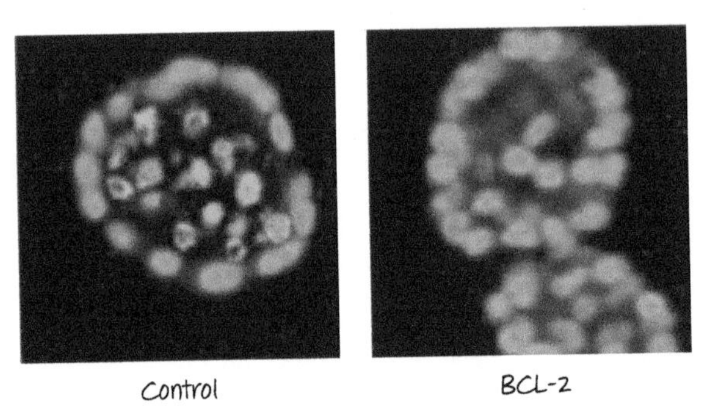

Figure 11.5. Breast epithelial cells resistant to apoptosis fill the acinus. Control cells (*left*) undergo apoptosis (green), whereas those expressing the inhibitor of apoptosis BCL-2 do not die and are able to populate the acinus.

One type of cancer appears to be caused predominantly by this anti-apoptotic effect. As B lymphocytes develop, rearrangements in the DNA occur that are essential for the generation of antibody diversity. Sometimes, the rearrangement goes awry, leading to constitutive expression of BCL-2. This can result in a cancer called follicular lymphoma. It appears that these cells fail to die, and they increase in number when they receive occasional signals to proliferate.[4]

Inhibition of apoptosis allows cells not only to proliferate but also to invade areas where normal cells would die. Adherent cells undergo anoikis if they lose contact with the basement membrane (see Chapter 5). However, cells that are resistant to apoptosis have no such requirement. An example can be found in the acinus of the breast (mentioned in Chapter 10). The center (lumen) is devoid of cells, because any cell that enters the lumen dies by anoikis. Nontransformed breast epithelial cells that express anti-apoptotic BCL-2 proteins, however, can fill the acinus (Fig. 11.5).

TUMOR-SUPPRESSOR PROTEINS PREVENT UNRESTRICTED PROLIFERATION

Signals that drive cells into the cell cycle are offset by tumor-suppressor proteins that prevent uncontrolled proliferation. Most tumor suppressors that have been identified are involved in signaling pathways affecting apoptosis, the cell cycle, DNA repair, or other processes necessary for cell survival, proliferation, and the maintenance of the genome.[5]

[4] BCL-2 (B-cell lymphoma-2) is named for its presence at the breakpoint of some genetic translocations in B-cell lymphomas. It is unlikely that the constitutive expression of BCL-2 is the only factor contributing to follicular lymphoma, and the example is therefore almost certainly an oversimplification.

[5] Generally, for a protein to be categorized as a tumor suppressor, it is insufficient to show that its expression can prevent oncogenesis; its elimination must be shown to promote cancer in some setting.

One tumor-suppressor protein appears to dominate over all others in terms of its importance—the transcription factor p53. About half of all known tumors have mutations in the gene that encodes p53 (*TP53*), and those that do not have mutations in other components of the pathway responsible for its function. As we will see below, p53 has a number of functions, including promotion of cell death.

ACTIVATION OF P53

The p53 protein is a transcription factor that induces the expression of a wide variety of genes, including genes involved in apoptosis. One target of p53 encodes a protein called mouse double minute 2 (MDM2). MDM2 is a ubiquitin ligase that binds p53 and modifies it for proteasomal degradation. Usually, p53 induces just enough MDM2 to ensure that neither is found at any detectable level in normal cells (Fig. 11.6).

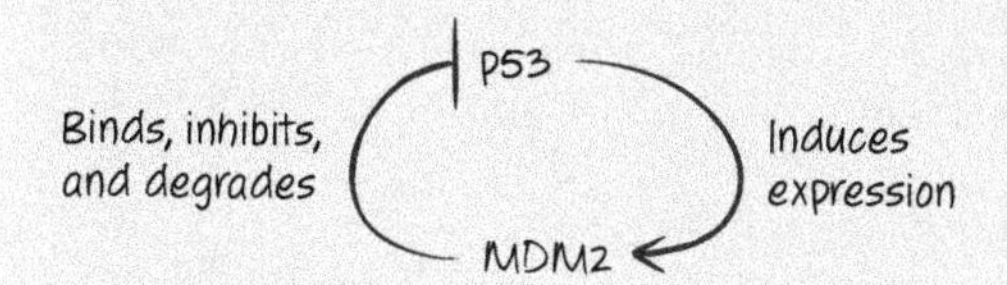

Figure 11.6. The p53–MDM2 feedback loop.

Anything that disrupts the interaction of MDM2 with p53 causes the accumulation of p53. The disruption of the inhibitory interaction between MDM2 and p53 occurs in two fundamentally different ways. One involves a competing protein, and the other involves modification of p53 itself.

Oncogenes capable of driving a cell into an unregulated cell cycle induce the expression of a protein called ARF.[6] ARF binds to MDM2 and thereby blocks its interaction with p53, allowing the latter to accumulate and perform its functions (Fig. 11.7). Tumors that have functional p53 very often have mutations in, or otherwise silence, the ARF locus. Thus, even if p53 is present, the absence of ARF prevents p53 from accumulating in response to the oncogenes, preventing a protective response.

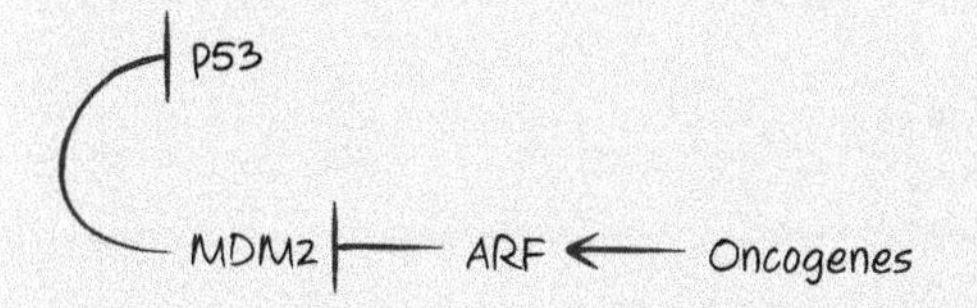

Figure 11.7. Oncogenes induce p53 via the ARF protein.

[6] ARF is an unusual protein in that its coding sequence overlaps that of another protein, INK4A, in a different reading frame (in fact, "ARF" stands for "alternative reading frame"). Interestingly, INK4A is a tumor suppressor that blocks the cell cycle.

(Continued)

(Continued from previous page)

ARF also controls another protein, an E3-ubiquitin ligase that targets p53 for degradation. Thus, ARF stabilizes p53 by not only blocking its interaction with MDM2 but also by blocking the function of this second ligase. Remarkably, this protein, ARF-BP1, turns out to be MULE,[7] one of the E3 ligases that target the anti-apoptotic BCL-2 protein MCL-1 for degradation (see Chapter 5). Therefore, the ARF-BP1–MULE E3-ubiquitin ligase seems to have opposing functions: promoting apoptosis by causing the degradation of MCL-1 and inhibiting apoptosis by promoting p53 degradation. How this fits into a broader scheme of cell regulation is not obvious, and clearly there are more-complex interactions around this cross talk between apoptosis and p53 control.

When active, p53 controls the expression of other proteins that regulate MULE activity, including "translationally controlled tumor protein" (TCTP), which is repressed by p53. Among its many functions, TCTP binds to MCL-1 and prevents its degradation by MULE. Therefore, when p53 is activated, the levels of TCTP decrease, leading to less MCL-1. Because MCL-1 is anti-apoptotic (see Chapter 5), this interaction can promote apoptosis. As we will see, the role of p53 in apoptosis is considerably more complicated even than this.

DNA DAMAGE INDUCES p53

Damage to DNA is a potent trigger for events that include DNA repair, cell cycle arrest, and/or apoptosis. Activation of p53 is an important component of this response, but this generally occurs independently of ARF. Instead, a second way in which p53 is activated is by phosphorylation of its amino-terminal region, to which MDM2 binds. Two related kinases, ATM and ATR, respond to damaged DNA either by directly phosphorylating p53 or by activating kinases Chk1 and Chk2 that phosphorylate p53 on a nearby site. These phosphorylation events interfere with the binding of MDM2 to p53, and hence the latter accumulates (Fig. 11.8).

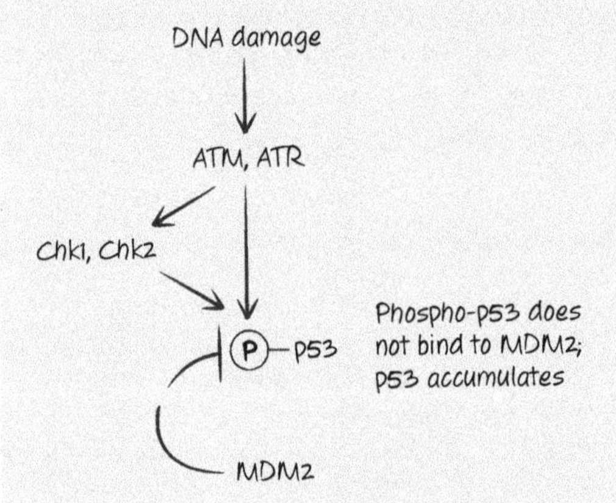

Figure 11.8. DNA damage induces p53.

[7] The proteins were identified and named independently but turned out to be the same protein.

(Continued)

(Continued from previous page)

Many other kinases (and phosphatases) target p53, and in most cases the effects of the phosphorylation are not known, but this might have effects on which genes p53 can activate. p53 is also modified in other ways, including by acetylation and sumoylation, and these probably influence its transcriptional activities.

RIBOSOMAL PROTEIN IMBALANCE INDUCES p53

Ribosomes are assembled in the nucleolus of the cell, and are essential for protein translation. Many stresses can disrupt proper ribosome assembly, including reactive oxygen, heat shock, hypoxia, growth factor deprivation, and DNA damage. As a result, the proteins that constitute the ribosome can accumulate without being associated with ribosomes. Many of these proteins have been shown to disrupt the interaction of MDM2 with p53, leading to p53 stabilization. It is possible that the action of ribosomal proteins to activate p53 in this way represents the most important mechanism of p53 engagement in the suppression of cancer. Tumors often have mutations in ribosomal proteins that result in improper ribosome assembly, and these, as well as other stresses that occur in tumors, result in a ribosome imbalance that can activate p53.

INDIVIDUALS WITH DEFECTIVE p53 ARE PRONE TO CANCER

People with Li–Fraumeni syndrome[8] carry a defective *TP53* allele (two defective alleles is unusual and generally embryonic lethal in humans), and such individuals are extremely prone to cancer, particularly breast cancer, sarcomas (soft tissue and bone), brain tumors, and acute leukemias. In the cancers that arise, the other p53 allele, which originally was wild type, has mutated, so the cells no longer have functional p53. It is not unusual for such individuals to have two or more different cancers in their lifetimes.

Similarly, mice lacking both alleles of *Tp53* always develop tumors. These tend to be tumors of thymic origin, but others also arise. Specific deletion of *Tp53* in particular tissues can also promote cancer, especially if DNA-damaging agents (such as γ-radiation) are used to promote mutagenesis.

Lack of even one allele of *Tp53* in mice usually greatly accelerates tumorigenesis induced by expression of an oncogene (such as Myc; this applies to many other oncogenes, but not all). Tumors that arise inevitably have lost expression or function of the remaining wild-type *Tp53* allele.

The p53 protein is activated in cells by two different mechanisms. The first responds to oncogenes; the second responds to DNA damage (see above). As tumors lose p53, they become genomically unstable, accumulating mutations and other DNA

[8] Named for Frederick Pei Li and Joseph F. Fraumeni, who identified this syndrome.

abnormalities, including chromosomal translocations as well as gain or loss of chromosomes. Presumably, this reflects a failure of the p53-defective cells to respond to DNA damage by engaging efficient repair mechanisms or undergoing apoptosis.

The function of p53 not only prevents cancer, it can also eliminate cancers once they have formed. This has been shown in two ways. In one system, the expression of p53 in animals was controlled by a transgenic short hairpin RNA (shRNA) that silenced p53. In the other, the endogenous p53 was fused to an element that made the protein functional only in the presence of a drug. In both cases, tumors that formed in the absence of functional p53 regressed when the function of p53 was restored.

NUCLEAR p53 INDUCES EXPRESSION OF A VARIETY OF GENES WITH DIFFERENT FUNCTIONS, INCLUDING APOPTOSIS

When p53 accumulates, it moves to the nucleus, where it regulates a variety of genes. Here, we focus on those that control cell death, but it is useful to survey the other consequences as well. These are illustrated in Figure 11.9. Several of these consequences can also be involved in inhibition of cancer.

When p53 is activated, choices are made among the different possible consequences. The cell can die or it can arrest in the cell cycle and repair damage to the DNA. Different tissues respond in their own way, and exactly what influences the decision is not well understood. It could well be that it is simply the level of p53 accumulation that determines the outcome; less p53 might engage repair, whereas more might lead to cell death.[9]

Several transcriptional targets of p53 promote cell death. Which of the pathways causes apoptosis triggered by p53 might depend on additional factors, including the

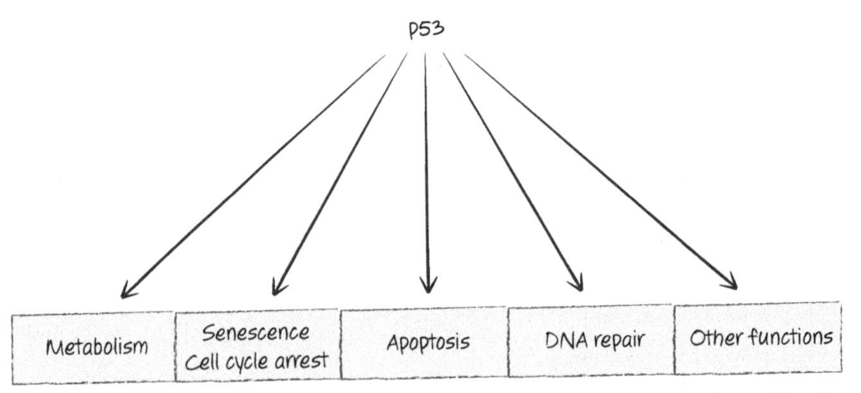

Figure 11.9. The p53 protein controls the expression of genes possessing a variety of functions.

[9] This is a bit simplistic; p53 is modified in many ways that affect its transcriptional activity (through alterations in the binding of additional transcriptional cofactors). Many studies have suggested that the different outcomes of p53 activation depend on specific modifications of this protein.

tissue that is involved. One important apoptosis target of p53 is the BH3-only protein PUMA (for "p53-up-regulated mediator of apoptosis"), which binds to and inactivates all of the anti-apoptotic BCL-2 proteins (e.g., BCL-2, BCL-xL, MCL-1, and A1; see Chapter 5). PUMA can activate BAX and BAK, although other BH3-only proteins do so more effectively (e.g., BID and BIM), and additional signals might be required for mitochondrial outer membrane permeabilization (MOMP) and the mitochondrial pathway of apoptosis when p53 is stabilized (see below).

Mice lacking PUMA are resistant to DNA-damage-induced p53-dependent apoptosis in the thymus (which is normally very sensitive to this). They show no increase in spontaneous cancer, although, if oncogenes are expressed, cancer is somewhat accelerated, but not to the extent seen in mice with defective p53. Other p53 targets therefore presumably contribute to tumor suppression.

Another p53 target is the BH3-only protein NOXA. NOXA neutralizes the anti-apoptotic BCL-2 proteins MCL-1 and A1, but not BCL-2 or BCL-xL. However, lack of NOXA does not generally have dramatic effects on p53-mediated apoptosis. Nevertheless, NOXA can influence apoptosis (and tumor suppression) in some settings.

Yet another target is the BH3-only protein BID. This protein, when activated by proteolytic cleavage, can activate BAX and BAK to cause MOMP and engage the mitochondrial pathway (see Chapter 5). As we have discussed, BID is efficiently activated by caspase-8 following ligation of death receptors. For this reason, activation of p53 sensitizes cells to death-receptor-induced apoptosis (see Chapter 6).

The gene encoding the pro-apoptotic BCL-2 effector BAX (see Chapter 5) is a direct target of p53 transcriptional activity in humans. In rodents, this is not the case, but stabilizing p53 in rodents nevertheless causes increased expression of BAX by an indirect mechanism. Although increased BAX expression does not itself cause MOMP, it does make cells more sensitive for engagement of this pathway. The activation of p53 therefore has a number of effects that promote the mitochondrial pathway of apoptosis (Fig. 11.10).

The p53 protein also induces expression of the death receptors CD95 and DR5 (one of the TRAIL receptors). Together with the up-regulation of BID, this makes cells more sensitive to death ligands when p53 is activated. However, the extent to which the death receptor pathway contributes to p53-induced apoptosis is unclear. Mice

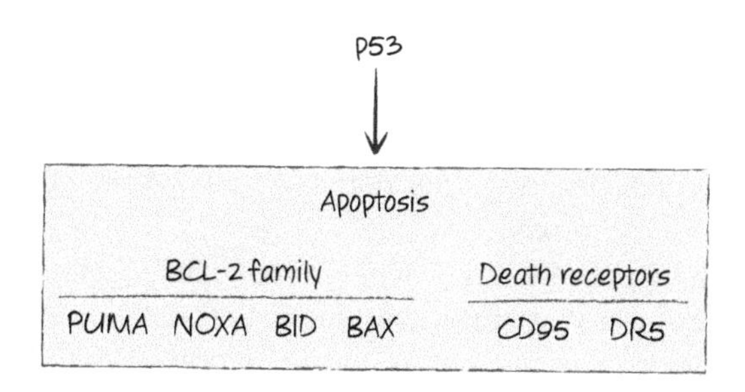

Figure 11.10. The p53 protein controls a number of genes encoding pro-apoptotic proteins.

lacking CD95 or its ligand CD95-L show no defects in p53-mediated cell death, although mice that have deficient TRAIL signaling show increased metastasis in a tumor model. It is therefore possible that loss of DR5 expression can contribute to the effects of loss of p53.

Which of the apoptotic pathways are most important for p53-induced cell death and how crucial is apoptosis for tumor suppression by p53? A hint comes from studies in which tumors are induced by the constitutive expression of Myc in B lymphocytes in mice. Lymphomas that arise in such animals appear slowly and generally have inactivating mutations, deletions, or loss of expression of p53. However, if lymphomas are induced by coexpressing Myc and BCL-2, which results in much more rapid tumorigenesis, the transformed cells do not develop mutations in p53. At least in these cells, inhibition of the mitochondrial pathway of apoptosis therefore appears to be sufficient to bypass the tumor-suppressive effects of p53. This further suggests that the mitochondrial pathway of apoptosis is a major mechanism of tumor suppression (again, at least in these cells).

In addition to apoptosis, p53 also sensitizes cells for ferroptosis (discussed in Chapter 8). In this case, p53 acts by repressing the expression of a cystine transporter that is a crucial component of system Xc^-. As a result, the activation of p53 causes a reduction in cystine uptake, depletion of glutathione, and cell death by ferroptosis. Tumor cells frequently increase their expression of system Xc^-, and this is seen even in cancers without mutations in p53 (suggesting that there are other mechanisms controlling the expression of system Xc^- as well).

Mice that lack MDM2 (see Box, above) die early in embryogenesis. This lethality is completely prevented by elimination of p53. Intriguingly, a particular mutant of p53 that fails to rescue MDM2-deficient mice does not induce the expression of any of the genes involved in apoptosis or cellular senescence. This mutant does, however, repress the expression of system Xc^-. Pharmacologic inhibition of ferroptosis in these developing embryos allowed them to progress further in development, suggesting that ferroptosis plays a role in this developmental phenomenon.

CYTOSOLIC p53 CAN ALSO FUNCTION IN APOPTOSIS

When p53 is stabilized and accumulates in cells, it can be found in both the nucleus and the cytosol. Cytosolic p53 can promote apoptosis by a mechanism other than regulation of transcription. Like some BH3-only proteins, it can directly activate the pro-apoptotic effector protein BAX (and, at higher concentrations, BAK) to cause MOMP. Moreover, as with BH3-only proteins, p53 is bound by anti-apoptotic BCL-2 proteins. Note that, despite these activities, p53 does not have a BH3 domain and shares no sequence similarity with BCL-2 proteins.

As cytosolic p53 accumulates, it can be sequestered by anti-apoptotic BCL-2 proteins. The binding of p53 to BCL-xL is not like that of BH3 regions in BH3-only proteins or BAX and BAK—instead, p53 binds near, but not in, the BH groove. When

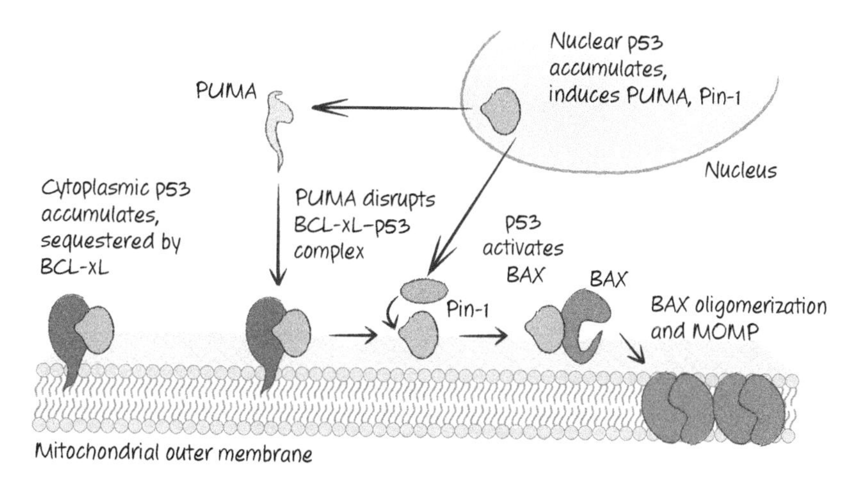

Figure 11.11. Nuclear and cytoplasmic p53 can cooperate in causing mitochondrial outer membrane permeabilization (MOMP).

nuclear p53 accumulates, however, we have already seen that it induces expression of PUMA and BAX. PUMA is remarkably good at displacing p53 from anti-apoptotic BCL-2 proteins, releasing p53 to activate BAX and promote MOMP (Fig. 11.11). This is because PUMA (and not other BH3-only proteins) induces a conformational change in BCL-xL that releases p53. Similarly, p53 engages BAX outside of its BH groove, but induces the same "unlatching" process to activate the effector. Interestingly, p53 only does this when two prolines in the protein are isomerized by an enzyme, Pin-1, whose expression is also induced by nuclear p53.

Although biochemical and cellular studies support this cytosolic role for p53 in apoptosis, no compelling genetic evidence unambiguously shows that this functions in tumor suppression. It is a complicated problem because, without transcription factor activity, p53 cannot induce expression of its regulator MDM2 or PUMA to promote the cytosolic activity. Ingenious experiments will be needed to resolve this controversial area of p53 research.

CASPASE-2 CAN ACTIVATE P53 WHEN CENTROSOMES ARE ABERRANT

In Chapter 7, we saw that PIDD associates with mature centrosomes and can activate caspase-2 when extra mature centrosomes appear in cells, such as when cytokinesis (the last step in mitosis) fails. In this way, caspase-2 can help to prevent chromosomal abnormalities that occur when centrosomes do not function correctly in mitosis.

This function of caspase-2 involves the activation of p53. MDM2 (see above) is a substrate for caspase-2, and upon caspase-2 activation, MDM2 is cleaved and ceases to inhibit p53. As a result, p53 is activated, leading to either apoptosis or cell cycle arrest (Fig. 11.12).

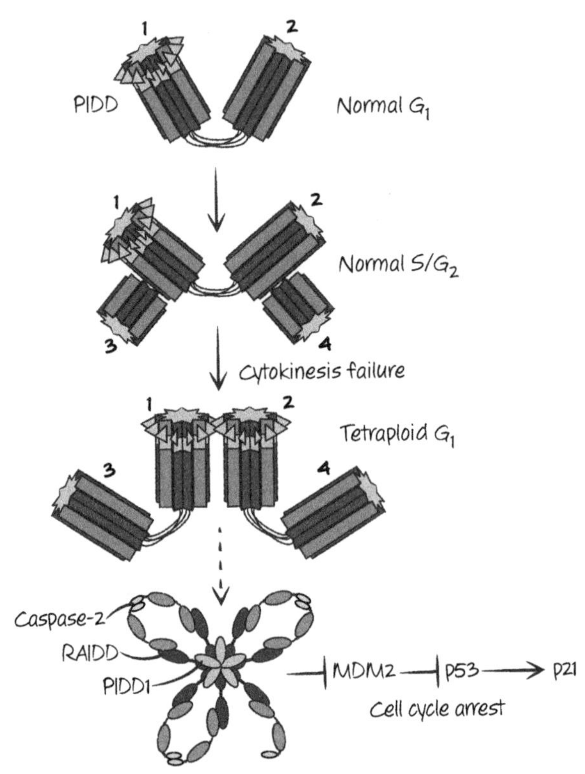

Figure 11.12. During passage through the cell cycle, the pair of centrosomes is duplicated. Errors in cytokinesis can lead to a cell possessing supernumerary mature centrosomes. These can lead to activation of a signaling pathway, involving RAIDD, PIDD1, and caspase-2, that results in destruction of the p53 inhibitor MDM2, thereby allowing p53 to accumulate and induce either apoptosis or cell cycle arrest (via p21).

Mice lacking PIDD or caspase-2 do not activate p53 in response to the presence of extra mature centrosomes. As a result, the proper assortment of chromosomes during mitosis can be disrupted, resulting in an increased propensity for cancers with chromosome abnormalities. This is therefore a potential explanation for how caspase-2 acts as a tumor suppressor in some cancers (i.e., lack of caspase-2 promotes cancer in some models). Although this has not formally been proven, $p53^{-/-}$ mice lacking caspase-2 do not show any increase in cancer in comparison with $p53^{-/-}$ mice that express caspase-2, suggesting that caspase-2 might act via p53 to suppress cancers in some settings.

p53-INDUCED APOPTOSIS IN OTHER ANIMALS

Proteins related to the mammalian p53 family have been found not only in vertebrates but also in several invertebrate phyla. In *Drosophila*, a p53-like protein is

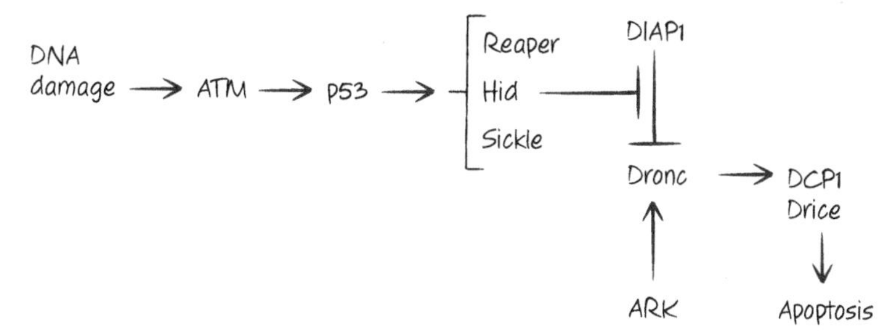

Figure 11.13. Pathway of induction of apoptosis via *Drosophila* p53.

activated in response to DNA damage. As in humans, *Drosophila* p53 acts as a transcription factor, and it leads to apoptosis. In the fly, it induces the expression of Reaper, Hid, and Sickle, all of which are direct transcriptional targets (see Fig. 11.13).

There are, however, fundamental differences between p53 in mammals and p53 in the fly. Although activation of p53 in mammalian cells leads to either cell cycle arrest or apoptosis (among other things), activation of p53 in the fly does not arrest cell division. In addition, the protein is regulated in a different way; there is no MDM2 homolog in insects. Intriguingly, *Drosophila* have a homolog of ATM, and this seems to regulate p53 in the fly.

Nematodes also have a p53 protein, which is called CEP-1. Activation of CEP-1 induces apoptosis, but as in flies, this does not cause cell cycle arrest. CEP-1 is activated following DNA damage and transcriptionally up-regulates the BH3-only protein EGL1, leading to apoptosis (Fig. 11.14). Like in the fly, *Caenorhabditis elegans* does not have an MDM2 homolog but does have a homolog of ATM that regulates p53.

Clearly, p53 in these organisms is linked to the core apoptotic pathways that function in them. However, it is not obvious that this has tumor-suppressor functions because the short life spans of such animals make cancer a minor selection force. It is likely that the role of p53 in these organisms is to preserve the genome in the germline, so that cells that have extensive DNA damage die, rather than contribute to the next generation. The role of p53 in longer-lived invertebrates might be a

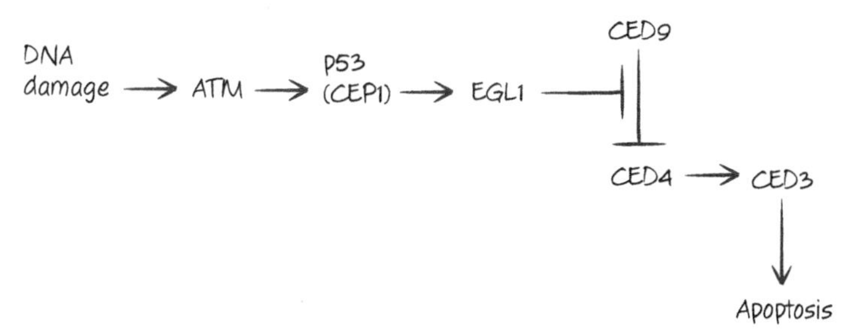

Figure 11.14. Pathway of induction of apoptosis via nematode transcription factor CEP-1 (p53).

different story, and p53-family members have been identified in other phyla. At present, we do not know what functions, if any, these have in tumor suppression.[10]

SOME COMPONENTS OF APOPTOTIC PATHWAYS ARE TUMOR SUPPRESSORS

Several proteins that we have already met are also tumor suppressors. Genetic ablation of BAX, BIM, or PUMA (see Chapter 5) promotes tumorigenesis in mice when a known oncogene, such as Myc, is expressed. In addition, genes encoding PUMA and another pro-apoptotic BCL-2 protein, BOK, are frequently deleted in human cancers, whereas BAX and BAK can be mutated to lose function. Conversely, human tumors often amplify the genes encoding the anti-apoptotic proteins MCL-1 and BCL-xL. Similarly, gene-translocation events can enforce the expression of BCL-2 so that it is constitutively expressed at elevated levels.

APAF1, caspase-9, and caspase-3 have not been observed to act as tumor suppressors in mouse models. This is intriguing because human cancers often have mutations in caspase-3, indicating that it might be a tumor suppressor in humans. As we mentioned in Chapter 8, caspase-2 acts as a tumor suppressor in some model systems, although we do not know whether this involves apoptosis. Another interesting example is caspase-8, which is often mutated in some types of human cancer, such as neuroblastoma.

OTHER TUMOR SUPPRESSORS ENGAGE APOPTOTIC PATHWAYS

Many tumor suppressors influence apoptosis indirectly. After p53, the most common tumor suppressor that is mutated in cancers is the "phosphatase and tensin homolog" PTEN. PTEN is an enzyme that removes a phosphate group from phosphatidylinositol (3,4,5)-trisphosphate (PIP_3), a lipid involved in activation of AKT. AKT promotes cell growth and inhibits apoptosis by several mechanisms (see Chapter 5). By blocking AKT activation, PTEN can inhibit cell proliferation and promote apoptosis. When PTEN is inactivated by mutation, AKT is constitutively active, and this promotes cancer.

There are many other well-defined tumor-suppressor pathways, and in several cases it is clear that they influence apoptosis. Their links to specific apoptotic pathways are less well established, however, and we have much to learn about how they control apoptosis.

[10] In fact, p53 is part of a family of related transcription factors that includes p63 and p73 (collectively, the "p53 family"). p53 members present in flies and nematodes more closely resemble p63, which in vertebrates has roles in stem cell function (more than it has roles in apoptosis).

APOPTOSIS AS AN ACHILLES' HEEL OF CANCER

At first glance, any disruption of apoptosis that occurs as a step in oncogenesis might seem to make killing a cancer an intractable problem. But this is not the case. Consider what it takes to make a tissue expand rather than regress. If the rate of cell division is greater than the rate of cell death, cells accumulate (the tissue/cancer expands); conversely, if the rate of cell death is greater than the rate of cell division, the tissue (or cancer) regresses. This means that anything that tips the balance toward cell division over cell death will result in the expansion of a tissue or cancer. Tumor suppressors, whether they act by inhibiting the cell cycle or promoting cell death, push the balance the other way, and in general, this works to prevent cancer.

But when cancer occurs, mutations only need to be sufficient to tip the scale toward expansion. Apoptosis mechanisms might be engaged, for example, in response to tumor suppressors that are activated or to apoptotic signals from oncogenes such as Myc, but these do not effectively tip the balance in our simple equation. Thus, in many cases, cancer cells are essentially "primed" for death but do not die in sufficient numbers to prevent the expansion of a tumor. The goal is therefore to devise therapies that take advantage of such priming and tip the balance toward regression.

Remember one way in which apoptotic pathways are controlled: Activation of BAX and/or BAK induces MOMP unless anti-apoptotic BCL-2 proteins prevent this by sequestering BH3-only proteins (such as BIM) or activated BAX and BAK themselves. If a drug could disrupt the interaction between pro-apoptotic BCL-2 proteins and anti-apoptotic BCL-2 proteins, apoptosis should proceed (Fig. 11.15). These are called BH3 mimetics because they act like derepressor BH3-only proteins. Indeed, one such BH3-mimetic drug (specific for BCL-2) has been approved for

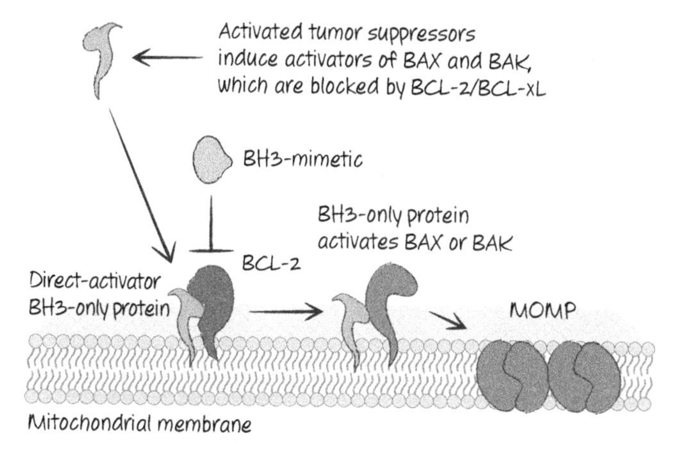

Figure 11.15. One way a BH3 mimetic can kill a cell.

the treatment of human cancer, and a number of drugs targeting BCL-xL and MCL-1 are currently in clinical trials (discussed in more detail in Chapter 12).

But what would happen if a normal cell were treated with a BH3 mimetic? Without tumor-suppressor signals to activate BAX and BAK (such signaling would not be occurring in normal cells), neutralizing the anti-apoptotic BCL-2 protein would not be sufficient to engage MOMP. In contrast, the primed cancer-causing cells would be more prone to engaging the mitochondrial pathway, and only they would die. One could therefore preferentially destroy the cancer cells.

THE "PRIMED FOR DEATH" STATUS OF MANY CANCERS

A cell is said to be "primed for death" if anti-apoptotic BCL-2 proteins are actively sequestering the active forms of BAX and/or BAK or the BH3-only proteins that can activate them. It is possible to determine whether a cell in a human cancer is thereby precariously poised to die when anti-apoptotic influences are disrupted. If the cell is permeabilized and the mitochondria kept intact, then upon addition of peptides corresponding to BH3-only proteins that bind to anti-apoptotic BCL-2 proteins (i.e., those BH3-only proteins that do not activate BAX and BAK), MOMP will occur, and this can be detected by the release of cytochrome *c* (or by the loss of the membrane potential $\Delta\Psi m$ that occurs when cytochrome *c* is released).[11] This approach is called "BH3 profiling" (Fig. 11.16).

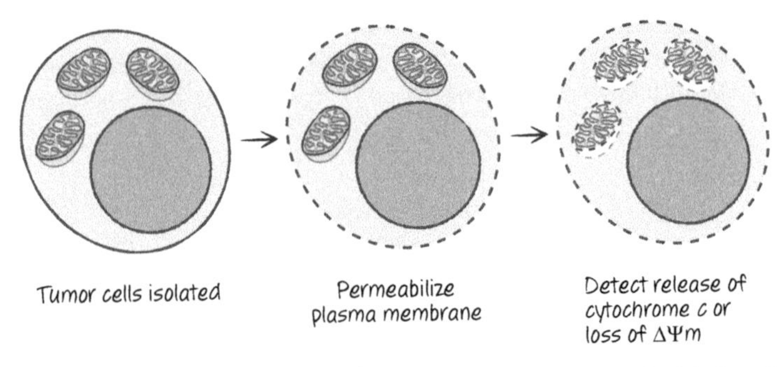

Figure 11.16. The principle of BH3 profiling. A tumor cell is permeabilized while keeping the mitochondria intact. Next, peptides corresponding to the BH3 regions of BH3-only proteins are added; these will bind to any anti-apoptotic BCL-2 proteins, allowing mitochondrial outer membrane permeabilization (MOMP) to occur, which can be detected by the release of cytochrome *c* or loss of the membrane potential.

[11] You may be wondering why it would not be simpler to see if the tumor cells undergo apoptosis in response to the agent of choice. While possible, such assessments are plagued by high background cell death in the control cultures (culture conditions are notoriously harsh), and the results are unreliable. In contrast, the assay described here is extremely rapid and can be performed using small numbers of cells.

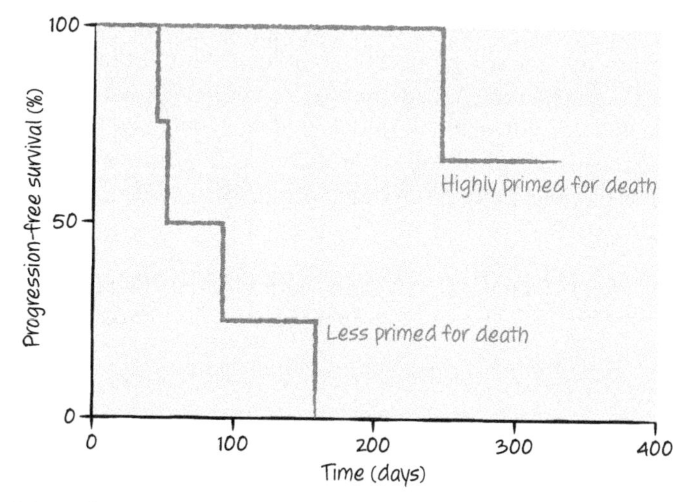

Figure 11.17. Priming for tumor therapy. BH3 profiling can be used to assay the degree to which different tumor types and controls are primed for cell death through mitochondrial outer membrane permeabilization (MOMP). Highly primed cancers (gray) show a marked clinical response to treatment with chemotherapeutic drugs, corresponding with good progression-free survival, compared with poorly primed cells (brown).

Remarkably, there is a strong correlation between the extent to which cells in a cancer are primed for death, and the likelihood that the cancer will respond to conventional therapies (Fig. 11.17). There are several important conclusions that arise based on this finding. First, it shows that the way in which most conventional cancer therapies work is by inducing the mitochondrial pathway of apoptosis. Second, it shows that the sensitivity of cancer cells to such treatments is due to this primed state in which anti-apoptotic proteins are required for the cell's survival. And third, it suggests that the "therapeutic window" in which a cancer therapy kills the cancer but not the patient is dictated by this status of being primed for death.

The last idea has been tested more directly. Acute myelogenous leukemia cells and normal hematopoietic stem cells from the same patient were compared for their "primed for death" status. If the cancer cells were more primed for death in this assay than the stem cells, the outcome was favorable. However, if the cancer cells and stem cells were similar in their sensitivity (primed for death status), this predicted a need for bone marrow transplantation, a procedure that is performed when conventional treatments fail.

BH3 profiling goes beyond whether a cell is primed for death or not and can give insights into which BCL-2 proteins act to keep it alive. As we saw in Chapter 5, some BH3-only proteins are specific for only one of the anti-apoptotic BCL-2 proteins; thus, BAD is specific for BCL-2, HRK for BCL-xL, and NOXA for MCL-1. By using peptides corresponding to these proteins, we can identify whether BCL-2, BCL-xL, or MCL-1 predominantly act to keep the cell alive. As it turns out, this cannot be simply determined by which of the anti-apoptotic BCL-2 proteins are most likely expressed; the interplay of the different pro- and anti-apoptotic BCL-2 proteins is complex, and one would have to know the relative levels of all of them and how they interact. Efforts to do that are under way (as we will see in Chapter 12), but for now the approach outlined here has the best predictive power.

MOST ADULT TISSUES ARE NOT PRIMED FOR DEATH

The fact that many tumors are primed for death (i.e., their mitochondria are "poised" to undergo MOMP) is an essential part of why many cancer therapies work (although not always as well as we might wish). Oncogenes engage tumor-suppressor mechanisms that promote apoptosis, and if these are only blunted, the cancer can survive until an additional stress (such as radiation or chemotherapy) is applied. But there is another factor as well—for such therapy to be effective, it is essential that indispensable host tissues be less sensitive than the tumor. Indeed, most tissues in an adult are not primed for death. The major exceptions are cells of the hematopoietic system, and, indeed, the extent to which leukemia cells are more primed for death than are the hematopoietic stem cells of the patient is predictive of the need for a bone marrow transplant (discussed above). Similarly, the cells of the adult intestine are somewhat primed for death (and we know that intestinal damage is often an unwanted side effect of cancer therapy). But other tissues, including heart, brain, liver, and kidney, are strikingly unprimed, and this is likely crucial for any successful cancer therapy.

In contrast, tissues of young individuals are more primed for death than are their adult counterparts. As a consequence, collateral damage during cancer therapy (e.g., to the heart and brain) is a problem in young cancer patients. It is likely that this reflects growing tissue; adult tissue that is undergoing regeneration similarly shows an increased priming for death. As we saw above, one factor that accounts for the difference in priming between young and adult tissues is the expression of c-Myc. It is sufficient to express c-Myc in any of several adult tissues to prime them for death.

CAN APOPTOTIC PATHWAYS CAUSE CANCER?

To this point, this chapter has been about the roles of apoptosis and the mechanisms that engage it in the suppression of tumorigenesis. However, there is also evidence that inefficient engagement of apoptotic pathways can actually promote cancer.

One way apoptosis can promote cancer is by extrinsic effects (the consequence of the dying cell for the behavior of other cells). As discussed in Chapter 9, apoptotic cells produce prostaglandin E_2 (PGE_2) when executioner caspases cleave and activate phospholipase A_2. PGE_2 promotes the proliferation of some stem cells, including cancer stem cells, and thus can promote cancer. Other factors produced by some dying cells (such as some cytokines and chemokines) can also promote cell proliferation as well as an increased blood supply to the region of the dying cells. Finally, as we also saw in Chapter 9, clearance of apoptotic cells can result in silencing of the adaptive immune response, which would otherwise help to destroy the cancer.

But the induction of apoptotic pathways can also have intrinsic effects to promote cancer if the affected cell does not actually die. As we have seen in Chapter 4, engagement of MOMP induces caspase activation and apoptosis. However, sublethal stress can induce MOMP in some but not all mitochondria (an effect called "minority MOMP"; Fig. 11.18). When minority MOMP occurs, cells often remain alive. Presumably, there is a threshold for the extent of MOMP needed to engage apoptosis (it is likely that this threshold depends not only on how many mitochondria are involved, but also on the levels of downstream components of the pathway, such as APAF1, caspase-9, and the executioner caspases). However, in cells that survive minimal MOMP, low levels of executioner caspase activation (dependent on APAF1) occur that cleave and activate low levels of iCAD (see Chapter 2). As a result, some caspase-activated DNase (CAD) is activated and cuts DNA, and, consequently, mutations are introduced as the DNA is repaired (remember, these cells survive). Some of these mutations can promote tumorigenesis or resistance to anticancer therapies. Sublethal engagement of the death receptor pathway of apoptosis (Chapter 6) can similarly promote oncogenic mutation because of low levels of CAD activation.

It has long been known that treatments for cancer can give rise to new ("iatrogenic") cancers as a consequence of the therapy. Although it has always been

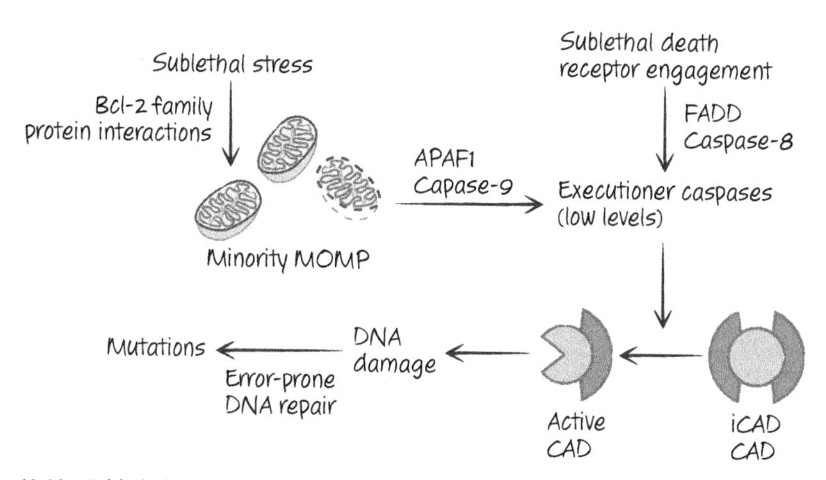

Figure 11.18. Sublethal engagement of apoptotic pathways can promote mutations in surviving cells. Cells subjected to sublethal stress or death receptor ligation can activate low (nonlethal) levels of executioner caspases and activated caspase-activated DNase (CAD). CAD cuts DNA, and error-prone DNA repair of the damage can lead to mutations that promote cancer.

assumed that such effects are a direct consequence of the therapy inducing DNA damage, the above findings suggest that the activation of CAD can contribute to iatrogenic cancers. Indeed, experimental treatment of cells with low levels of a BH3-mimetic drug (see above) can increase the frequency of oncogenic mutations.

This raises an interesting possibility. It might be that pharmacologic inhibition of executioner caspases (which do not prevent cell death following MOMP; see Chapters 4 and 8) will actually improve cancer outcomes in response to therapies. Several experimental studies have shown that caspase inhibitors (see Chapter 3) do not impair the effects of cancer therapeutics, and can even be beneficial, but the mechanisms were obscure until now.

OTHER FORMS OF CELL DEATH IN CANCER

Cancers are not well-organized tissues, and as they grow, they expand beyond their ability to provide nutrients and oxygen to all of the cells in the tumor. As a result, cancers often have central regions of extensive cell death that are necrotic (see Chapter 8). Although such cell death certainly fits into our simple equation above, necrosis in a cancer is often associated with a poor prognosis. This may be because it activates inflammatory responses (see Chapter 9) that produce cytokines that stimulate growth of the cancer. Nevertheless, tipping the balance in favor of any form of cell death will produce regression, and there are therapies that aim for such an outcome.

Autophagy is usually a cell survival mechanism, and inhibition of autophagy is another goal in tumor therapy. But beclin-1, which initiates autophagy (see Chapter 8), is a tumor suppressor that is sometimes deleted in human cancer. Mice that lack one allele of the gene encoding beclin-1 display spontaneous cancer and show accelerated tumor formation when an oncogene is expressed. Cells that have defective autophagy accumulate unfolded proteins, and these induce the production of reactive oxygen species that can promote additional mutations that contribute to oncogenesis.

Induction of autophagic cell death (see Chapter 9) is also being explored as a therapy. Chloroquine (which has a long history in the treatment of malaria infection) causes autophagic cell death (i.e., nonapoptotic cell death associated with, but not dependent on, autophagy), and several trials with this drug in combination with other therapeutics are under way.

Necroptosis (see Chapter 8) plays an unexpected role in cancer. Many cancers express APP, which is a ligand for the death receptor DR6, expressed on endothelial cells. The ligation of DR6 by APP on the cancer cell surface induces necroptosis in endothelial cells, allowing a tumor cell that is in the circulation to escape into the tissue, producing metastases. In two tumor models, neutralization of APP or DR6, or endothelium-specific deletion of RIPK3, dramatically prevented metastasis.

Ferroptosis might also function in tumor suppression. In addition to the mechanisms discussed above, activation of p53 can also promote ferroptosis, by inhibiting the expression of a component of system Xc^-. When cells lose p53 function during oncogenesis, expression of system Xc^- increases, and inhibitors of either system Xc^- or glutathione synthesis could prove to be promising therapeutics, especially for cancers that have high levels of iron (remember that iron participates in ferroptosis). Such cancers include some renal and some hepatic tumors.

Entosis (see Chapter 9) is frequently observed in epithelial cancers; however, it is not clear whether this is a pro- or anticancer mechanism. As we discussed, oncogenes such as *KRAS* and c-*Myc* can promote entotic "winners," and cancer cells can potentially gain nutrients by the digestion of normal, "loser" cells, and therefore entosis could potentially promote cancer. However, as we also discussed, cells that lose attachment do not increase in number as long as entosis occurs, as sister cells will eat each other and one of the two will die by this process. In this way, entosis can function to restrict cancer. In any case, the appearance of cell-in-cell structures, indicative of entosis, generally corresponds with a poor prognosis.

The regulation of cell death in cancer and cancer therapy is varied and complex. Tumor-suppressor mechanisms can specify cells for death, much as do developmental cues (Chapter 10), and oncogenesis necessarily involves mechanisms to avoid such cell death. An important step forward is to identify ways to counteract these avoidance mechanisms.

Of course, cancer is not the only disease process in which cell death pathways have a prominent role. The approaches to controlling cell death and survival will have applications in the treatment of autoimmunity, neurodegenerative diseases, and aging. In Chapter 12, we consider how our knowledge of cell death processes can be tested by formal and practical approaches.

IMMUNOGENIC CELL DEATH

The induction of an adaptive immune response against tumors is a crucial factor in the efficacy of therapies against cancer. In animal models, cancer therapies that prevent growth of a tumor often do not work if the immune system is compromised, and human studies have shown that the presence of an immune response is associated with positive outcomes of therapies.

In Chapter 9, we discussed the general principle that whereas necrotic cell death can promote innate and adaptive immunity, apoptosis tends to silence immune responses. Although this is generally true in most contexts, it turns out to be more complex; some agents that induce apoptosis in tumor cells can subvert this principle to make the immune system respond in a manner that promotes anticancer immunity. For example, apoptosis induced by the chemotherapeutic agent cisplatin does not promote adaptive T-cell responses to tumors, whereas its derivative oxaliplatin does. At least part of the reason for this appears to be an oxaliplatin-induced production of DAMPs (ATP and HMGB1; see Chapter 9) and the surface exposure of calreticulin on the dying tumor cells. The class of cancer therapeutics called anthracyclines also have this property.

It might also be possible to combine other drugs to ensure that cell death, induced by a therapeutic, will be more immunogenic. Cardiac glycosides, long used for treatment of heart disease, promote the anticancer effects of drugs that do not themselves induce immunogenic cell death, by altering the cell death to engage immune responses. Studies have shown that people who take such drugs for heart conditions have lower rates of some cancers, and perhaps this relates to the promotion of immunogenic cell death in the cancers that do arise. The beneficial effects of cardiac glycosides in animal models of cancer depend on the presence of an intact immune response.

Dying cells, taken up by dendritic cells, can engage a process called "cross-priming," allowing some novel proteins in the dying cells (such as might be produced by mutations in a cancer) to be processed and presented to cytotoxic T lymphocytes for their activation and expansion. As it turns out, the form of cell death matters. Although apoptotic cells can cross-prime T-cell responses to some extent, cells undergoing necroptosis do so much more efficiently. But this effect is dependent not on the form of death, per se, but on the engagement of a nuclear factor-κB (NF-κB) response, elicited by RIPK1 (see Chapters 5 and 8). If NF-κB is inhibited, the dying cells cross-prime much less efficiently.

Immunogenic cell death has crucial implications for cancer therapy. There are several mechanisms that restrict T-cell responses (very probably to prevent autoimmunity); collectively these are called "immune checkpoints." There are therapeutics that block these immune checkpoints to unleash anticancer immunity to fight (and frequently eradicate) the tumor. But, to work, it is necessary that the tumor is recognized by the immune system, and therefore chemotherapeutics that engage immunogenic forms of cell death are more likely to work in combination with immune checkpoint blockade strategies.

CHAPTER 12

The Future of Death

TESTING CELL DEATH MODELS

In this book, we have surveyed a great deal of information, organizing it into superficially complete signaling pathways that lead from a variety of stresses and developmental cues to the death of a cell. We understand, in principle, how a cell is specified for death, how this engages cell death pathways, how these are regulated, how death is performed, how the cell is removed, and the consequences for the cells that remain. It is tempting to think that what remains to be discovered are merely the details, mainly of interest to specialists. But is this "big picture" really complete, and are there approaches that we can use to determine this?

In this final chapter, we consider two such approaches to this problem, one formal and one informal (but very important). The first involves taking what we know, modeling it mathematically, and asking whether the models we generate conform to what we observe. This approach does not explicitly tell us if we are right or wrong but can reveal areas of ignorance and provide a different way to look at the pathways that we have considered.

The second approach is a practical one—can we use the information that we have gleaned to successfully engineer cell death? Again, this cannot tell us whether we are right or wrong but, instead, where our ignorance lies and whether our knowledge is useful.

CELLULAR STATES AND MODELS OF DEATH

Living cells can be in many states, including rest, proliferation, and differentiation, and can transition between such states. But any transition to the cell death state can lead to an irreversible outcome; the consequences can differ, but from the point of view of the cell, dead is dead.

Apoptosis (as well as most other forms of cell death) is a balancing act between molecular mechanisms that determines whether a cell survives or dies. The outcome depends on the relative amounts of specific molecules in the cell and the interactions between them. A simple experiment shows that cell-to-cell variation in these molecules is stochastic. If cells are cultured under conditions conducive to apoptosis and monitored (e.g., by time-lapse microscopy), we see that some cells die at different times—some relatively quickly, some only after more time has passed. And if the signal was not too overwhelming, some survive. Which cells do what appears to be random. However, if a cell divides and one of the daughters dies, the other daughter is more likely to die in a given time frame than are other cells that are less closely related (Fig. 12.1). This correlation is strongest immediately after division and wanes with time. Undoubtedly, this is because the quantities of key molecules in the daughter cells are more similar than the quantities of such molecules in more distantly related cells, but as it happens, no single molecule accounts for the variation. Even when all of the cells used are derived from a single cell, the divergence over generations is remarkably rapid. This means that there is significant "noise" in the system.

There is another interesting result that comes from this type of analysis. Unless the death-inducing signal is very strong, some cells in a clonal population always survive the treatment. When these "persister" cells are allowed to expand, the population responds in the same manner to the same treatment—that is, the same proportions of dying and surviving cells are observed (therefore, this is not selection for a genetically resistant population). However, when persisters are examined shortly after the original treatment, this state of relative resistance continues for a period of up to a few days. This suggests that cells can stochastically enter a state in which apoptotic

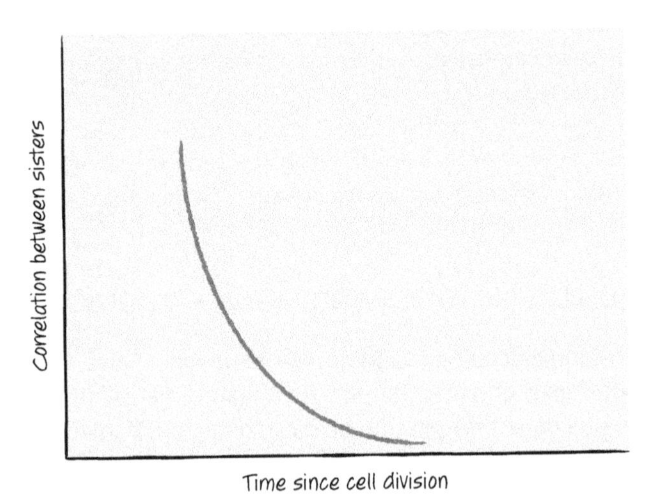

Figure 12.1. Cell division and stochastic death. If, following cell division, sister cells are monitored for their response to an apoptotic stimulus, MOMP is closely correlated shortly after division, but this relationship quickly decays with time.

pathways are "off," but this state does not stably pass to the progeny. At this point, we do not know how persisters differ from the susceptible cells in the population.

The simple idea that levels of key molecules randomly vary between cells and that the outcome of a signaling event depends on these levels applies to many cellular responses, not just cell death. But cell death is a special case, as we have said, because it is irreversible. Given a constant level of stress (and our cells are always under some stress) and random variation in the levels of different key molecules, all cells would eventually die (for some cells, this does not actually occur; neurons and cardiac myocytes, for example, are postmitotic, and most persist throughout the life span of the individual). In other cases, random cell death can be offset by proliferation, both in normal tissue (see Chapter 10) and in neoplastic tissue (see Chapter 11). The trick for complex biological systems is to control this balance and limit the variability through regulation. One way to look at this is to consider the probability that a given cell will die in response to a given level of stress (or any pro-death signal) during a fixed period. In the absence of built-in regulatory mechanisms, the relationship would be expected to be largely linear (Fig. 12.2).

But the actual, observed relationship between stress and cell death is more similar to that shown in Figure 12.3. This type of relationship, which approximates a "step function," is all around us. It occurs in biological systems of all types, but also in many other systems. Often, it is an effect of negative-feedback regulation.[1] But cell death pathways (those that we understand) do not have any obvious built-in negative-

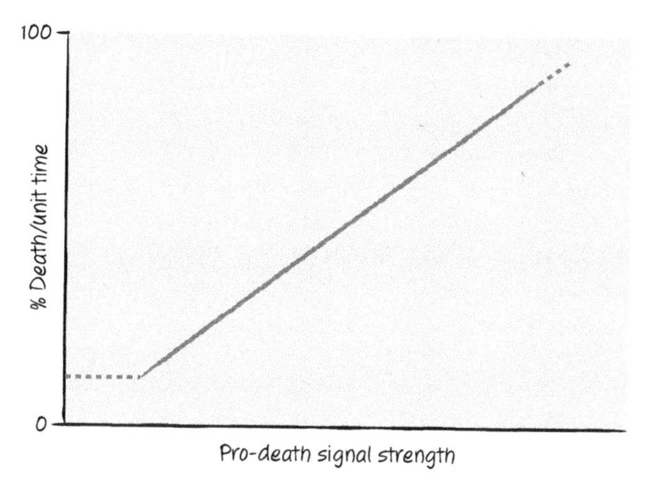

Figure 12.2. Expected relationship between the strength of a pro-death signal and outcome, if regulatory mechanisms are absent.

[1] The conception and application of negative-feedback regulation is often attributed to James Watt in creating the steam engine, but goes back at least as far as Ktesibios of Alexandria in the third century BCE who used it to refine the water clock.

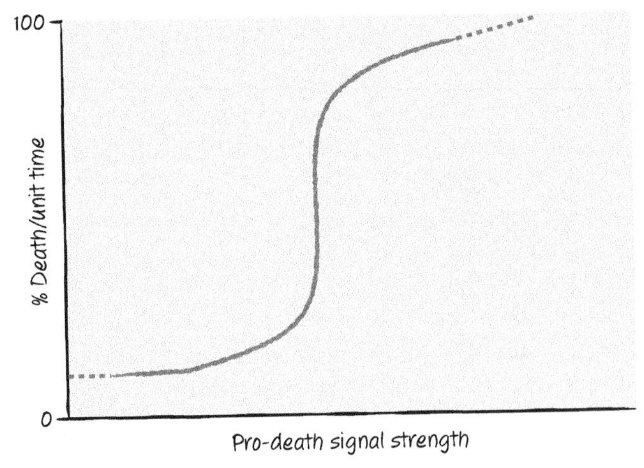

Figure 12.3. Observable relationship between the strength of a pro-death signal and outcome.

feedback mechanisms. If we look back on the pathways that we have covered, despite the existence of many inhibitory interactions, there appear to be no points at which the induction of a later event inhibits an earlier event in the pathway. How then is cell death controlled to produce this step function?

BISTABILITY IN CELL DEATH PATHWAYS

A system that can exist in two stable states (e.g., survival vs. death, inactive vs. active caspases, or intact mitochondria vs. permeabilized mitochondria) is said to be bistable.[2] We can analyze such systems by drawing a pathway, assigning concentrations (real or estimated) for all of the components, and then generating mass-action formulas. A stimulus is introduced (simulated by the increase of an upstream component over a range) and the resulting curves can be generated (usually in silico). We can then see whether our model behaves in a bistable manner, displaying two states. A robust bistable system is one that is relatively resistant to small changes in concentrations of the different components (noise).

Let us begin with an example discussed in Chapter 5. Recall that there are two models for how interactions among the BCL-2 proteins lead to mitochondrial outer membrane permeabilization (MOMP) in vertebrate cells (Fig. 12.4): (1) the neutralization

[2] We use the term bistable to mean "having two states that can stably exist as input changes." Technically, this is incorrect: formally, "bistability" is a consequence of a mathematical representation having two solutions. In its use here, it is more correctly called transcritical bifurcation, which can be bistable or not, in the formal sense. Our admittedly sloppy use of "bistability" in this context is taken from literature in this area. As for our use of the term "stable," we refer to a state that can persist despite noise in the system—for example, death is a persistent state (of course), but so too must cell survival be if we are to function as organisms.

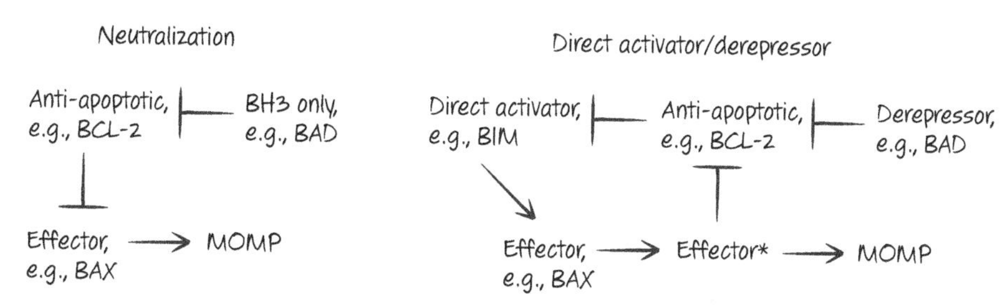

Figure 12.4. Schematic representation of two models of BCL-2-family protein interactions in MOMP. *, activated effector.

model, in which inhibition of anti-apoptotic function by BH3-only proteins is necessary and sufficient to cause MOMP, which is mediated by the pro-apoptotic BCL-2 effector proteins, and (2) the more complex direct activator/derepressor model, in which direct activators for the effectors are sequestered by the anti-apoptotic BCL-2 proteins and then released to trigger the effectors, leading to MOMP.

When tested as described above, the simpler neutralization model is unstable. In contrast, the direct activator/derepressor model is bistable, but it has a problem—it is not very robust. That is, when the system is subjected to noise (more than a very small amount of variation in the component molecules), it is unstable.

Does this mean that neither model is correct? Not necessarily. The starting concentrations used for the components might have been wrong in our model, or the model might simply need tweaking. It turns out that a small tweak to the second model makes it work in a robustly bistable manner (showing much more resistance to noise)—the tweak is the positive-feedback loop shown in Figure 12.5.

Now, the activated pro-apoptotic BCL-2 effectors (BAX and BAK) can activate inactive effectors, without relying on the appearance of more direct activator proteins. Indeed, this is supported by experimental evidence. Intuitively, one might have thought that this addition to the system would have made it even less resistant to random fluctuations, but the results show otherwise.

This does not show that the direct activator/derepressor model is correct. But the pathway is formally consistent with robust bistability; it is an on–off switch that is resistant to random noise, such as small variations in the amounts of the main components.

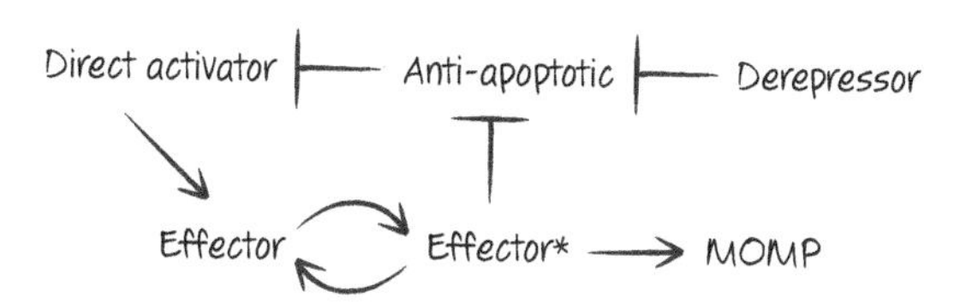

Figure 12.5. Schematic representation of the direct activator/derepressor model of BCL-2 protein interactions in MOMP. Included is a feed-forward mechanism, whereby activated effector proteins (*) can activate other effector proteins.

The model itself still might not be correct or only partially so. For example, it does not include inhibition of the activated pro-apoptotic effectors BAX and BAK by the anti-apoptotic BCL-2 proteins, and we have good reason to believe that this occurs. Would adding such interactions make the system more or less robust?

By considering both ways in which anti-apoptotic proteins work in the model (i.e., the "unified model" of BCL-2 protein interactions), it turns out that not only is robustness sustained, but there is a gain in predictive power. A computational model built on these principles was applied to a set of samples from patients, in which the average levels of BCL-2, BCL-xL, MCL-1, BAX, and BAK for the cells in each tumor were determined. The model separated the patients into two groups: those predicted to require lower doses or those predicted to require higher doses of a therapeutic to induce apoptosis. This turned out to correlate very well with the actual outcome of therapy (Fig. 12.6). We will return to the application of such models to cancer therapy when we consider the development of activators and inhibitors of the BCL-2 family proteins and their applications.

MOMP is not the only source of bistability in apoptosis, and the mitochondrial pathway is only one route to cell death. Bistability is also found in caspase activation and inhibition by inhibitor of apoptosis proteins (IAPs; see Chapter 3). Bistability

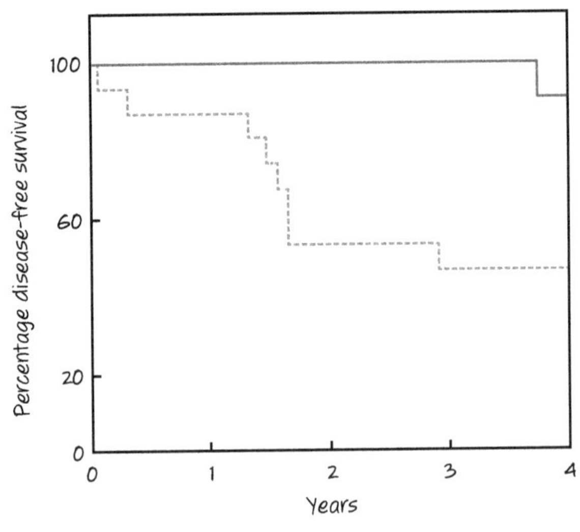

Figure 12.6. A model of BCL-2 protein interactions in MOMP predicts outcome in cancer patients. Tumors from patients with colorectal cancer were analyzed for levels of BCL-2, BCL-xL, MCL-1, BAX, and BAK, which were then input into a computational model of BCL-2 interactions. The model outputs a minimal value, η, of stress-induced BH3-only proteins needed for MOMP, based on the considerations of the unified model of BCL-2 protein interactions and the activation of BAX and BAK. The values of η for each patient (solid line, $\eta < 0.3$ µmol/L; dotted line, $\eta > 0.3$ µmol/L) were then correlated with clinical outcome. Patients needing the high-dose treatment to effect MOMP (dotted line) had poorer disease-free outcomes over the 4-year study period. $P = 0.0169$. Solid line, $N = 11$; dotted line, $N = 15$.

manifests in the model if we assume that, in addition to the inhibition of caspases by IAPs, caspases inhibit IAPs—that is, the bound caspase prevents the IAP from inhibiting any other caspases, which is a reasonable assumption.

Another source of bistability in cell death occurs at the level of transcription. As discussed in Chapter 11, p53 induces the expression of a number of proteins with various functions, including apoptosis. Among the proteins induced by p53 is its inhibitor MDM2, and the p53–MDM2 interaction and the regulation of this interaction have been shown to be bistable under some conditions. Indeed, models based on observations in single cells have revealed different outcomes for this interaction that can predict not only cell survival versus cell death, but also other effects of p53 (such as cell cycle arrest).

IDENTIFYING OTHER LIFE–DEATH DECISIONS

Some of the cell death pathways that we have discussed are not inherently bistable, even though the result (e.g., life vs. death) clearly achieves this. An example is the apoptotic pathway in the neurosecretory motor neuron (NMN) sister cells in *Caenorhabditis elegans* development, discussed in Chapter 10 and shown again in Figure 12.7.

Although this has not been formally tested for bistability, the pathway, as we understand it, is linear and probably insufficient for robust bistability. If so, then from where does bistability in the system arise? The same query applies to the other nematode cell deaths that occur during development and in response to genotoxic stress in germ cells (see Chapters 10 and 11). It is likely that bistability arises from upstream transcriptional control mechanisms, whereas the pathway itself (composed of several "inhibitor of inhibitor" connections) functions to dampen the effects of noise in the system. If this proves to be correct (following formal analysis of the pathway), with this information we might set out to identify the transcriptional control mechanisms that convert the pathways to bistable systems.

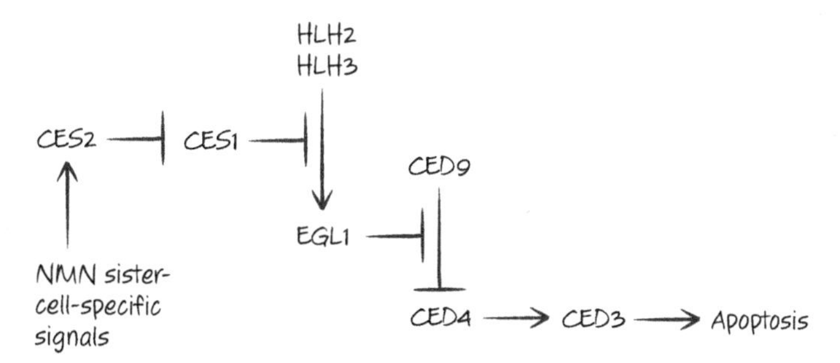

Figure 12.7. The neurosecretory motor neuron (NMN) pathway of apoptosis in nematodes.

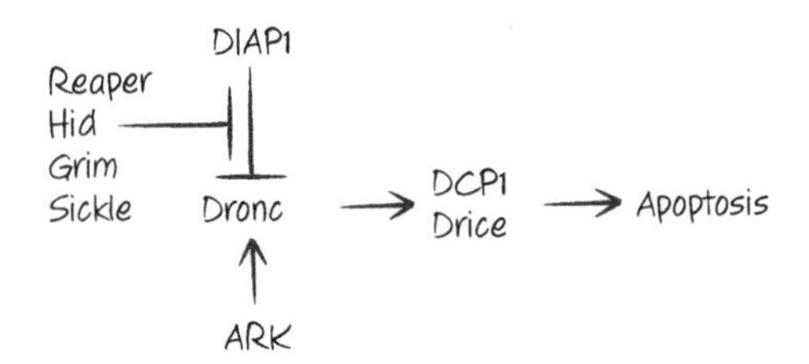

Figure 12.8. Relationships between effectors of developmental apoptosis in *Drosophila*.

This fresh perspective can be brought to other apoptotic pathways as well. Developmental cell death in *Drosophila* (see Chapter 10) includes the pathway shown in Figure 12.8. This pathway has been formally tested for bistability, and one source resides within the IAP–caspase interactions, as discussed above. However, as with developmental cell death in nematodes, it depends on transcriptional control of the antagonists of IAPs—Reaper, Hid, and Grim. We saw in Chapter 10 that, during metamorphosis, such transcriptional control can be triggered by the binding of ecdysone to its nuclear receptors. Bistability might therefore also emerge from tissue-specific regulatory interactions governing the expression and function of the ecdysone receptor and/or other regulatory steps in the pathway that remain to be uncovered.

CELL DEATH BY DESIGN

The other approach to evaluating our understanding of cell death—one perhaps more in keeping with the mandate placed on us by the public and private entities that support the research enterprise—is to ask whether we can manipulate the system for desired ends. There is obviously great therapeutic potential to controlling the survival and death of specific cells at will. But, to do this, we must probe each step in the cell death mechanisms that we have detailed to determine where the effective points of manipulation lie. When we consider such manipulations, we are especially concerned with two approaches—pharmacology and genetic engineering. Both hold promise. But, as we will see, there is a surprising amount that we do not yet know.

MAKING CELLS DIE

It is easy to kill a cell, but the trick is to kill only those cells we wish to, with as little collateral damage as possible to other cells. A good place to start is the generation of "suicide switches."

Gene therapy, in principle, holds great potential for treatment of disease, but any introduction of foreign DNA into an individual's cells carries with it the risk of

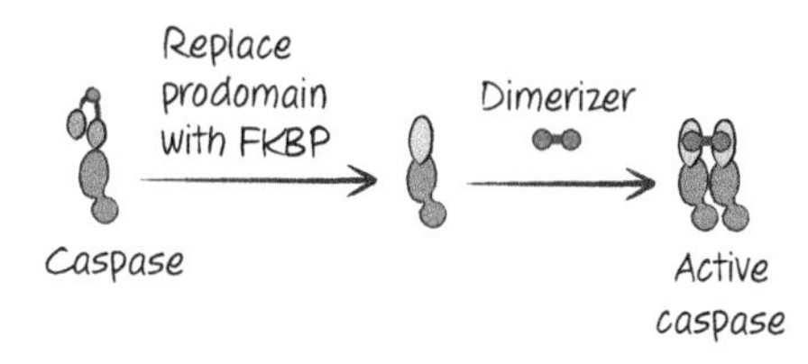

Figure 12.9. The principle behind FKBP-mediated dimerization.

mutations that could lead to cancer.[3] One strategy is to include a switch that can trigger death in engineered cells at will. Some such suicide switches exploit the cell death pathways already present in our cells.

Several proteins have been identified that bind to the immunosuppressive drug FK506, collectively called FK-binding proteins (FKBPs). Sequences derived from one of these, FKBP12, can be fused onto a protein of choice, and a derivative of the drug (without immunosuppressive activity) can be safely used to dimerize chimeric molecules in cells (Fig. 12.9).

Creating a suicide switch this way is straightforward (at least in principle). Because initiator caspases are activated by induced proximity by adapter molecules (see Chapter 3), we can replace the prodomain of an initiator caspase with the FKBP motif and express the chimeric protein in the cells that we wish to control. Addition of the FK506 derivative would then kill the cell. Figure 12.10 shows this for a caspase-9-based switch.

This approach can also be extended to the proteins of the necroptosis pathway. FKBP-fusion proteins of RIPK3 or MLKL cause rapid necroptosis when the dimerizing agent is added.

Whether such suicide switches are sufficiently resistant to the effects of noise to be of practical value is unclear—that is, we do not know whether they behave as robust bistable systems. If not, additional elements will have to be incorporated until they do. Alternatively, this might not be the ideal point in the pathway to exercise control.

There are other potential therapeutic applications for drugs that promote cell death. Probably the most pressing one relates to cancer therapy. In Chapter 11, we

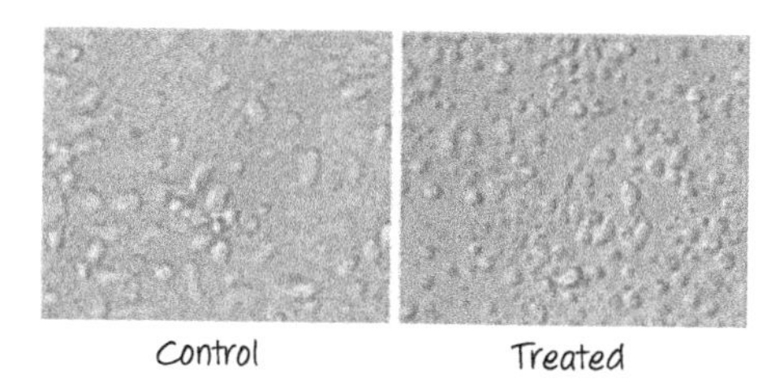

Figure 12.10. A caspase-9 suicide switch in engineered T cells. Addition of the dimerizer agent ("treated" group) induces rapid apoptosis in most of the cells.

[3] Unfortunately, this is only one of many challenges facing gene therapy; the rest lie outside the scope of our discussion.

discussed how cell death pathways can be thwarted in cancer, but we developed the idea that this can produce an "Achilles' heel" when these pathways are engaged but held in check (i.e., cancer cells are often poised to die in a way that normal cells are not). The challenge is to disrupt the resistance mechanism and thereby kill the transformed cells.

Anti-apoptotic BCL-2 proteins block apoptosis by sequestering activated pro-apoptotic effectors (BAX and BAK) and the proteins that activated them. BH3-only proteins can disrupt this sequestration, leading to death (see Chapter 5 and above). We can mimic this effect by using drugs—the BH3 mimetics (mentioned in Chapter 11)—that bind to the anti-apoptotic proteins in the same manner. One such drug is ABT-737 (or its orally available form, Navitoclax), which binds to BCL-2 and BCL-xL, but not to MCL-1 (Fig. 12.11).

ABT-737 and Navitoclax do not trigger death in most cells, with the exception of some lymphocytes and platelets (remarkably, the latter return to ~70% of normal levels even during the course of treatment; this is because it is the aging platelets that die owing to inhibition of BCL-xL). In contrast, some tumor cells are very sensitive to the drug, rapidly undergoing apoptosis. Others are highly responsive to Navitoclax in combination with other chemotherapeutic agents.

Based on the success with Navitoclax, other BH3-mimetic drugs have been developed. One of these, Venetoclax, binds specifically to BCL-2 and not to the other anti-apoptotic BCL-2 proteins. It has been approved for the treatment of a particularly aggressive lymphoma, 17p-deleted chronic lymphoblastic leukemia, and is showing promise in the treatment of other lymphomas and leukemias. BH3 mimetics that specifically target BCL-xL or MCL-1 have also been developed, and are currently in clinical trials.

As we discussed in Chapter 11, BH3 profiling is currently the most efficient way of determining whether a cancer cell is dependent on one or more anti-apoptotic BCL-2 proteins for its survival. However, as the computational approaches discussed in the last chapter improve, it might be possible to make this determination by assessing the levels of key proteins (at this point, however, we do not know which ones will

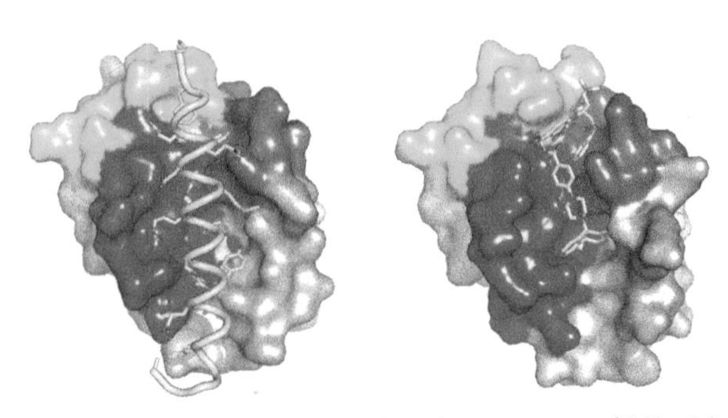

Figure 12.11. The BH3-mimetic drug ABT-737 binds to the BH groove of BCL-xL in the same manner as do BH3-only proteins.

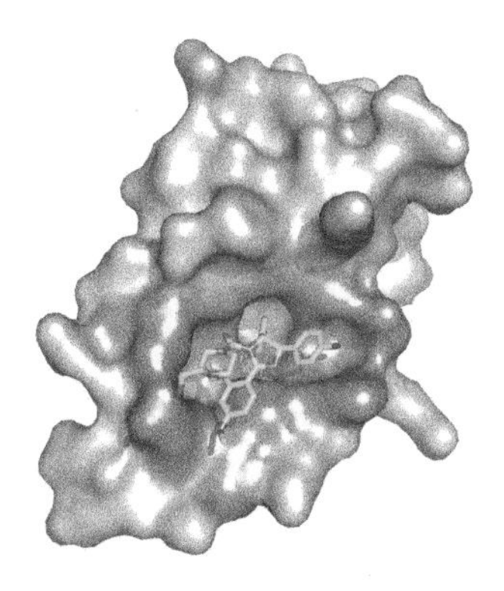

Figure 12.12. Nutlin-3 (green) bound to the binding pocket (brown) of MDM2 that binds to the tumor suppressor protein p53.

be key; presumably we will need to know the levels of not only the anti-apoptotic proteins, but also both the effectors and the BH3-only proteins). By knowing which anti-apoptotic BCL-2 proteins are required for the survival of a cancer cell, it might be possible to specifically inhibit them as an approach to cancer therapy.

Inhibition of the anti-apoptotic proteins is not the only way to promote apoptosis by the mitochondrial pathway. Drugs capable of directly activating BAX or BAK are also in development. At this point, we do not know whether these will have the necessary "therapeutic window" to allow their use in the treatment of cancers.

Other survival mechanisms can also be targeted pharmacologically. In Chapter 11, we discussed how p53 is stabilized by disruption of the p53–MDM2 interaction, leading to apoptosis (and additional effects of p53). Although many tumors mutate p53 itself to bypass this control mechanism, many others subvert the p53 pathway by losing ARF (which normally blocks the p53–MDM2 interaction). In the latter case, p53 can be stabilized by drugs that disrupt the binding of p53 to MDM2. One such drug is nutlin-3 (Fig. 12.12). At this point, we do not know whether this is effective as an anticancer therapy, but the prospect is intriguing.

As we discussed in Chapter 6, Smac and Omi are released upon MOMP, and these act to antagonize XIAP. Pharmacologic Smac-mimetic drugs induce cell death in some cells and sensitize others. As it turns out, however, this is not because they inhibit XIAP, but rather because of their action on cIAP-1 and cIAP-2. This sensitizes cells to the induction of necroptosis, and in some cells, also induces the production of tumor necrosis factor (TNF, which ligates the TNF receptor to trigger necroptosis).[4] Addition of a Smac mimetic plus a caspase inhibitor can potently induce necroptosis, and this approach is being explored as a route to cancer therapy.

Ferroptosis can also be induced pharmacologically (indeed, this is how it was first described). An inhibitor of system Xc^-, called erastin, is remarkably efficient in

[4] This effect is due to the inhibitory effect of cIAPs on one of the NF-κB pathways; inhibiting the cIAPs induces this pathway, leading to production of TNF.

inducing ferroptosis. Other promising drugs to induce ferroptosis do so by inhibiting GPX4, the lipid peroxidase that detoxifies lipid peroxides (see Chapter 8). It is possible that some cancers are more susceptible to ferroptosis than are normal tissues, and if so, drugs that induce ferroptosis could have therapeutic application.

These are only a few examples of pharmacological approaches to making cells die. Others involve blocking survival signals or engaging death pathways upstream of the core molecular mechanisms. Indeed, most cancer therapies are designed to kill tumor cells, and to do so they must engage one or more of the cell death pathways that we have discussed.

MAKING CELLS LIVE

Just as it can be desirable to make some cells die, it can be beneficial to limit cell death in some cases. The approach taken depends on the mode of cell death and the pathways with which it is engaged.

Caspases (like many proteases) are amenable to pharmacological inhibition (see Chapter 2). However, inhibition of caspases does not necessarily prevent cell death. Remember, two pathways of caspase activation and apoptosis result in cell death even when caspases are blocked: (1) death-receptor-mediated engagement of necroptosis and (2) MOMP-dependent caspase-independent cell death. Nevertheless, in some cases, inhibition of caspases does, indeed, preserve the survival of the cell. As discussed in Chapter 6, caspase inhibitors preserve hepatocytes and result in the survival of animals injected with CD95-ligand or antibodies that stimulate the death receptor CD95 (Fig. 12.13). This is because the necroptosis pathway might not be active in these cells.

This has potential value in the treatment of conditions that cause liver pathology through CD95-ligand-mediated apoptosis. These include chronic hepatitis infection and alcoholic cirrhosis. As it turns out, all caspase inhibitors tested in humans so far tend to accumulate in the liver, which may be useful for these conditions (but more problematic for other target tissues). That said, the long-term consequences of caspase inhibition remain unknown.

Other examples of cell death that depend on caspase activity include cell death caused by caspase-1 activity and, potentially, death that is dependent on the activation of caspase-2 (see Chapter 7). Potentially lethal bacterial sepsis in experimental animals that involves the former has been effectively treated with caspase inhibitors. Whether such approaches will have therapeutic value will depend on the development of new caspase inhibitors with better availability in vivo.

Necroptosis can be engaged by death receptors and other mechanisms (see Chapter 8). RIPK1 has been shown, in principle, to be an effective target for pharmacological inhibition by necrostatins, and whether RIPK3 and MLKL (also required for this death pathway) can be effectively targeted remains to be seen. As we discussed, kinase-inactive RIPK3 promotes apoptosis in animals, and some RIPK3 inhibitors do

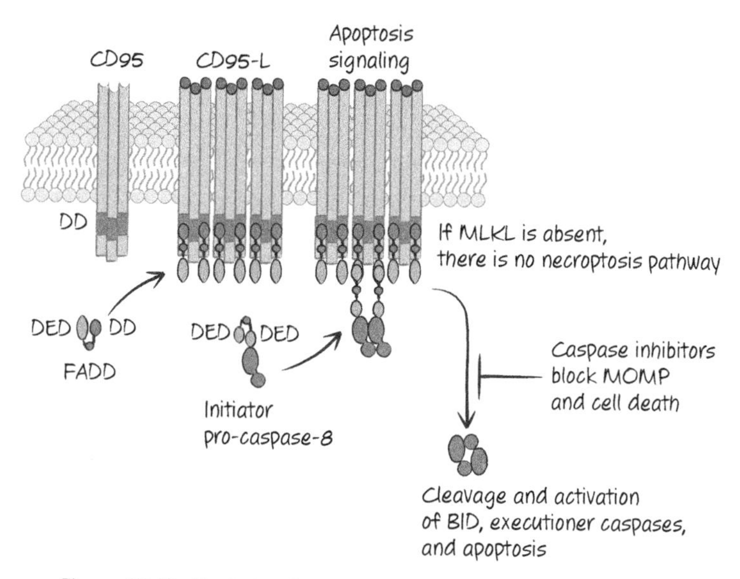

Figure 12.13. Fas-L signaling and protection by caspase blockade.

so as well. However, some mutants of RIPK3 that do not promote apoptosis (or permit necroptosis) have been identified, making the development of an effective inhibitor a possibility. Human MLKL can be inhibited by a compound called necrosulfonamide (although this is not specific for MLKL), and therefore MLKL might be amenable to pharmacologic inhibition. Necrostatins have been developed for human use and have entered clinical trials.

Inhibitors of ferroptosis have also been developed, called ferrostatins. A ferrostatin that was developed for use in vivo showed striking protection in a model of ischemia–reperfusion kidney damage.

In Chapter 8, we discussed the role of the mitochondrial permeability transition (MPT) in necrosis—for example, in ischemia–reperfusion injury. Mice lacking cyclophilin D display a defective MPT and are resistant to such injury. The drug cyclosporin A blocks the MPT by inhibiting cyclophilin D. Although this drug has other effects (it is widely used as an immunosuppressive drug in organ transplant, at doses well below those needed to inhibit MPT), other compounds that more specifically inhibit cyclophilin D have been described. These drugs could have promise in the treatment of stroke and myocardial infarction as well as other diseases involving ischemia–reperfusion injury.[5]

As another approach to keeping cells alive, we would ideally like to have drugs that inhibit the activation of BAX and BAK to prevent MOMP and all forms of

[5] Although cyclosporin A has protective effects in animal models of ischemia–reperfusion injury, neurotoxic effects have been observed in up to 60% of organ transplant patients who receive this drug as an immunosuppressant. The high doses needed to block cyclophilin D are therefore problematic for this drug.

MOMP-dependent cell death. Such pharmacologic agents are in development and are remarkably efficient in blocking the mitochondrial pathway of apoptosis. Other approaches are also possible; a drug has been identified that prevents MOMP, apparently without inhibiting the activation of BAX and BAK, although how it works is currently obscure. Nevertheless, the progress so far shows that inhibition of the mitochondrial pathway of apoptosis is possible. We can envision using such drugs to limit damage to tissues such as the gut during chemotherapy for cancer, provided the agents can be delivered to normal tissue without compromising their antitumor effects.

BACK TO THE FUTURE

As we have seen above, there are ways to challenge our knowledge of cell death by probing models of the pathways in silico and by engineering the systems for applications in vivo. Beyond these, it could be tempting simply to stand back and admire the pool of understanding that we have gained. But, when we do, it is only necessary to extend our gaze further to view a sea of ignorance, of which our knowledge only skims the surface.

An elegant way to illustrate this is by considering the "button experiment."[6] Imagine a table covered with buttons. Choose, at random, two buttons and connect them with a piece of thread. Then choose another button, again at random, pick it up, and count how many buttons are collected with it. Now repeat the entire process and score again. If we do this many times (this is more practically performed in a computer model) something interesting happens (Fig. 12.14).

Early on, one or at most a few buttons come up with the one we pick up. Later, the entire system undergoes a phase transition such that any chosen button brings with it a large number of other buttons. Quickly, nearly any chosen button pulls with it most of the others. Astonishingly, this transition occurs near the point at which the number of threads approximately equals half the number of buttons. Chosen randomly, some buttons act as "nodes," tied to many threads, and some remain attached to only one or two. But, at this phase transition, the system has been altered dramatically.

The human genome comprises approximately 28,000 genes encoding proteins. We take it as a given that all of these do something. The majority are acted on by and/or act on at least one other protein. But proteins are not the only "buttons." Lipids, carbohydrates, nucleic acids (and their sequences), and metabolites are all buttons, similarly connected by threads. Viruses, bacteria, and other passengers in our bodies carry their own sets of buttons and threads, and many of these connect to our buttons as well. It is a dizzying thought that picking up almost any button will drag along large numbers of other buttons.

[6] Conceived by Stephan Kaufmann, a theoretical biologist and geneticist.

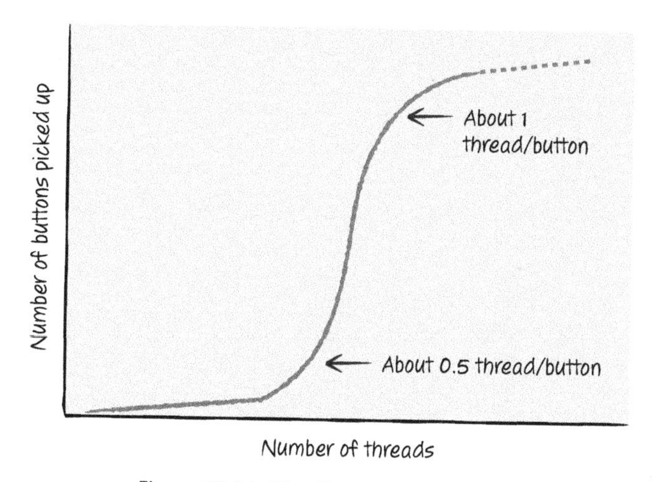

Figure 12.14. The "button experiment."

Clearly, any button that we choose is likely to be connected to the pathways controlling life and death. Perturb a protein, alter a metabolite, mutate a sequence, or add a drug, and we can impact cell death pathways. Because the cell death pathways are robustly bistable, most perturbations will not shift the state dramatically, but they have the potential to do so. Selection has acted over eons to favor those buttons and threads that impart fitness to the pathways for cell death and survival. Moreover, the molecular events that control energy, movement, replication, repair, and all other important cellular processes, connect to this machinery. The devil is in the details—how are they all connected? This is the future to which we will return.

Darwin concluded *On the Origin of Species* with an evocative image—an "entangled bank" on which varieties of life forms comingled, competed, and survived. We now know that evolution acts not only at the level of individuals, but also at the level of cells engaged in development, homeostasis, and disease. Cell survival and death are the outcomes of battles on the entangled banks within all of us.

Figure Credits

Chapter 1: **1.1**, Images provided by Douglas R. Green. **1.2**, Reprinted from Maclean KH, et al. 2008. *J Clin Invest* **118:** 79–88, ©2008 with permission from the American Society for Clinical Investigation. **1.3**, Images courtesy of Dr. Nigel Waterhouse, Mater Medical Research Institute, Brisbane, Australia.

Chapter 2: **2.8, 2.10, 2.11**, Reprinted from Mahrus S, et al. 2008. *Cell* **134:** 866–876, ©2008 with permission from Elsevier. **2.12**, Image courtesy of Dr. Yufang Shi, Institutes for Translational Medicine, Soochow University, Suzhou, China; structure (*inset*), PDB 1GQF (Riedl S, Bode W, Fuentes-Prior P. 2001. *Proc Natl Acad Sci* **98:** 14790). **2.13**, Courtesy of Michael Olson, Beatson Institute. **2.15**, Reprinted from Ellis HM, Horvitz HR. 1986. *Cell* **44:** 817–829, ©1986 with permission from Elsevier. **2.16**, Reprinted from Quinn LM, et al. 2000. *J Biol Chem* **275:** 40416–40424, ©2000 with permission from the American Society for Biochemistry and Molecular Biology. **2.17**, Reprinted with permission from Macmillan Publishers Ltd.: Kuida K, et al. 1996. *Nature* **384:** 368–372, ©1996.

Chapter 3: **3.1, left**, PDB 1K86 (Chai J, Wu Q, Shiozaki E, Srinivasula SM, Alnemri ES, Shi Y. 2001. *Cell* **107:** 299–407); **right**, PDB 1GQF (Riedl S, Bode W, Fuentes-Prior P. 2001. *Proc Natl Acad Sci* **98:** 14790). **3.2, left**, *see* 3.1a; **center**, PDB 1F1J (Wei Y, Fox T, Chambers SP, Sintchak J, Coll JT, Golec JM, Swenson L, Wilson KP, Charifson PS. 2000. *Chem Biol* **7:** 423–432); **right**, *see* 3.1b. **3.3**, Reproduced from Paul WE. 2008. *Fundamental immunology*, 6th Edition, Bleakley: Cytotoxic T lymphocytes, ©2008 with permission from Lippincott, Williams & Wilkins. Image kindly provided by Dr. I.S. Goping, University of Alberta. **3.7, left to right**, PDB 1DDF (Huang B, Eberstadt M, Olejniczak ET, Meadows RP, Fesik SW. 1996. *Nature* **384:** 638–641); PDB 1CWW (Day CL, Dupont C, Lackmann M, Vaux DL, Hinds MG. 1999. *Cell Death Differ* **6:** 1125–1132); PDB 1PN5 (Hiller S, Kohl A, Fiorito F, Herrmann T, Wider G, Tschopp J, Grutter MG, Wuthrich K. 2003. *Structure* **11:** 1199–1205); PDB 1A1Z (Eberstadt M, Huang B, Chen Z, Meadows RP, Ng SC, Zheng L, Lenardo MJ, Fesik SW. 1998. *Nature* **392:** 941–945). **3.8**, Reproduced with permission from Fuentes-Prior P, Salvesen GS. 2004. *Biochem J* **384:** 201–232, ©The Biochemical Society. **3.9**, Reprinted by permission from Macmillan Publishers Ltd.: Fernandes-Alnemri T, et al. 2007. *Cell Death Differ* **14:** 1590–1604, ©2007. **3.12**, PDB 1CP3 (Mittl PR, Di Marco S, Krebs JF, Bai X, Karanewsky DS, Priestle JP, Tomaselli KJ, Grutter MG. 1997. *J Biol Chem* **272:** 6539–6547). **3.13**, Images courtesy of Dr. Nigel Waterhouse, Mater Medical Research Institute, Brisbane, Australia. **3.14**, Reprinted

---

Structures created using data from Research Collaboratory for Structural Bioinformatics Protein Data Bank (RCSB PDB; Berman HM et al. 2000. *Nucleic Acids Res* **28:** 235–242; http://www.pdb.org) and PyMOL (http://www.pymol.org/).

(AAAS) American Association for the Advancement of Science.

from Wang SL, Hawkins CJ, Yoo SJ, Müller HA, Bay A. 1999. *Cell* **98:** 453–463, ©1999 with permission from Elsevier. **3.16**, PDB 1I3O (Riedl SJ, Renatus M, Schwarzenbacher R, Zhou Q, Sun C, Fesik SW, Liddington RC, Salvesen GS. 2001. *Cell* **104:** 791–800).

Chapter 4: **4.2**, PDB 3YGS (Qin H, Srinivasula SM, Wu G, Fernandes-Alnemri T, Alnemri ES, Shi Y. 1999. *Nature* **399:** 549–557). **4.5**, Images courtesy of Dr. Christopher Dillon, St. Jude Children's Research Hospital, Memphis, Tennessee. **4.6**, Images provided by Dr. Stephen Tait, St. Jude Children's Research Hospital, Memphis, Tennessee. **4.8**, Reprinted from Chautan M, Chazal G, Cecconi F, Gruss P, Golstein P. 1999. *Curr Biol* **9:** 967–970, ©1999 with permission from Elsevier.

Chapter 5: **5.2**, Courtesy of Dr. Stephen Tait, St. Jude Children Research Hospital, Memphis, Tennessee. **5.4, left**, PDB 1MAZ (Muchmore SW, Sattler M, Liang H, Meadows RP, Harlan JE, Yoon HS, Nettesheim D, Chang BS, Thompson CB, Wong SL, et al. 1996. *Nature* **381:** 335–341); **right**, PDB 1BXL (Sattler M, Liang H, Nettesheim D, Meadows RP, Harlan JE, Eberstadt M, Yoon HS, Shuker SB, Chang BS, Minn AJ, et al. 1997. *Science* **275:** 983–986). **5.5, left**, PDB 1F16 (Suzuki M, Youle RJ, Tjandra N. 2000. *Cell* **103:** 645–654); **right**, PDB 2IMS (Moldoveanu T, Liu Q, Tocilj A, Watson M, Shore G, Gehring K. 2006. *Mol. Cell* **24:** 677–688). **5.13**, *see* 5.5, left; **overlay**, PDB 2K7W (Gavathiotis E, Suzuki M, Davis ML, Pitter K, Bird GH, Katz SG, Tu HC, Kim H, Cheng EH, Tjandra N, et al. 2008. *Nature* **455:** 1076–1081). **5.19**, PDB 2BID (Chou JJ, Li H, Salvesen GS, Yuan J, Wagner G. 1999. *Cell* **96:** 615–624). **5.28**, Reprinted from Hardwick JM, Youle RJ. 2009. *Cell* **138:** 404, ©2009 with permission from Elsevier. **5.29, left**, *see* 5.3, left; **right**, PDB 1K3K (Huang Q, Petros AM, Virgin HW, Fesik SW, Olejniczak ET. 2002. *Proc Natl Acad Sci* **99:** 3428–3433). **5.30, top left**, *see* 5.4, left; **top right**, PDB 1F0L (Choe S, Bennett M, Fujii G, Curmi PM, Kantardjieff KA, Collier RJ, Eisenberg D. 1992. *Nature* **357:** 216–222); **lower left**, PDB 1CII (Wiener M, Freymann D, Ghosh P, Stroud RM. 1997. *Nature* **385:** 461–464); **lower right**, PDB 2C9K (Boonserm P, Mo M, Angsuthanasombat C, Lescar J. 2006. *J Bacteriol* **188:** 3391). **5.31**, Image courtesy of Paul A. Ney, MD, St. Jude Children's Research Hospital, Memphis, Tennessee. **5.33**, PDB 2P1L (Oberstein A, Jeffrey PD, Shi Y. 2007. *J Biol Chem* **282:** 13123–13132).

Chapter 6: **6.2**, PDB 1TNR (Banner DW, D'Arcy A, Janes W, Gentz R, Schoenfeld HJ, Broger C, Loetscher H, Lesslauer W. 1993. *Cell* **73:** 431–445). **6.4** and **6.5**, Redrawn from Wang L, Yang JK, Kabaleeswaran V, Rice AJ, Cruz AC, Park AY, Yin Q, Damko E, Jang SB, Raunser S, et al. 2010. The Fas-FADD death domain complex structure reveals the basis of DISC assembly and disease mutations. *Nat Struct Biol* **17:** 1324–1329. PDB kindly provided by Dr. Hao Wu. **6.9**, Image courtesy of Dr. Ralph C. Budd, University of Vermont, Burlington.

Chapter 7: **7.9, bottom left**, Reprinted from Gillray J. 1799. *The Gout*. **7.16, left**, PDB 2OF5 (Park HH, Logette E, Raunser S, Cuenin S, Walz T, Tschopp J, Wu H. 2007. *Cell* **128:** 533–546); **right**, Reprinted from Park H, et al. 2007. *Cell* **128:** 533–546, ©2007 with permission from Elsevier. **7.17**, Photo courtesy of Dr. Luca Fava and Dr. Andreas Villunger, University of Innsbruck.

Chapter 8: **8.1**, Reprinted from Edinger AL, Thompson CB. 2004. *Curr Opin Cell Biol* **16:** 663–669, ©2004 with permission from Elsevier. **8.2**, Reprinted with permission from Springer Science+Business Media: Silva M, et al. 2008. *Apoptosis* **13:** 463–482, ©2008. **8.5B**, Model made by Dr. Scott Brown, St. Jude Children's Research Institute. **8.6B**, 3D scanning electron micrographs by Dr. Giovanni Quarato and Dr. Sharon Frase, St. Jude Children's Research Institute. **8.9**, Reprinted from He, et al. 2009. **Cell** *137:* 1100–1111, ©2009 with permission from Elsevier. **8.14**, Reprinted from Eskelinen E-L. 2008. *Intl Rev Cell Mol Biol* **266:** 207–247, ©2008 with permission from Elsevier. **8.19**, Reprinted from Lum J, Bauer DE, Kong M, Harris MH, Li C, Lindsten T, Thompson CB. 2005. *Cell* **120:** 237–248, ©2005 with permission from Elsevier. **8.21**, Courtesy of Ynjie Wei and Beth Levine. **8.23**, Reprinted from Kirsch D, Santiago PM, di Tomaso E, Sullivan JM, Hou WS, Dayton T, Jeffords LB, Sodha P, Mercer KL, Cohen R, et al. 2010. *Science* **327:** 593–596, ©2010 with permission from AAAS.

Chapter 9: **9.2**, Reprinted from Brumatti G, et al. 2008. *Methods* **44:** 235–240, ©2008 with permission from Elsevier. **9.5**, Reprinted from Hanayama R, Tanaka M, Miyasaka K, Aozasa K, Koike M,

Uchiyama Y, Nagata S. 2004. *Science* **304:** 1147–1150, ©2004 with permission from AAAS. **9.12**, Franc NC, Heitzler P, Ezekowitz RA, White K. 1999. *Science* **284:** 1991–1994. **9.21**, Reprinted from Lämmermann T, Afonso PV, Angermann BR, Wang JM, Kastenmüller W, Parent CA, Germain RN. 2013. *Nature* **98:** 371–375 (Fig. 1B). **9.25**, Provided by Jennifer Martinez and Clifford Guy, Department of Immunology, St. Jude Children's Research Hospital. **9.26**, Reprinted from Holman S. 1951. *J Clin Pathol* **4:** 290–295, with permission from BMJ Publishing Group Ltd. **9.27**, Michael Overholter, unpublished.

Chapter 10: **10.1**, Reprinted from Sulston JE, Schierenberg E, White JG, Thomson JN. 1983. *Dev Biol* **100:** 64–119, ©1983 with permission from Elsevier. **10.2**, Reprinted from Abraham M, Lu Y, Shaham S. 2007. *Dev Cell* **12:** 73–86, ©2007 with permission from Elsevier. **10.5**, Reprinted from Maurer CW, Chiorazzi M, Shaham S. 2007. *Development* **134:** 1357–1368, ©2007 with permission from The Company of Biologists. **10.7**, Reprinted from Lee C-Y, Baehrecke EH. 2001. *Development* **128:** 1443–1455, ©2001 with permission from The Company of Biologists. **10.8**, Reprinted from Neufeld T, Baehrecke E. 2008. *Autophagy* **4:** 557–562, ©Landes Bioscience. **10.9, left**, Reproduced with permission from Zuzarte-Luís V, Hurlé JM. 2002. *Int J Dev Biol* **46:** 871–876; **right**, Reprinted from Gañan Y, Macías D, Basco RD, Merino R, Hurle JM. 1998. *Dev Biol* **196:** 33–41, ©1998 with permission from Elsevier. **10.10, left two panels**, Reprinted from Lindsten T, Ross AJ, King A, Zong WX, Rathmell JC, Shiels HA, Ulrich E, Waymire KG, Mahar P, Frauwirth K, et al. 2000. *Mol Cell* **6:** 1389–1399, ©2000 with permission from Elsevier; **right two panels**, Reprinted from Hubner A, Cavanagh-Kyros J, Rincon M, Flavell RA, Davis RJ. 2010. *Mol Cell Biol* **30:** 98–105, ©2010 with permission from American Society for Microbiology. **10.11**, Adapted from Coucouvanis E, Martin G. 1995. *Cell* **83:** 279–287, ©1995 with permission from Elsevier. **10.12**, Reprinted from Kerr JFR, Harmon B, Searle J. 1974. *J Cell Sci* **14:** 571–585, ©1974 with permission from The Company of Biologists.

Chapter 11: **11.3** and **11.4**, Modified from Pelengaris S, Khan M, Evan GI. 2002. *Cell* **109:** 321–334, ©2002 with permission from Elsevier. **11.5**, Modified from Debnath J, Mills KR, Collins NL, Reginato MJ, Muthuswamy SK, Brugge JS. 2002. *Cell* **111:** 29–40, ©2002 with permission from Elsevier. **11.17**, From Ni Chonghaile, Sarosiek KA, Vo TT, Ryan JA, Tammareddi A, Moore Vdel G, Deng J, Anderson KC, Richardson P, Tai YT, et al. 2011. *Science* **2011:** 1129–1131.

Chapter 12: **12.1**, Redrawn with permission from Macmillan Publishers Ltd.: Spencer SL, Gaudet S, Albeck JG, Burke JM, Sorger PK. 2009. *Nature* **459:** 428–432, ©2009. **12.6**, Redrawn from Lindner AU, Concannon CG, Boukes GJ, Cannon MD, Llambi F, Ryan D, Boland K, Kehoe J, McNamara DA, Murray F, et al. 2013. *Cancer Res* **73:** 519–528. **12.10**, Reprinted with permission from the American Society of Hematology: Straathof KC, Pulè MA, Yotnda P, Dotti G, Vanin EF, Brenner MK, Heslop HE, Spencer DM, Rooney CM. 2005. *Blood* **105:** 4247–4254, ©2005. **12.11, left**, PDB2BZW; **right**, PDB 2YXJ. **12.12**, PDB 1RV1 (Vassilev LT, Vu BT, Graves B, Carvajal D, Podlaski F, Filipovic Z, Kong N, Kammlott U, Lukacs C, Klein C, et al. 2004. *Science* **303:** 844–848).

Additional Reading

CHAPTER 1

Cell Death Nomenclature

Andre N. 2003. Hippocrates of Cos and apoptosis. *Lancet* **361:** 1306.

A note on an early use of "apoptosis" in medicine.

Degli-Esposti M. 1998. Apoptosis: Who was first? *Cell Death Differ* **5:** 719.

A hunt for the origins of the word "apoptosis."

Galluzzi L, Bravo-San Pedro JM, Vitale I, Aaronson SA, Abrams JM, Adam D, Alnemri ES, Altucci L, Andrews D, Annicchiarico-Petruzelli M, et al. 2015. Essential versus accessory aspects of cell death: Recommendations of the NCCD 2015. *Cell Death Differ* **22:** 58–73.

More than simply nomenclature, this review outlines the different types of cell death and problems with their classification.

Green DR, Victor B. 2012. The pantheon of the fallen: Why are there so many forms of cell death? *Trends Cell Biol* **22:** 555–556.

A potentially useful way to think about different types of cell death.

History of Apoptosis

Diamantis A, Magiorkinis E, Sakorafas GH, Androutsos G. 2008. A brief history of apoptosis: From ancient to modern times. *Onkologie* **31:** 702–706.

Not essential reading, but an interesting overview of the study of apoptosis.

Horvitz HR. 2003. Nobel lecture. Worms, life and death. *Biosci Rep* **23:** 239–303.

The modern study of apoptosis was ignited by the identification of genes that control cell death during development in the nematode Caenorhabditis elegans. *This is the Nobel lecture that describes those studies.*

Kerr JF, Wyllie AH, Currie AR. 1972. Apoptosis: A basic biological phenomenon with wide-ranging implications in tissue kinetics. *Br J Cancer* **26:** 239–257.

This is the original paper formally describing apoptosis.

Saunders JW Jr. 1966. Death in embryonic systems. *Science* **154:** 604–612.

A classic early overview of the role of cell death in development.

Wyllie AH, Kerr JF, Currie AR. 1980. Cell death: The significance of apoptosis. *Int Rev Cytol* **68:** 251–306.

An early review that details the state of knowledge on apoptosis before molecular mechanisms became understood.

Lockshin RA. 2016. Programmed cell death 50 (and beyond). *Cell Death Differ* **23:** 10–17.

History and perspectives on cell death from the author who coined the term "programmed cell death."

Green DR. 2016. The cell's dilemma, or the story of cell death: An entertainment in three acts. *FEBS J* **283:** 2568–2576.

Just for fun, a play about the study of cell death; where we've been, where we are, and where it might be going.

CHAPTER 2

Caspases

Lamkanfi M, Festjens N, Declercq W, Vanden Berghe T, Vandenabeele P. 2007. Caspases in cell survival, proliferation and differentiation. *Cell Death Differ* **14:** 44–55.

A review of the caspases and their functions in life and death.

Yuan J, Shaham S, Ledoux S, Ellis HM, Horvitz HR. 1993. The *C. elegans* cell death gene *ced-3* encodes a protein similar to mammalian interleukin-1 β-converting enzyme. *Cell* **75:** 641–652.

The original landmark paper that identified caspases as being important in apoptosis.

Fernandes-Alnemri T, Litwack G, Alnemri ES. 1994. CPP32, a novel human apoptotic protein with homology to *Caenorhabditis elegans* cell death protein Ced-3 and mammalian interleukin-1 β-converting enzyme. *J Biol Chem* **269:** 30761–30764.

The first description of mammalian caspase-3, originally called CPP32.

Alnemri ES, Livingston DJ, Nicholson DW, Salvesen G, Thornberry NA, Wong WW, Yuan J. 1996. Human ICE/CED-3 protease nomenclature. *Cell* **87:** 171.

Eckhart L, Ballaun C, Hermann M, VandeBerg JL, Sipos W, Uthman A, Fischer H. Tschachler E. 2008. Identification of novel mammalian caspases reveals an important role of gene loss in shaping the human caspase repertoire. *Mol Biol Evol* **25:** 831–841.

An overview of the evolutionary relationships among mammalian caspases, including caspases-15, -16, -17, and -18.

Kuida K, Zheng TS, Na S, Kuan C, Yang D, Karasuyama H, Rakic P, Flavell RA. 1996. Decreased apoptosis in the brain and premature lethality in CPP32-deficient mice. *Nature* **384:** 368–372.

The first knockout of caspase-3 (called CPP32) and effects on development.

Caspase Specificities

Timmer JC, Salvesen GS. 2007. Caspase substrates. *Cell Death Differ* **14:** 66–72.

Thornberry NA, et al. 1997. A combinatorial approach defines specificities of members of the caspase family and granzyme B. Functional relationships established for key mediators of apoptosis. *J Biol Chem* **272:** 17907–17911.

The identification of caspase preferences through the use of a combinatorial peptide library.

Mahrus S, Trinidad JC, Barkan DT, Sali A, Burlingame AL, Wells JA. 2008. Global sequencing of proteolytic cleavage sites in apoptosis by specific labeling of protein N termini. *Cell* **134:** 866–876.

Another approach to characterization of caspase substrates.

Poreba M, Strozyk A, Salvesen GS, Drag M. 2013. Caspase substrates and inhibitors. *Cold Spring Harb Perspect Biol* **5:** a008680.

Julien O, Wells JA. 2017. Caspases and their substrates. *Cell Death Differ* **24:** 1380–1389.

A useful review of the specificities and functions of caspases, and how they cause cell death.

Functional Caspase Substrates

Luthi AU, Martin SJ. 2007. The CASBAH: A searchable database of caspase substrates. *Cell Death Differ* **14:** 641–650.

An introduction to a searchable database of all known caspase substrates.

Kumar S, van Raam B., Salvesen GS, Cieplak P. 2014. Caspase cleavage sites in the human proteome: CaspDB, a database of predicted substrates. *PLoS One* **9:** e110539.

Another searchable database of caspase substrates.

Sakahira H, Enari M, Nagata S. 1998. Cleavage of CAD inhibitor in CAD activation and DNA degradation during apoptosis. *Nature* **391:** 96–99.

One of the original descriptions of the iCAD–CAD system and its role in DNA fragmentation during apoptosis.

Zhang J, Liu X, Scherer DC, van Kaer L, Wang X, Xu M. 1998. Resistance to DNA fragmentation and chromatin condensation in mice lacking the DNA fragmentation factor 45. *Proc Natl Acad Sci* **95:** 12480–12485.

A clear demonstration of the role of iCAD/DFF45 in DNA fragmentation versus cell death.

Sebbagh M, Renvoize C, Hamelin J, Riche N, Bertoglio J, Breard J. 2001. Caspase-3-mediated cleavage of ROCK I induces MLC phosphorylation and apoptotic membrane blebbing. *Nat Cell Biol* **3:** 346–352.

One of the original descriptions of ROCK as a caspase substrate and its role in blebbing during apoptosis.

Coleman ML, Sahai EA, Yeo M, Bosch M, Dewar A, Olson MF. 2001. Membrane blebbing during apoptosis results from caspase-mediated activation of ROCK I. *Nat Cell Biol* **3:** 339–345.

See above.

Kothakota S, Azuma T, Reinhard C, Klippel A, Tang A, Chu K, McGarry TJ, Kirschner MW, Koths K, Kwiatkowski DJ, et al. 1997. Caspase-3-generated fragment of gelsolin: Effector of morphological change in apoptosis. *Science* **278:** 294–298.

The role of gelsolin in blebbing during apoptosis.

Ricci JE, Munoz-Pinedo C, Fitzgerald P, Bailly-Maitre B, Perkins GA, Yadava N, Scheffler IE, Ellisman MH, Green DR. 2004. Disruption of mitochondrial function during apoptosis is mediated by caspase cleavage of the p75 subunit of complex I of the electron transport chain. *Cell* **117:** 773–786.

The original description of NDUFS1 as a mitochondrial caspase substrate.

Suzuki J, Denning DP, Imanishi E, Horvitz HR, Nagata S. 2013. Xk-related protein 8 and CED-8 promote phosphatidylserine exposure in apoptotic cells. *Science.* **341:** 403–406.

The original description of the caspase-activated phospholipid scramblase Xkr8.

Suzuki J, Imanishi E, Nagata S. 2016. Xkr8 phospholipid scrambling complex in apoptotic phosphatidylserine exposure. *Proc Natl Acad Sci* **113:** 9509–9514.

More insights into the function of Xkr8 during apoptosis.

Segawa K, Kurata S, Yanagihashi Y, Brummelkamp TR, Matsuda F, Nagata S. 2014. Caspase-mediated cleavage of phospholipid flippase for apoptotic phosphatidylserine exposure. *Science* **344:** 1164–1168.

The original description of how executioner caspases inhibit the lipid translocase responsible for keeping phosphatidylserine sequestered to the inner leaflet of the plasma membrane.

CHAPTER 3

Mechanisms of Caspase Activation

Fuentes-Prior P, Salvesen GS. 2004. The protein structures that shape caspase activity, specificity, activation and inhibition. *Biochem J* **384:** 201–232.

Pop C, Salvesen GS. 2009. Human caspases: Activation, specificity, and regulation. *J Biol Chem* **284:** 21777–21781.

Salvesen GS, Riedl SJ. 2008. Caspase mechanisms. *Adv Exp Med Biol* **615:** 13–23.

Park HH, Lo YC, Lin SC, Wang L, Yang JK, Wu H. 2007. The death domain superfamily in intracellular signaling of apoptosis and inflammation. *Annu Rev Immunol* **25:** 561–586.

An extensive overview of the structures involved in caspase activation. Several of the mechanisms discussed are covered in later chapters.

Induced Proximity

Muzio M, Stockwell BR, Stennicke HR, Salvesen GS, Dixit VM. 1998. An induced proximity model for caspase-8 activation. *J Biol Chem* **273:** 2926–2930.

The original description of induced proximity, applied to caspase-8.

Stennicke HR, Deveraux QL, Humke EW, Reed JC, Dixit VM, Salvesen GS. 1999. Caspase-9 can be activated without proteolytic processing. *J Biol Chem* **274:** 8359–8362.

Boatright KM, Renatus M, Scott FL, Sperandio S, Shin H, Pedersen IM, Ricci JE, Edris WA, Sutherlin DP, Green DR, et al. 2003. A unified model for apical caspase activation. *Mol Cell* **11:** 529–541.

Caspase Activation Cascades

Fernandes-Alnemri T, Armstrong RC, Krebs J, Srinivasula SM, Wang L, Bullrich F, Fritz LC, Trapani JA, Tomaselli KJ, Litwack G, et al. 1996. In vitro activation of CPP32 and Mch3 by Mch4, a novel human apoptotic cysteine protease containing two FADD-like domains. *Proc Natl Acad Sci* **93:** 7464–7469.

One of the first papers to show a caspase activation cascade, with caspase-8 (called Mch4) cleaving and activating caspase-3 (called CPP32) and caspase-7 (called Mch3).

Slee EA, Adrain C, Martin SJ. 1999. Serial killers: Ordering caspase activation events in apoptosis. *Cell Death Differ* **6:** 1067–1074.

Slee EA, Harte MT, Kluck RM, Wolf BB, Casiano CA, Newmeyer DD, Wang HG, Reed JC, Nicholson DW, Alnemri ES, et al. 1999. Ordering the cytochrome *c*–initiated caspase cascade: Hierarchical activation of caspases-2, -3, -6, -7, -8, and -10 in a caspase-9-dependent manner. *J Cell Biol* **144:** 281–292.

The order of caspase cleavage following activation of caspase-9 in a cell extract (see Chapter 4 for the initial activation mechanism).

IAPs

Salvesen GS, Riedl SJ. 2007. Caspase inhibition, specifically. *Structure* **15:** 513–514.

An overview of IAPs and other intracellular caspase inhibitors.

Silke J, Vaux DL. 2001. Two kinds of BIR-containing protein: Inhibitors of apoptosis, or required for mitosis. *J Cell Sci* **114:** 1821–1827.

An important distinction between different IAPs and their functions.

Vaux DL, Silke J. 2005. IAPs, RINGs and ubiquitylation. *Nat Rev Mol Cell Biol* **6:** 287–297.

Orme M, Meier P. 2009. Inhibitor of apoptosis proteins in *Drosophila*: Gatekeepers of death. *Apoptosis* **14:** 950–960.

Hay BA, Wassarman DA, Rubin GM. 1995. *Drosophila* homologs of baculovirus inhibitor of apoptosis proteins function to block cell death. *Cell* **83:** 1253–1262.

The discovery of IAPs.

Deveraux QL, Takahashi R, Salvesen GS, Reed JC. 1997. X-linked IAP is a direct inhibitor of cell death proteases. *Nature* **388:** 300–304.

The first paper to show that XIAP is a caspase inhibitor.

Riedl SJ, Renatus M, Schwarzenbacher R, Zhou Q, Sun C, Fesik SW, Liddington RC, Salvesen GS. 2001. Structural basis for the inhibition of caspase-3 by XIAP. *Cell* **104:** 791–800.

CHAPTER 4

Mitochondria and Cell Death

Tait SWG, Green DR. 2010. Mitochondria and cell death: Outer membrane permeabilization and beyond. *Nat Rev Cell Mol Biol* **11:** 621–632.

A detailed review of MOMP and its consequences.

Green DR, Kroemer G. 2004. The pathophysiology of mitochondrial cell death. *Science* **305:** 626–629.

Wallace DC, Fan W. 2009. The pathophysiology of mitochondrial disease as modeled in the mouse. *Genes Dev* **23:** 1714–1736.

This review discusses how defects in mitochondrial function affect cell survival and death.

Green DR, Fitzgerald P. 2016. Just so stories about the evolution of apoptosis. *Curr Biol* **26:** R620–R627.

Cytochrome *c* and the Apoptosome

Green DR. 1998. Apoptotic pathways: The roads to ruin. *Cell* **94:** 695–698.

An early review describing the mitochondrial pathway of apoptosis.

Green DR. 2005. Apoptotic pathways: Ten minutes to dead. *Cell* **121:** 671–674.

A later review that incorporates MOMP and additional players into the mitochondrial pathway of apoptosis.

Shakeri R, Kheirollahi A, Davoodi, J. 2017. Apaf-1: Regulation and function in cell death. *Biochimie* **135:** 111–125.

Liu X, Kim CN, Yang J, Jemmerson R, Wang X. 1996. Induction of apoptotic program in cell-free extracts: Requirement for dATP and cytochrome *c*. *Cell* **86:** 147–157.

The original paper showing that cytochrome c *induces caspase activation.*

Zou H, Henzel WJ, Liu X, Lutschg A, Wang X. 1997. Apaf-1, a human protein homologous to *C. elegans* CED-4, participates in cytochrome *c*–dependent activation of caspase-3. *Cell* **90:** 405–413.

The identification and characterization of APAF1.

Li P, Nijhawan D, Budihardjo I, Srinivasula SM, Ahmad M, Alnemri ES, Wang X. 1997. Cytochrome *c* and dATP-dependent formation of Apaf-1/caspase-9 complex initiates an apoptotic protease cascade. *Cell* **91:** 479–489.

The interaction between APAF1 and caspase-9 leading to caspase activation was first described in this paper.

Cheng TC, Hong C, Akey IV, Yuan S, Akey CW. 2016. A near atomic structure of the active human apoptosome. *Elife* **5:** 17755.

Li Y, Zhou M, Hu Q, Bai XC, Huang W, Scheres SH, Shi Y. 2017. Mechanistic insights into caspase-9 activation by the structure of the apoptosome holoenzyme. *Proc Natl Acad Sci* **114:** 1542–1547.

The above two papers provide insights into the structure of the apoptosome and its activity.

MOMP

Goldstein JC, Waterhouse NJ, Juin P, Evan GI, Green DR. 2000. The coordinate release of cytochrome c during apoptosis is rapid, complete and kinetically invariant. *Nat Cell Biol* **2:** 156–162.

The original paper demonstrating MOMP in cells using fluorescent cytochrome c.

Waterhouse NJ, Goldstein JC, von Ahsen O, Schuler M, Newmeyer DD, Green DR. 2001. Cytochrome c maintains mitochondrial transmembrane potential and ATP generation after outer mitochondrial membrane permeabilization during the apoptotic process. *J Cell Biol* **153:** 319–328.

The immediate consequences of MOMP for mitochondria are explored.

Smac/DIABLO and Omi

Verhagen AM, Kratina TK, Hawkins CJ, Silke J, Ekert PG, Vaux DL. 2007. Identification of mammalian mitochondrial proteins that interact with IAPs via N-terminal IAP binding motifs. *Cell Death Differ* **14:** 348–357.

Green DR. 2000. Apoptotic pathways: Paper wraps stone blunts scissors. *Cell* **102:** 1–4.

Du C, Fang M, Li Y, Li L, Wang X. 2000. Smac, a mitochondrial protein that promotes cytochrome c–dependent caspase activation by eliminating IAP inhibition. *Cell* **102:** 33–42.

The original description of Smac, published with the original description of the mouse homolog DIABLO.

Verhagen AM, Ekert PG, Pakusch M, Silke J, Connolly LM, Reid GE, Moritz RL, Simpson RJ, Vaux DL. 2000. Identification of DIABLO, a mammalian protein that promotes apoptosis by binding to and antagonizing IAP proteins. *Cell* **102:** 43–53.

See above.

Hegde R, Srinivasula SM, Zhang Z, Wassell R, Mukattash R, Cilenti L, DuBois G, Lazebnik Y, Zervos AS, Fernandes-Alnemri T, et al. 2002. Identification of Omi/HtrA2 as a mitochondrial apoptotic serine protease that disrupts inhibitor of apoptosis protein–caspase interaction. *J Biol Chem* **277:** 432–438.

One of several papers initially describing Omi and its role in apoptosis.

Caspase-Independent Cell Death

Chipuk JE, Green DR. 2005. Do inducers of apoptosis trigger caspase-independent cell death? *Nat Rev Mol Cell Biol* **6:** 268–275.

Lartigue L, Kushnareva Y, Seong Y, Lin H, Faustin B, Newmeyer DD. 2009. Caspase-independent mitochondrial cell death results from loss of respiration, not cytotoxic protein release. Mol Biol Cell **20:** 4871–4884.

A study showing that MOMP-induced, caspase-independent cell death is a consequence of loss of mitochondrial function.

Susin SA, Lorenzo HK, Zamzami N, Marzo I, Snow BE, Brothers GM, Mangion J, Jacotot E, Costantini P, Loeffler M, et al. 1999. Molecular characterization of mitochondrial apoptosis-inducing factor. *Nature* **397:** 441–446.

The initial characterization of AIF.

Hangen E, Féraud O, Lachkar S, Mou H, Doti N, Fimia GM, Lam NV, Zhu C, Godin I, Muller K, et al. 2015. Interaction between AIF and CHCHD4 regulates respiratory chain biogenesis. *Mol Cell* **58:** 1001–1014.

Characterization of the role of AIF in mitochondrial function.

Milasta S, et al. 2016. Apoptosis-inducing-factor-dependent mitochondrial function is required for T cell but not B cell function. *Immunity* **44:** 88–102.

Further insights into AIF function, and evidence that it may not be directly involved in cell death.

Li LY, Luo X, Wang X. 2001. Endonuclease G is an apoptotic DNase when released from mitochondria. *Nature* **412:** 95–99.

Initial description of endonuclease G and its proposed role in caspase-independent cell death.

Mitochondrial Permeability Transition

Lemasters JJ, Theruvath TP, Zhong Z, Nieminen AL. 2009. Mitochondrial calcium and the permeability transition in cell death. *Biochim Biophys Acta* **1787:** 1395–1401.

An overview of the permeability transition in cell death, but one that suggests roles in apoptosis.

Bernardi P, Rasola A, Forte M, Lippe G. 2015. The mitochondrial permeability transition pore: Channel formation by F-ATP synthase, integration in signal transduction, and role in pathophysiology. *Physiol Rev.* **95:** 1111–1155.

He J, Ford HC, Carroll J, Ding S, Fearnley IM, Walker JE. 2017. Persistence of the mitochondrial permeability transition in the absence of subunit c of human ATP synthase. *Proc Natl Acad Sci* **114:** 3409–3414.

The above two papers highlight continued controversies in the nature of the MPTP.

Izzo V, Bravo-San Pedro JM, Sica V, Kroemer G, Galluzzi L. 2016. Mitochondrial permeability transition: New findings and persisting uncertainties. *Trends Cell Biol.* **26:** 655–667.

Nakagawa T, Shimizu S, Watanabe T, Yamaguchi O, Otsu K, Yamagata H, Inohara H, Kubo T, Tsujimoto Y. 2005. Cyclophilin D–dependent mitochondrial permeability transition regulates some necrotic but not apoptotic cell death. *Nature* **434:** 652–658.

One of several papers showing that MPT has roles in necrosis but not apoptosis.

Apoptosomes in Worms, Flies, and Other Beasts

Bao Q, Shi Y. 2007. Apoptosome: A platform for the activation of initiator caspases. *Cell Death Differ* **14:** 56–65.

Qi S, Pang Y, Hu Q, Liu Q, Li H, Zhou Y, He T, Liang Q, Liu Y, Yuan X, et al. 2010. Crystal structure of the *Caenorhabditis elegans* apoptosome reveals an octameric assembly of CED-4. *Cell* **141:** 446–457.

Cheng TC, Akey IV, Yuan S, Yu Z, Ludtke SJ, Akey CW. 2017. A near-atomic structure of the Dark apoptosome provides insight into assembly and activation. *Structure* **25:** 40–52.

Rodriguez A, Chen P, Oliver H, Abrams JM. 2002. Unrestrained caspase-dependent cell death caused by loss of Diap1 function requires the *Drosophila* Apaf-1 homolog, Dark. *EMBO J* **21:** 2189–2197.

The role of DIAP1 in restraining ARK (called "Dark")-induced caspase activity in flies.

Rodriguez A, Oliver H, Zou H, Chen P, Wang X, Abrams JM. 1999. Dark is a *Drosophila* homologue of Apaf-1/CED-4 and functions in an evolutionarily conserved death pathway. *Nat Cell Biol* **1:** 272–279.

The identification of ARK (called "Dark").

Yuan J, Horvitz HR. 1992. The *Caenorhabditis elegans* cell death gene *ced-4* encodes a novel protein and is expressed during the period of extensive programmed cell death. *Development* **116:** 309–320.

The original characterization of CED4.

Chinnaiyan AM, Chaudhary D, O'Rourke K, Koonin EV, Dixit VM. 1997. Role of CED-4 in the activation of CED-3. *Nature* **388:** 728–729.

One of the original papers showing that CED4 biochemically activates CED3.

Bender CE, Fitzgerald P, Tait SW, Llambi F, McStay GP, Tupper DO, Pellettieri J, Sanchez Alvarado A, Salvesen GS, Green, DR. 2012. Mitochondrial pathway of apoptosis is ancestral in metazoans. *Proc Natl Acad Sci* **109:** 4904–4909.

Evidence for the mitochondrial pathway of apoptosis, including cytochrome c activation of apoptosome function, in platyhelminths (flatworms) and echinoderms.

CHAPTER 5

BCL-2 Proteins and MOMP

Birkinshaw RW, Czabotar PE. 2017. The BCL-2 family of proteins and mitochondrial outer membrane permeabilisation. *Semin Cell Dev Biol* **72:** 152–162.

Kalkavan H, Green DR. 2018. MOMP, cell suicide as a BCL-2 family business. *Cell Death Differ* **25:** 46–55.

Kale J, Osterlund EJ, Andrews DW. 2018. BCL-2 family proteins: Changing partners in the dance towards death. *Cell Death Differ* **25:** 65–80.

Suhaili SH, Karimian H, Stellato M, Lee TH, Aguilar MI. 2017. Mitochondrial outer membrane permeabilization: A focus on the role of mitochondrial membrane structural organization. *Biophys Rev* **9:** 443–457.

Anti-Apoptotic BCL-2 Proteins

Vaux DL, Cory S, Adams JM. 1988. *Bcl-2* gene promotes haemopoietic cell survival and cooperates with c-myc to immortalize pre-B cells. *Nature* **335:** 440–442.

The first paper to show that BCL-2 preserves cell survival in primary and transformed cells.

Hockenbery D, Nunez G, Milliman C, Schreiber RD, Korsmeyer SJ. 1990. Bcl-2 is an inner mitochondrial membrane protein that blocks programmed cell death. *Nature* **348:** 334–336.

The first paper to show that BCL-2 is associated with mitochondria (the original assertion that it is on the inner membrane was due to an artifact) and that BCL-2 inhibits apoptosis.

Kluck RM, Bossy-Wetzel E, Green DR, Newmeyer DD. 1997. The release of cytochrome c from mitochondria: A primary site for Bcl-2 regulation of apoptosis. Science **275:** 1132–1136.

One of two papers (see below) identifying the role of BCL-2 in blocking the release of cytochrome c *from mitochondria.*

Yang J, Liu X, Bhalla K, Kim CN, Ibrado AM, Cai J, Peng TI, Jones DP, Wang X. 1997. Prevention of apoptosis by Bcl-2: Release of cytochrome c from mitochondria blocked. *Science* **275:** 1129–1132.

See above.

Vaux DL, Weissman IL, Kim SK. 1992. Prevention of programmed cell death in *Caenorhabditis elegans* by human bcl-2. *Science* **258:** 1955–1957.

One of three papers that showed that CED9 and BCL-2 are functionally equivalent in nematodes. Although BCL-2 does not bind to CED4, it can sequester EGL1, which was not shown until later.

Hengartner MO, Horvitz HR. 1994. *C. elegans* cell survival gene *ced-9* encodes a functional homolog of the mammalian proto-oncogene *bcl-2*. *Cell* **76:** 665–676.

See above.

Jabbour AM, Puryer MA, Yu JY, Lithgow T, Riffkin CD, Ashley DM, Vaux DL, Ekert PG, Hawkins CJ. 2006. Human Bcl-2 cannot directly inhibit the *Caenorhabditis elegans* Apaf-1 homologue CED-4, but can interact with EGL-1. *J Cell Sci* **119:** 2572–2582.

See above.

Pro-Apoptotic BCL-2 Effectors

Oltvai ZN, Milliman CL, Korsmeyer SJ. 1993. Bcl-2 heterodimerizes in vivo with a conserved homolog, Bax, that accelerates programmed cell death. *Cell* **74:** 609–619.

The discovery of BAX.

Chittenden T, Harrington EA, O'Connor R, Flemington C, Lutz RJ, Evan GI, Guild BC. 1995. Induction of apoptosis by the Bcl-2 homologue Bak. *Nature* **374:** 733–736.

The discovery of BAK.

Wei MC, Zong WX, Cheng EH, Lindsten T, Panoutsakopoulou V, Ross AJ, Roth KA, MacGregor GR, Thompson CB, Korsmeyer SJ. 2001. Proapoptotic BAX and BAK: A requisite gateway to mitochondrial dysfunction and death. *Science* **292:** 727–730.

This paper showed that BAX and BAK are necessary for MOMP and apoptosis via the mitochondrial pathway.

Llambi F, Victor B, Wang Y, Yang M, Schneider DM, Gingras S, Zheng J, Parsons MJ, Brown SA, Pelletier S, et al. 2016. BOK is a non-canonical BCL-2 family effector of apoptosis regulated by ER-associated degradation. *Cell* **165:** 421–433.

This paper shows that Bok can also promote MOMP in a manner that can be independent of other Bcl-2 family proteins.

BH3-Only Proteins

Giam M, Huang DC, Bouillet P. 2008. BH3-only proteins and their roles in programmed cell death. *Oncogene* (Suppl 1) **27:** S128–S136.

Stoka V, Turk B, Schendel SL, Kim TH, Cirman T, Snipas SJ, Ellerby LM, Bredesen D, Freeze H, Abrahamson M, et al. 2001. Lysosomal protease pathways to apoptosis. Cleavage of bid, not pro-caspases, is the most likely route. *J Biol Chem* **276:** 3149–3157.

BID as a protease sensor for engaging the mitochondrial pathway of apoptosis.

Datta SR, Ranger AM, Lin MZ, Sturgill JF, Ma YC, Cowan CW, Dikkes P, Korsmeyer SJ, Greenberg ME. 2002. Survival factor–mediated BAD phosphorylation raises the mitochondrial threshold for apoptosis. *Dev Cell* **3:** 631–643.

The regulation of BAD by survival factor signaling.

Bouillet P, Metcalf D, Huang DC, Tarlinton DM, Kay TW, Kontgen F, Adams JM, Strasser A. 1999. Proapoptotic Bcl-2 relative Bim required for certain apoptotic responses, leukocyte homeostasis, and to preclude autoimmunity. *Science* **286:** 1735–1738.

The BIM knockout mouse illustrates roles for this protein in apoptosis.

Puthalakath H, Huang DC, O'Reilly LA, King SM, Strasser A. 1999. The proapoptotic activity of the Bcl-2 family member Bim is regulated by interaction with the dynein motor complex. *Mol Cell* **3:** 287–296.

The association of BIM with the cytoskeleton.

Conradt B, Horvitz HR. 1998. The *C. elegans* protein EGL-1 is required for programmed cell death and interacts with the Bcl-2-like protein CED-9. *Cell* **93:** 519–529.

EGL1 is a BH3-only protein in nematodes.

BCL-2 Protein Interactions

Letai A, Bassik MC, Walensky LD, Sorcinelli MD, Weiler S, Korsmeyer SJ. 2002. Distinct BH3 domains either sensitize or activate mitochondrial apoptosis, serving as prototype cancer therapeutics. *Cancer Cell* **2:** 183–192.

The proposal of the "direct activator/derepressor" (called "sensitizer") model of BH3-only protein function.

Chen L, Willis SN, Wei A, Smith BJ, Fletcher JI, Hinds MG, Colman PM, Day CL, Adams JM, Huang DC. 2005. Differential targeting of prosurvival Bcl-2 proteins by their BH3-only ligands allows complementary apoptotic function. *Mol Cell* **17:** 393–403.

The "neutralization" model of BH3-only protein function and specificities of BH3-only proteins for different anti-apoptotic proteins.

Willis SN, Chen L, Dewson G, Wei A, Naik E, Fletcher JI, Adams JM, Huang DC. 2005. Proapoptotic Bak is sequestered by Mcl-1 and Bcl-xL, but not Bcl-2, until displaced by BH3-only proteins. *Genes Dev* **19:** 1294–1305.

The "neutralization" model of BH3-only protein function.

Kuwana T, Mackey MR, Perkins G, Ellisman MH, Latterich M, Schneiter R, Green DR, Newmeyer DD. 2002. Bid, Bax, and lipids cooperate to form supramolecular openings in the outer mitochondrial membrane. *Cell* **111:** 331–342.

Biochemical characterization of BAX activation and membrane permeabilization.

Kim H, Tu HC, Ren D, Takeuchi O, Jeffers JR, Zambetti GP, Hsieh JJ, Cheng EH. 2009. Stepwise activation of BAX and BAK by tBID, BIM, and PUMA initiates mitochondrial apoptosis. *Mol Cell* **36:** 487–499.

Further analysis of the "direct activator/derepressor" model of BH3-only protein function.

Lovell JF, Billen LP, Bindner S, Shamas-Din A, Fradin C, Leber B, Andrews DW. 2008. Membrane binding by tBid initiates an ordered series of events culminating in membrane permeabilization by Bax. *Cell* **135:** 1074–1084.

Use of FRET technology to characterize derepression and direct activation of BAX.

Leber B, Lin J, Andrews DW. 2007. Embedded together: The life and death consequences of interaction of the Bcl-2 family with membranes. *Apoptosis* **12:** 897–911.

One of two approaches (with below) to unify the neutralization and sensitizer/depression models of BH3-only protein action.

Llambi F, Moldoveanu, T, Tait SW, Bouchier-Hayes L, Temirov J, McCormick LL, Dillon CP, Green, DR. 2011. A unified model of mammalian BCL-2 protein family interactions at the mitochondria. *Mol Cell* **44:** 517–531.

See above.

Gavathiotis E, Suzuki M, Davis ML, Pitter K, Bird GH, Katz SG, Tu HC, Kim H, Cheng EH, Tjandra N, et al. 2008. BAX activation is initiated at a novel interaction site. *Nature* **455:** 1076–1081.

Structural analysis of the binding of the BIM BH3 region to BAX and early steps in BAX activation.

Czabotar PE, Westphal D, Dewson G, Ma S, Hockings C, Fairlie WD, Lee EF, Yao S, Robin AY, Smith BJ, et al. 2013. Bax crystal structures reveal how BH3 domains activate Bax and nucleate its oligomerization to induce apoptosis. *Cell* **152:** 519–531.

Insights into the activation of BAX to form oligomers.

Moldoveanu T, Grace CR, Llambi F, Nourse A, Fitzgerald P, Gehring K, Kriwacki RW, Green DR. 2013. BID-induced structural changes in BAK promote apoptosis. *Nat Struct Mol Biol* **20:** 589–597.

Insights into the activation of BAK.

Other Functions of the BCL-2 Family

Karbowski M, Lee YJ, Gaume B, Jeong SY, Frank S, Nechushtan A, Santel A, Fuller M, Smith CL, Youle RJ. 2002. Spatial and temporal association of Bax with mitochondrial fission sites, Drp1, and Mfn2 during apoptosis. *J Cell Biol* **159:** 931–938.

Characterization of interactions between BAX and proteins involved in mitochondrial fission and fusion.

Sheridan C, Delivani P, Cullen SP, Martin SJ. 2008. Bax- or Bak-induced mitochondrial fission can be uncoupled from cytochrome c release. *Mol Cell* **31:** 570–585.

Evidence that mitochondrial fission occurs at the time of MOMP but is not required for permeabilization.

Rong Y, Distelhorst CW. 2008. Bcl-2 protein family members: Versatile regulators of calcium signaling in cell survival and apoptosis. *Annu Rev Physiol* **70:** 73–91.

Chen R, Valencia I, Zhong F, McColl KS, Roderick HL, Bootman MD, Berridge MJ, Conway SJ, Holmes AB, Mignery GA, et al. 2004. Bcl-2 functionally interacts with inositol 1,4,5-trisphosphate receptors to regulate calcium release from the ER in response to inositol 1,4,5-trisphosphate. *J Cell Biol* **166:** 193–203.

Characterization of the regulation of calcium homeostasis by BCL-2.

Levine B, Sinha S, Kroemer G. 2008. Bcl-2 family members: Dual regulators of apoptosis and autophagy. *Autophagy* **4:** 600–606.

Pattingre S, Tassa A, Qu X, Garuti R, Liang XH, Mizushima N, Packer M, Schneider MD, Levine B. 2005. Bcl-2 antiapoptotic proteins inhibit Beclin 1–dependent autophagy. *Cell* **122:** 927–939.

A mechanism for the regulation of autophagy by BCL-2. Autophagy is covered in more detail in Chapter 8.

Cuconati A, White E. 2002. Viral homologs of BCL-2: Role of apoptosis in the regulation of virus infection. *Genes Dev* **16:** 2465–2478.

CHAPTER 6

Death Receptors

Banner DW, D'Arcy A, Janes W, Gentz R, Schoenfeld HJ, Broger C, Loetscher H, Lesslauer W. 1993. Crystal structure of the soluble human 55 kd TNF receptor-human TNF β complex: Implications for TNF receptor activation. *Cell* **73:** 431–445.

Hymowitz SG, Christinger HW, Fuh G, Ultsch M, O'Connell M, Kelley RF, Ashkenazi A, de Vos AM. 1999. Triggering cell death: The crystal structure of Apo2L/TRAIL in a complex with death receptor 5. *Mol Cell* **4:** 563–571.

The structure of TRAIL shows a role for zinc that is not seen in other TNF-family ligands.

Quistad SD, Traylor-Knowles N. 2016. Precambrian origins of the TNFR superfamily. *Cell Death Discov* **2:** 16058.

Evolution of the TNFR family, including death receptors.

Death Receptor Signaling for Apoptosis

Tummers B, Green DR. 2017. Caspase-8: Regulating life and death. *Immunol Rev* **277:** 76–89.

Muzio M, Stockwell BR, Stennicke HR, Salvesen GS, Dixit VM. 1998. An induced proximity model for caspase-8 activation. *J Biol Chem* **273:** 2926–2930.

The activation of caspase-8 by FADD-mediated induced by ligation of death receptors.

Dickens LS, Boyd RS, Jukes-Jones R, Hughes MA, Robinson GL, Fairall L, Schwabe JW, Cain K, Macfarlane M. 2012. A death effector domain chain DISC model reveals a crucial role for caspase-8 chain assembly in mediating apoptotic cell death. *Mol Cell* **47:** 291–305.

Fu TM, Li Y, Lu A, Li Z, Vajjhala PR, Cruz AC, Srivastava DB, DiMaio F, Penczek PA, Siegel RM, et al. 2016. Cryo-EM structure of caspase-8 tandem DED filament reveals assembly and regulation mechanisms of the death-inducing signaling complex. *Mol Cell* **64:** 236–250.

Insights into the activation of caspase-8 to form filaments.

Hughes MA, Powley IR, Jukes-Jones R, Horn S, Feoktistova M, Fairall L, Schwabe JW, Leverkus M, Cain K, MacFarlane, M. 2016. Co-operative and hierarchical binding of c-FLIP and caspase-8: A unified model defines how c-FLIP isoforms differentially control cell fate. *Mol Cell* **61:** 834–849.

Lu M, Lawrence DA, Marsters S, Acosta-Alvear D, Kimmig P, Mendez AS, Paton AW, Paton JC, Walter P, Ashkenazi, A. 2014. Opposing unfolded-protein-response signals converge on death receptor 5 to control apoptosis. *Science* **345:** 98–101.

Describes how endoplasmic reticulum stress can cause TRAIL receptor activation and apoptosis in the absence of its ligand, TRAIL.

Defects in the CD95 Pathway Cause ALPS

Bidere N, Su HC, Lenardo MJ. 2006. Genetic disorders of programmed cell death in the immune system. *Annu Rev Immunol* **24:** 321–352.

An overview of diseases in mice and humans caused by defects in CD95, CD95L, and associated caspases.

Rieux-Laucat, F. 2017. What's up in the ALPS. *Curr Opin Immunol* **49:** 79–86.
Further insights into ALPS.

Watanabe-Fukunaga R, Brannan CI, Copeland NG, Jenkins NA, Nagata S. 1992. Lymphoproliferation disorder in mice explained by defects in Fas antigen that mediates apoptosis. *Nature* **356:** 314–317.
Defects in CD95 (called "Fas antigen") are responsible for a lymphoaccumulative disease in mice (originally thought to be a lymphoproliferation defect).

Lynch DH, Watson ML, Alderson MR, Baum PR, Miller RE, Tough T, Gibson M, Davis-Smith T, Smith CA, Hunter K, et al. 1994. The mouse Fas-ligand gene is mutated in *gld* mice and is part of a TNF family gene cluster. *Immunity* **1:** 131–136.
Defects in CD95L (called "Fas ligand") are responsible for a lymphoaccumulative disease in mice.

TNFR Signaling

Brenner D, Blaser H, Mak TW. 2015. Regulation of tumour necrosis factor signalling: Live or let die. *Nat Rev Immunol* **15:** 362–374.
A useful review of TNFR signaling, including NF-κB activation, apoptosis, and the process of necroptosis (discussed in Chapter 8).

Death Receptors and the Mitochondrial Pathway

Li H, Zhu H, Xu CJ, Yuan J. 1998. Cleavage of BID by caspase 8 mediates the mitochondrial damage in the Fas pathway of apoptosis. *Cell* **94:** 491–501.
Two papers identified BID as a target for caspase-8, linking the death receptor and mitochondrial pathways of apoptosis.

Luo X, Budihardjo I, Zou H, Slaughter C, Wang X. 1998. Bid, a Bcl2 interacting protein, mediates cytochrome c release from mitochondria in response to activation of cell surface death receptors. *Cell* **94:** 481–490.
See above.

Scaffidi C, Fulda S, Srinivasan A, Friesen C, Li F, Tomaselli KJ, Debatin KM, Krammer PH, Peter ME. 1998. Two CD95 (APO-1/Fas) signaling pathways. *EMBO J* **17:** 1675–1687.
The description of two cell types ("type I and type II") in which the mitochondrial pathway is dispensable (type I) or required (type II) for death receptor–induced apoptosis.

Jost PJ, Grabow S, Gray D, McKenzie MD, Nachbur U, Huang DC, Bouillet P, Thomas HE, Borner C, Silke J, et al. 2009. XIAP discriminates between type I and type II FAS-induced apoptosis. *Nature* **460:** 1035–1039.
Inhibition of XIAP is required for apoptosis induced by CD95 (called "Fas") in type II cells in mice.

CHAPTER 7

The Inflammatory Caspases

Shi J, Zhao Y, Wang Y, Gao W, Ding J, Li P, Hu L, Shao F. 2014. Inflammatory caspases are innate immune receptors for intracellular LPS. *Nature* **514:** 187–192.
The activation of caspases-4, -5, and -11 by lipopolysaccharide.

Shi J, Zhao Y, Wang K, Shi X, Wang Y, Huang H, Zhuang Y, Cai T, Wang F, Shao F. 2015. Cleavage of GSDMD by inflammatory caspases determines pyroptotic cell death. *Nature* **526:** 660–665.
One of the two original papers (with below) on the role of gasdermin D in pyroptosis.

Kayagaki N, Stowe IB, Lee BL, O'Rourke K, Anderson K, Warming S, Cuellar T, Haley B, Roose-Girma M, Phung QT, et al. 2015. Caspase-11 cleaves gasdermin D for non-canonical inflammasome signalling. *Nature* **526:** 666–671.

See above.

Kersse K, Vanden Berghe T, Lamkanfi M, Vandenabeele P. 2007. A phylogenetic and functional overview of inflammatory caspases and caspase-1-related CARD-only proteins. *Biochem Soc Trans* **35:** 1508–1511.

An overview of the inflammatory caspases and the molecules that interact with them.

Cerretti DP, Kozlosky CJ, Mosley B, Nelson N, Van Ness K, Greenstreet TA, March CJ, Kronheim SR, Druck T, Cannizzaro LA, et al. 1992. Molecular cloning of the interleukin-1 β converting enzyme. *Science* **256:** 97–100.

Two papers describing the identification of caspase-1, the first caspase discovered.

Thornberry NA, Bull HG, Calaycay JR, Chapman KT, Howard AD, Kostura MJ, Miller DK, Molineaux SM, Weidner JR, Aunins J, et al. 1992. A novel heterodimeric cysteine protease is required for interleukin-1 β processing in monocytes. *Nature* **356:** 768–774.

See above.

Dinarello CA. 2018. Overview of the IL-1 family in innate inflammation and acquired immunity. *Immunol Rev* **281:** 8–27.

Interleukin-1 is a major target of caspase-1, and an understanding of its biological roles provides a context for appreciating the significance of caspase-1 function.

Keller M, Ruegg A, Werner S, Beer HD. 2008. Active caspase-1 is a regulator of unconventional protein secretion. *Cell* **132:** 818–831.

The role of caspase-1 in secretion of proteins that lack conventional secretory signal sequences.

TLRs and NLRs

Gao D, Li W. 2017. Structures and recognition modes of toll-like receptors. *Proteins* **85:** 3–9.

Meunier E, Broz P. 2017. Evolutionary convergence and divergence in NLR function and structure. *Trends Immunol* **38:** 744–757.

Medzhitov R, Preston-Hurlburt P, Janeway CA Jr. 1997. A human homologue of the *Drosophila* Toll protein signals activation of adaptive immunity. *Nature* **388:** 394–397.

The original paper describing mammalian Toll-like receptors and their importance in the immune response.

Inflammasomes

Malik A, Kanneganti TD. 2017. Inflammasome activation and assembly at a glance. *J Cell Sci* **130:** 3955–3963.

Lu A, Wu H. 2015. Structural mechanisms of inflammasome assembly. *FEBS J* **282:** 435–444.

Duncan JA, Canna SW. 2018. The NLRC4 inflammasome. *Immunol Rev* **281:** 115–123.

Lugrin J. Martinon F. 2018. The AIM2 inflammasome: Sensor of pathogens and cellular perturbations. *Immunol Rev* **281:** 99–114.

Martinon F, Burns K, Tschopp J. 2002. The inflammasome: A molecular platform triggering activation of inflammatory caspases and processing of proIL-β. *Mol Cell* **10:** 417–426.

The first paper describing an inflammasome, which in this case involved NLRP1 (called "Nalp1"). Although this inflammasome was suggested to include caspase-5, all others described do not, and the details in this paper may not be entirely correct.

Agostini L, Martinon F, Burns K, McDermott MF, Hawkins PN, Tschopp J. 2004. NALP3 forms an IL-1β-processing inflammasome with increased activity in Muckle–Wells autoinflammatory disorder. *Immunity* **20:** 319–325.

A description of the NLRP3 (called "NALP3") inflammasome.

Mariathasan S, Newton K, Monack DM, Vucic D, French DM, Lee WP, Roose-Girma M, Erickson S, Dixit VM. 2004. Differential activation of the inflammasome by caspase-1 adaptors ASC and Ipaf. *Nature* **430:** 213–218.

A description of the NLRC4 (called IPAF) inflammasome.

Youm YH, Grant RW, McCabe LR, Albarado DC, Nguyen KY, Ravussin A, Pistell P, Newman S, Carter R, Laque A, et al. 2013. Canonical Nlrp3 inflammasome links systemic low-grade inflammation to functional decline in aging. *Cell Metab* **18:** 519–532.

A role for NLRP3 in aging.

Cell Death and Caspase-1 Activation

Gaidt, MM, Hornung V. 2017. The NLRP3 inflammasome renders cell death pro-inflammatory. *J Mol Biol* **430:** 133–141.

A useful overview of the interplay of cell death pathways.

Kovacs SB, Miao EA. 2017. Gasdermins: Effectors of pyroptosis. *Trends Cell Biol* **27:** 673–684.

Lamkanfi M, Moreira LO, Makena P, Spierings DC, Boyd K, Murray PJ, Green DR, Kanneganti TD. 2009. Caspase-7 deficiency protects from endotoxin-induced lymphocyte apoptosis and improves survival. *Blood* **113:** 2742–2745.

One of two papers (see below) suggesting that the activation of caspase-7 by caspase-1 results in cell death. Note that this work preceded the discovery of gasdermin D.

Lamkanfi M, Kanneganti TD, Van Damme P, Vanden Berghe T, Vanoverberghe I, Vandekerckhove J, Vandenabeele P, Gevaert K, Nunez G. 2008. Targeted peptidecentric proteomics reveals caspase-7 as a substrate of the caspase-1 inflammasomes. *Mol Cell Proteomics* **7:** 2350–2363.

See above.

Caspase-2

Sladky V, Schuler F, Fava LL, Villunger, A. 2017. The resurrection of the PIDDosome—Emerging roles in the DNA-damage response and centrosome surveillance. *J Cell Sci* **130:** 3779–3787.

Roles for caspase-2 in apoptosis and other phenomena.

Forsberg J, Zhivotovsky B, Olsson M. 2017. Caspase-2: An orphan enzyme out of the shadows. *Oncogene* **36:** 5441–5444.

Tinel A, Tschopp J. 2004. The PIDDosome, a protein complex implicated in activation of caspase-2 in response to genotoxic stress. *Science* **304:** 843–846.

The first characterization of a complex responsible for activating caspase-2.

Park HH, Logette E, Raunser S, Cuenin S, Walz T, Tschopp J, Wu H. 2007. Death domain assembly mechanism revealed by crystal structure of the oligomeric PIDDosome core complex. *Cell* **128:** 533–546.

Bouchier-Hayes L, Oberst A, McStay GP, Connell S, Tait SW, Dillon CP, Flanagan JM, Beere HM, Green DR. 2009. Characterization of cytoplasmic caspase-2 activation by induced proximity. *Mol Cell* **35:** 830–840.

The activation of caspase-2 by several stressors, analyzed using a live-cell-imaging technique.

Fava LL, Schuler F, Sladky V, Haschka MD, Soratroi C, Eiterer L, Demetz E, Weiss G, Geley S, Nigg EA, et al. 2017. The PIDDosome activates p53 in response to supernumerary centrosomes. *Genes Dev* **31:** 34–45.

Ando K, Parsons MJ, Shah RB, Charendoff CI, Paris SL, Liu PH, Fassio SR, Rohrman BA, Thompson R, Oberst A, et al. 2017. NPM1 directs PIDDosome-dependent caspase-2 activation in the nucleolus. *J Cell Biol* **216:** 1795–1810.

CHAPTER 8

Tonnus W, Linkermann A. 2017. The in vivo evidence for regulated necrosis. *Immunol Rev* **277:** 128–149.

Conrad M, Angeli JP, Vandenabeele P, Stockwell BR. 2016. Regulated necrosis: Disease relevance and therapeutic opportunities. *Nat Rev Drug Discov* **15:** 348–366.

Secondary Necrosis

Silva MT, do Vale A, dos Santos NM. 2008. Secondary necrosis in multicellular animals: An outcome of apoptosis with pathogenic implications. *Apoptosis* **13:** 463–482.

Rogers C, Fernandes-Alnemri T, Mayes L, Alnemri D, Cingolani G, Alnemri ES. 2017. Cleavage of DFNA5 by caspase-3 during apoptosis mediates progression to secondary necrotic/pyroptotic cell death. *Nat Commun* **8:** 14128.

One of two papers showing that caspase-3 induces secondary necrosis via cleavage of DFNA5/gasdermin E.

Wang Y, Gao W, Shi X, Ding J, Liu W, He H, Wang K, Shao, F. 2017. Chemotherapy drugs induce pyroptosis through caspase-3 cleavage of a gasdermin. *Nature* **547:** 99–103.

See above.

Necroptosis

Weinlich R, Oberst A, Beere HM, Green DR. 2017. Necroptosis in development, inflammation and disease. *Nat Rev Mol Cell Biol* **18:** 127–136.

Vanden Berghe T, Hassannia B, Vandenabeele P. 2016. An outline of necrosome triggers. *Cell Mol Life Sci* **73:** 2137–2152.

Brault M, Oberst, A. 2017. Controlled detonation: Evolution of necroptosis in pathogen defense. *Immunol Cell Biol* **95:** 131–136.

Holler N, Zaru R, Micheau O, Thome M, Attinger A, Valitutti S, Bodmer JL, Schneider P, Seed B, Tschopp J. 2000. Fas triggers an alternative, caspase-8-independent cell death pathway using the kinase RIP as effector molecule. *Nat Immunol* **1:** 489–495.

The first description of RIPK-dependent cell death, in this case induced by CD95 (called "Fas").

Xie T, Peng W, Liu Y, Yan C, Maki J, Degterev A, Yuan J, Shi Y. 2013. Structural basis of RIP1 inhibition by necrostatins. *Structure* **21:** 493–499.

He S, Wang L, Miao L, Wang T, Du F, Zhao L, Wang X. 2009. Receptor interacting protein kinase-3 determines cellular necrotic response to TNF-α. *Cell* **137:** 1100–1111.

One of three papers that initially defined the role of RIPK3 in necroptosis.

Cho YS, Challa S, Moquin D, Genga R, Ray TD, Guildford M, Chan FK. 2009. Phosphorylation-driven assembly of the RIP1-RIP3 complex regulates programmed necrosis and virus-induced inflammation. *Cell* **137:** 1112–1123.

See above.

Zhang DW, Shao J, Lin J, Zhang N, Lu BJ, Lin SC, Dong MQ, Han J. 2009. RIP3, an energy metabolism regulator that switches TNF-induced cell death from apoptosis to necrosis. *Science* **325:** 332–336.

See above. The proposed mechanism of action of RIPK3 to promote necroptosis was ultimately shown to be incorrect, but the role of RIPK3 was demonstrated.

Sun L, Wang H, Wang Z, He S, Chen S, Liao D, Wang L, Yan J, Liu W, Lei X, et al. 2012. Mixed lineage kinase domain-like protein mediates necrosis signaling downstream of RIP3 kinase. *Cell* **148:** 213–227.

The first paper to define MLKL as the target of RIPK3 in necroptosis.

Murphy JM, Czabotar PE, Hildebrand JM, Lucet IS, Zhang JG., Alvarez-Diaz S, Lewis R, Lalaoui N, Metcalf D, Webb AI, et al. 2013. The pseudokinase MLKL mediates necroptosis via a molecular switch mechanism. *Immunity* **39:** 443–453.

Oberst A, Dillon CP, Weinlich R, McCormick LL, Fitzgerald P, Pop C, Hakem R, Salvesen GS, Green DR. 2011. Catalytic activity of the caspase-8-FLIP(L) complex inhibits RIPK3-dependent necrosis. *Nature* **471:** 363–367.

One of two papers that initially showed that ablation of RIPK3 prevents embryonic lethality in caspase-8-deficient mice.

Kaiser WJ, Upton JW, Long AB, Livingston-Rosanoff D, Daley-Bauer LP, Hakem R, Caspary T, Mocarski ES. 2011. RIP3 mediates the embryonic lethality of caspase-8-deficient mice. *Nature* **471:** 368–372.

See above.

Dillon CP, Tummers B, Baran K, Green DR. 2016. Developmental checkpoints guarded by regulated necrosis. *Cell Mol Life Sci* **73:** 2125–2136.

Weinlich R, Green DR. 2014. The two faces of receptor interacting protein kinase-1. *Mol Cell* **56:** 469–480.

Ferroptosis

Stockwell BR, Friedmann Angeli JP, Bayir H, Bush AI, Conrad M, Dixon SJ, Fulda S, Gascon S, Hatzios SK, Kagan VE, et al. 2017. Ferroptosis: A regulated cell death nexus linking metabolism, redox biology, and disease. *Cell* **171:** 273–285.

Other Forms of Regulated Necrosis

Bouchard VJ, Rouleau M, Poirier GG. 2003. PARP-1, a determinant of cell survival in response to DNA damage. *Exp Hematol* **31:** 446–454.

Szydlowska K, Tymianski M. 2010. Calcium, ischemia and excitotoxicity. *Cell Calcium* **47:** 122–129.

Eliasson MJ, Sampei K, Mandir AS, Hurn PD, Traystman RJ, Bao J, Pieper A, Wang ZQ, Dawson TM, Snyder SH, et al. 1997. Poly(ADP-ribose) polymerase gene disruption renders mice resistant to cerebral ischemia. *Nat Med* **3:** 1089–1095.

A connection between ischemia/reperfusion injury and PARP.

Autophagy

Galluzzi L, Baehrecke EH, Ballabio A, Boya P, Bravo-San Pedro JM, Cecconi F, Choi AM, Chu CT, Codogno P, Colombo MI, et al. 2017. Molecular definitions of autophagy and related processes. *EMBO J* **36:** 1811–1836.

Green DR, Levine B. 2014. To be or not to be? How selective autophagy and cell death govern cell fate. *Cell* **157:** 65–75.

Mitophagy

Zimmermann M, Reichert AS. 2017. How to get rid of mitochondria: Crosstalk and regulation of multiple mitophagy pathways. *Biol Chem* **399:** 29–45.

Nguyen TN, Padman BS, Lazarou M. 2016. Deciphering the molecular signals of PINK1/Parkin mitophagy. *Trends Cell Biol* **26:** 733–744.

Autophagic Cell Death

Fulda S, Kogel D. 2015. Cell death by autophagy: Emerging molecular mechanisms and implications for cancer therapy. *Oncogene* **34:** 5105–5113.

Zhang H, Baehrecke EH. 2015. Eaten alive: Novel insights into autophagy from multicellular model systems. *Trends Cell Biol* **25:** 376–387.

Yu L, Alva A, Su H, Dutt P, Freundt E, Welsh S, Baehrecke EH, Lenardo MJ. 2004. Regulation of an ATG7-beclin 1 program of autophagic cell death by caspase-8. *Science* **304:** 1500–1502.

The first paper suggesting that caspase-8 inhibits autophagy-dependent cell death, which ultimately turned out to be RIPK-dependent necroptosis. The relationships between these modes of cell death remain unresolved.

Mitotic Catastrophe

Vakifahmetoglu H, Olsson M, Zhivotovsky B. 2008. Death through a tragedy: Mitotic catastrophe. *Cell Death Differ* **15:** 1153–1162.

Castedo M, Perfettini JL, Roumier T, Andreau K, Medema R, Kroemer G. 2004. Cell death by mitotic catastrophe: A molecular definition. *Oncogene* **23:** 2825–2837.

Kirsch DG, Santiago PM, di Tomaso E, Sullivan JM, Hou WS, Dayton T, Jeffords LB, Sodha P, Mercer KL, Cohen R, et al. p53 controls radiation-induced gastrointestinal syndrome in mice independent of apoptosis. *Science* **327:** 593–596.

Cell death in the intestines of BAX–BAK double-deficient mice may be a form of mitotic catastrophe. p53 is discussed in Chapter 11.

CHAPTER 9

Engulfment of Dying Cells

Elliott MR, Ravichandran KS. 2016. The dynamics of apoptotic cell clearance. *Dev Cell* **38:** 147–160.

Find-Me Signals

Medina CB, Ravichandran KS. 2016. Do not let death do us part: 'Find-me' signals in communication between dying cells and the phagocytes. *Cell Death Differ* **23:** 979–989.

Lauber K, Bohn E, Krober SM, Xiao YJ, Blumenthal SG, Lindemann RK, Marini P, Wiedig C, Zobywalski A, Baksh S, et al. 2003. Apoptotic cells induce migration of phagocytes via caspase-3-mediated release of a lipid attraction signal. *Cell* **113:** 717–730.

Identification of lysophosphatidylcholine as a "find-me" signal and one way in which it is produced following caspase activation.

Elliott MR, Chekeni FB, Trampont PC, Lazarowski ER, Kadl A, Walk SF, Park D, Woodson RI, Ostankovich M, Sharma P, et al. 2009. Nucleotides released by apoptotic cells act as a find-me signal to promote phagocytic clearance. *Nature* **461:** 282–286.

Identification of ATP as a "find-me" signal.

Eat-Me Signals and Engulfment

Nagata S, Suzuki J, Segawa K, Fujii T. 2016. Exposure of phosphatidylserine on the cell surface. *Cell Death Differ* **23:** 952–961.

Martin SJ, Reutelingsperger CP, McGahon AJ, Rader JA, van Schie RC, LaFace DM, Green DR. 1995. Early redistribution of plasma membrane phosphatidylserine is a general feature of apoptosis regardless of the initiating stimulus: Inhibition by overexpression of Bcl-2 and Abl. *J Exp Med* **182:** 1545–1556.

An early paper describing the use of annexin V to detect apoptosis.

Wu YC, Horvitz HR. 1998. The *C. elegans* cell corpse engulfment gene *ced-7* encodes a protein similar to ABC transporters. *Cell* **93:** 951–960.

The original characterization of CED7.

Scott RS, McMahon EJ, Pop SM, Reap EA, Caricchio R, Cohen PL, Earp HS, Matsushima GK. 2001. Phagocytosis and clearance of apoptotic cells is mediated by MER. *Nature* **411:** 207–211.

Hanayama R, Tanaka M, Miwa K, Shinohara A, Iwamatsu A, Nagata S. 2002. Identification of a factor that links apoptotic cells to phagocytes. *Nature* **417:** 182–187.

The identification of MFG-E8 as a bridge molecule recognizing phosphatidylserine and its role in engulfment of dying cells.

Miyanishi M, Tada K, Koike M, Uchiyama Y, Kitamura T, Nagata S. 2007. Identification of Tim4 as a phosphatidylserine receptor. *Nature* **450:** 435–439.

Park D, Tosello-Trampont AC, Elliott MR, Lu M, Haney LB, Ma Z, Klibanov AL, Mandell JW, Ravichandran KS. 2007. BAI1 is an engulfment receptor for apoptotic cells upstream of the ELMO/DOCK180/ RAC module. *Nature* **450:** 430–434.

Franc NC, Heitzler P, Ezekowitz RA, White K. 1999. Requirement for croquemort in phagocytosis of apoptotic cells in *Drosophila*. Science **284:** 1991–1994.

Reddien PW, Horvitz HR. 2004. The engulfment process of programmed cell death in *Caenorhabditis elegans*. *Annu Rev Cell Dev Biol* **20:** 193–221.

A detailed review of the engulfment process in nematodes.

Park SY, Kang KB, Thapa N, Kim SY, Lee SJ, Kim IS. 2008. Requirement of adaptor protein GULP during stabilin-2-mediated cell corpse engulfment. *J Biol Chem* **283:** 10593–10600.

Waste Management

Kiss RS, Elliott MR, Ma Z, Marcel YL, Ravichandran KS. 2006. Apoptotic cells induce a phosphatidylserine-dependent homeostatic response from phagocytes. *Curr Biol* **16:** 2252–2258.

A-Gonzalez N, Bensinger SJ, Hong C, Beceiro S, Bradley MN, Zelcer N, Deniz J, Ramirez C, Diaz M, Gallardo G, et al. 2009. Apoptotic cells promote their own clearance and immune tolerance through activation of the nuclear receptor LXR. *Immunity* **31:** 245–258.

In addition to illustrating the role of LXR in clearance of dying cells, this paper shows that activation of LXR may suppress disease consequences that arise due to defective apoptosis.

Dying Cells and Immunity

Elliott MR, Koster KM, Murphy PS. 2017. Efferocytosis signaling in the regulation of macrophage inflammatory responses. *J Immunol* **198:** 1387–1394.

Fond AM, Ravichandran KS. 2016. Clearance of dying cells by phagocytes: Mechanisms and implications for disease pathogenesis. *Adv Exp Med Biol* **930:** 25–49.

Rock KL, Latz E, Ontiveros F, Kono H. 2010. The sterile inflammatory response. *Annu Rev Immunol* **28:** 321–342.

Green DR, Ferguson T, Zitvogel L, Kroemer G. 2009. Immunogenic and tolerogenic cell death. *Nat Rev Immunol* **9:** 353–363.

A survey of the effects of dying cells on the adaptive immune system.

Gaipl US, Munoz LE, Grossmayer G, Lauber K, Franz S, Sarter K, Voll RE, Winkler T, Kuhn A, Kalden J, et al. 2007. Clearance deficiency and systemic lupus erythematosus (SLE). *J Autoimmun* **28:** 114–121.

An overview of the possible role of engulfment defects in autoimmune disease.

Matzinger P. 1994. Tolerance, danger, and the extended family. *Annu Rev Immunol* **12:** 991–1045.

The original review that triggered the idea that dying cells influence the adaptive immune response.

Martin SJ. 2016. Cell death and inflammation: The case for IL-1 family cytokines as the canonical DAMPs of the immune system. *FEBS J* **283:** 2599–2615.

A perspective on how cell death engages the adaptive immune response.

Entosis

Krishna S, Overholtzer M. 2016. Mechanisms and consequences of entosis. *Cell Mol Life Sci* **73:** 2379–2386.

Overholtzer M, Mailleux AA, Mouneimne G, Normand G, Schnitt SJ, King RW, Cibas ES, Brugge JS. 2007. A nonapoptotic cell death process, entosis, that occurs by cell-in-cell invasion. *Cell* **131:** 966–979.

The original paper describing entosis.

LC3-Associated Phagocytosis

Heckmann BL, Boada-Romero E, Cunha LD, Magne J, Green DR. 2017. LC3-associated phagocytosis and inflammation. *J Mol Biol* **429:** 3561–3576.

Martinez J, Almendinger J, Oberst A, Ness R, Dillon CP, Fitzgerald P, Hengartner MO, Green DR. 2011. Microtubule-associated protein 1 light chain 3 α (LC3)-associated phagocytosis is required for the efficient clearance of dead cells. *Proc Natl Acad Sci* **108:** 17396–17401.

The first paper showing LAP in the clearance of dying cells.

Compensatory Proliferation

Fan Y, Bergmann A. 2008. Apoptosis-induced compensatory proliferation. The cell is dead. Long live the cell! *Trends Cell Biol* **18:** 467–473.

Ryoo HD, Gorenc T, Steller H. 2004. Apoptotic cells can induce compensatory cell proliferation through the JNK and the Wingless signaling pathways. *Dev Cell* **7:** 491–501.

Two pathways involved in compensatory proliferation in flies.

Weavers H, Evans IR, Martin P, Wood W. 2016. Corpse engulfment generates a molecular memory that primes the macrophage inflammatory response. *Cell* **165:** 1658–1671.

A role for engulfment of apoptotic cells in wound healing in flies.

Li F, Huang Q, Chen J, Peng Y, Roop DR, Bedford JS, Li CY. 2010. Apoptotic cells activate the "phoenix rising" pathway to promote wound healing and tissue regeneration. *Sci Signal* **3:** ra13.

A possible mechanism of compensatory proliferation in response to apoptosis in mammals.

Laplante P, Brillant-Marquis F, Brissette MJ, Joannette-Pilon B, Cayrol R, Kokta V, Cailhier JF. 2017. MFG-E8 reprogramming of macrophages promotes wound healing by increased bFGF production and fibroblast functions. *J Invest Dermatol* **137:** 2005–2013.

Another mechanism linking clearance of dying cells to wound repair in mammals.

CHAPTER 10

Cell Death in Development

Conradt B. 2009. Genetic control of programmed cell death during animal development. *Annu Rev Genet* **43:** 493–523.

Baehrecke EH. 2002. How death shapes life during development. *Nat Rev Mol Cell Biol* **3:** 779–787.

Meier P, Finch A, Evan G. 2000. Apoptosis in development. *Nature* **407:** 796–801.

Opferman JT, Kothari A. 2018. Anti-apoptotic BCL-2 family members in development. *Cell Death Differ* **25:** 37–45.

Kutscher LM, Shaham S. 2017. Non-apoptotic cell death in animal development. *Cell Death Differ* **24:** 1326–1336.

Cell Death in Nematode Development

Ellis RE, Yuan JY, Horvitz HR. 1991. Mechanisms and functions of cell death. *Annu Rev Cell Biol* **7:** 663–698.

An early survey of cell death in nematode development.

Ellis HM, Horvitz HR. 1986. Genetic control of programmed cell death in the nematode *C. elegans*. *Cell* **44:** 817–829.

The original paper defining the genetic basis for cell death in nematode development.

Maurer CW, Chiorazzi M, Shaham S. 2007. Timing of the onset of a developmental cell death is controlled by transcriptional induction of the *C. elegans* ced-3 caspase-encoding gene. *Development* **134:** 1357–1368.

Abraham MC, Lu Y, Shaham S. 2007. A morphologically conserved nonapoptotic program promotes linker cell death in *Caenorhabditis elegans*. *Dev Cell* **12:** 73–86.

Cell death in the tail-spike cell.

Ellis RE, Horvitz HR. 1991. Two *C. elegans* genes control the programmed deaths of specific cells in the pharynx. *Development* **112:** 591–603.

The identification of CES1 and CES2 in specifying cell death in nematode development.

Metzstein MM, Hengartner MO, Tsung N, Ellis RE, Horvitz HR. 1996. Transcriptional regulator of programmed cell death encoded by *Caenorhabditis elegans* gene *ces-2*. *Nature* **382:** 545–547.

Zarkower D, Hodgkin J. 1992. Molecular analysis of the *C. elegans* sex-determining gene *tra-1*: A gene encoding two zinc finger proteins. *Cell* **70:** 237–249.

Cell Death in Fly Development

Hay BA, Guo M. 2006. Caspase-dependent cell death in *Drosophila*. *Annu Rev Cell Dev Biol* **22:** 623–650.

Baehrecke EH. 2000. Steroid regulation of programmed cell death during *Drosophila* development. *Cell Death Differ* **7:** 1057–1062.

Rusconi JC, Hays R, Cagan RL. 2000. Programmed cell death and patterning in *Drosophila*. *Cell Death Differ* **7:** 1063–1070.

Berry DL, Baehrecke EH. 2007. Growth arrest and autophagy are required for salivary gland cell degradation in *Drosophila*. *Cell* **131:** 1137–1148.

The first clear-cut description of autophagy-dependent cell death in development.

Denton D, Shravage B, Simin R, Mills K, Berry DL, Baehrecke EH, Kumar S. 2009. Autophagy, not apoptosis, is essential for midgut cell death in *Drosophila*. *Curr Biol* **19:** 1741–1746.

Another example of developmental cell death dependent on autophagy.

Cell Death in Vertebrate Development

Montero JA, Hurlé JM. 2010. Sculpturing digit shape by cell death. *Apoptosis* **15:** 365–375.

Coucouvanis E, Martin GR. 1995. Signals for death and survival: A two-step mechanism for cavitation in the vertebrate embryo. *Cell* **83:** 279–287.

Ishizuya-Oka A, Hasebe T, Shi YB. 2010. Apoptosis in amphibian organs during metamorphosis. *Apoptosis* **15:** 350–364.

Cell Death and Selection

Yuen EC, Howe CL, Li Y, Holtzman DM, Mobley WC. 1996. Nerve growth factor and the neurotrophic factor hypothesis. *Brain Dev* **18:** 362–368.

O'Leary DD, Fawcett JW, Cowan WM. 1986. Topographic targeting errors in the retinocollicular projection and their elimination by selective ganglion cell death. *J Neurosci* **6:** 3692–3705.

An example of how cell death can select for proper neuronal connections.

Daley SR, Teh C, Hu DY, Strasser A, Gray DHD. 2017. Cell death and thymic tolerance. *Immunol Rev* **277:** 9–20.

Li KP, Shanmuganad S, Carroll K, Katz JD, Jordan MB, Hildeman DA. 2017. Dying to protect: Cell death and the control of T-cell homeostasis. *Immunol Rev* **277:** 21–43.

Bouillet P, Purton JF, Godfrey DI, Zhang LC, Coultas L, Puthalakath H, Pellegrini M, Cory S, Adams JM, Strasser A. 2002. BH3-only Bcl-2 family member Bim is required for apoptosis of autoreactive thymocytes. *Nature* **415:** 922–926.

The role of BIM in the negative selection of immature T cells.

CHAPTER 11

Cancer and Apoptosis

Hanahan D, Weinberg RA. 2000. The hallmarks of cancer. *Cell* **100:** 57–70.

An influential overview of how cancer occurs, including the role of apoptosis evasion.

Hanahan D, Weinberg RA. 2011. Hallmarks of cancer: The next generation. *Cell* **144:** 646–674.

Dillon CP, Green DR. 2016. Molecular cell biology of apoptosis and necroptosis in cancer. *Adv Exp Med Biol* **930:** 1–23.

Green DR, Evan GI. 2002. A matter of life and death. *Cancer Cell* **1:** 19–30.

A proposal that cancer is largely a consequence of unregulated proliferation combined with evasion of apoptosis.

Letai AG. 2008. Diagnosing and exploiting cancer's addiction to blocks in apoptosis. *Nat Rev Cancer* **8:** 121–132.

Harnessing our knowledge of cell death in treating cancer.

Pelengaris S, Khan M, Evan GI. 2002. Suppression of Myc-induced apoptosis in β cells exposes multiple oncogenic properties of Myc and triggers carcinogenic progression. *Cell* **109:** 321–334.

An important paper that illustrates the relationships between apoptosis and the requirements of c-Myc-induced cancer.

p53

Kastenhuber ER, Lowe SW. 2017. Putting p53 in context. *Cell* **170:** 1062–1078.

Mello SS, Attardi LD. 2017. Deciphering p53 signaling in tumor suppression. *Curr Opin Cell Biol* **51:** 65–72.

Kaiser AM, Attardi LD. 2018. Deconstructing networks of p53-mediated tumor suppression in vivo. *Cell Death Differ* **25:** 93–103.

Kruiswijk F, Labuschagne CF, Vousden KH. 2015. p53 in survival, death and metabolic health: A lifeguard with a licence to kill. *Nat Rev Mol Cell Biol* **16:** 393–405.

Green DR, Kroemer G. 2009. Cytoplasmic functions of the tumour suppressor p53. *Nature* **458:** 1127–1130.

Oren M, Levine AJ. 1983. Molecular cloning of a cDNA specific for the murine p53 cellular tumor antigen. *Proc Natl Acad Sci* **80:** 56–59.

One of two early papers describing a role for p53 in cancer.

Lane DP. 1984. Cell immortalization and transformation by the *p53* gene. *Nature* **312:** 596–597.

See above. Interestingly, the p53 cloned was a dominant-negative mutant that promoted (rather than suppressed) cellular transformation.

p53 and Apoptosis

Lowe SW, Schmitt EM, Smith SW, Osborne BA, Jacks T. 1993. p53 is required for radiation-induced apoptosis in mouse thymocytes. *Nature* **362:** 847–849.

One of two original papers showing that p53 is required for apoptosis in some settings.

Clarke AR, Purdie CA, Harrison DJ, Morris RG, Bird CC, Hooper ML, Wyllie AH. 1993. Thymocyte apoptosis induced by p53-dependent and independent pathways. *Nature* **362:** 849–852.

See above.

Nakano K, Vousden KH. 2001. *PUMA*, a novel proapoptotic gene, is induced by p53. *Mol Cell* **7:** 683–694.

The discovery of PUMA as a BH3-only protein, induced by p53 (with below).

Yu J, Zhang L, Hwang PM, Kinzler KW, Vogelstein B. 2001. PUMA induces the rapid apoptosis of colorectal cancer cells. *Mol Cell* **7:** 673–682.

See above.

Villunger A, Michalak EM, Coultas L, Mullauer F, Bock G, Ausserlechner MJ, Adams JM, Strasser A. 2003. p53- and drug-induced apoptotic responses mediated by BH3-only proteins PUMA and NOXA. *Science* **302:** 1036–1038.

PUMA and NOXA are required for p53-induced apoptosis.

Follis AV, Llambi F, Merritt P, Chipuk JE, Green DR, Kriwacki RW. 2015. Pin1-induced proline isomerization in cytosolic p53 mediates BAX activation and apoptosis. *Mol Cell* **59:** 677–684.

Structural basis for cytosolic p53-induced activation of Bax.

Lowe SW, Cepero E, Evan G. 2004. Intrinsic tumour suppression. *Nature* **432:** 307–315.

Christophorou MA, Ringshausen I, Finch AJ, Swigart LB, Evan GI. 2006. The pathological response to DNA damage does not contribute to p53-mediated tumour suppression. *Nature* **443:** 214–217.

An exploration of how p53 suppresses oncogenesis.

Aubrey BJ, Kelly GL, Janic A, Herold MJ, Strasser A. 2018. How does p53 induce apoptosis and how does this relate to p53-mediated tumour suppression? *Cell Death Differ* **25:** 104–113.

Autophagy and Cancer

Levy JMM, Towers CG, Thorburn A. 2017. Targeting autophagy in cancer. *Nat Rev Cancer* **17:** 528–542.

Kimmelman AC, White E. 2017. Autophagy and tumor metabolism. *Cell Metab* **25:** 1037–1043.

Cadwell K, Debnath J. 2017. Beyond self-eating: The control of nonautophagic functions and signaling pathways by autophagy-related proteins. *J Cell Biol.* **217:** 813–822.

Problems in the interpretation of roles for autophagy in cancer.

Degenhardt K, Mathew R, Beaudoin B, Bray K, Anderson D, Chen G, Mukherjee C, Shi Y, Gelinas C, Fan Y, et al. 2006. Autophagy promotes tumor cell survival and restricts necrosis, inflammation, and tumorigenesis. *Cancer Cell* **10:** 51–64.

Evidence that autophagy acts as a tumor suppressor.

Maclean KH, Dorsey FC, Cleveland JL, Kastan MB. 2008. Targeting lysosomal degradation induces p53-dependent cell death and prevents cancer in mouse models of lymphomagenesis. *J Clin Invest* **118:** 79–88.

Induction of cell death, associated with autophagy, by chloroquine, and applied to cancer therapy.

Apoptosis as a Driver of Cancer

Ichim G, Tait SW. 2016. A fate worse than death: Apoptosis as an oncogenic process. *Nat Rev Cancer* **16:** 539–548.

Weigert A, Mora J, Sekar D, Syed S, Brune B. 2016. Killing is not enough: How apoptosis hijacks tumor-associated macrophages to promote cancer progression. *Adv Exp Med Biol* **930:** 205–239.

Immunogenic Cell Death and Cancer

Galluzzi L, Buque A, Kepp O, Zitvogel L, Kroemer G. 2017. Immunogenic cell death in cancer and infectious disease. *Nat Rev Immunol* **17:** 97–111.

Pitt JM, Kroemer G, Zitvogel, L. 2017. Immunogenic and non-immunogenic cell death in the tumor microenvironment. *Adv Exp Med Biol* **1036:** 65–79.

CHAPTER 12

Modeling Apoptosis

Flusberg DA, Sorger PK. 2015. Surviving apoptosis: Life–death signaling in single cells. *Trends Cell Biol* **25:** 446–458.

Spencer SL, Sorger PK. 2011. Measuring and modeling apoptosis in single cells. *Cell* **144:** 926–939.

Lavrik IN. 2014. Systems biology of death receptor networks: Live and let die. *Cell Death Dis* **5:** e1259.

Lindner AU, Concannon CG, Boukes GJ, Cannon MD, Llambi F, Ryan D, Boland K, Kehoe J, McNamara DA, Murray F, et al. 2013. Systems analysis of BCL2 protein family interactions establishes a model to predict responses to chemotherapy. *Cancer Res* **73:** 519–528.

Lindner AU, Salvucci M, Morgan C, Monsefi N, Resler AJ, Cremona M, Curry S, Toomey S, O'Byrne R, Bacon O, et al. 2017. BCL-2 system analysis identifies high-risk colorectal cancer patients. *Gut* **66:** 2141–2148.

Spencer SL, Gaudet S, Albeck JG, Burke JM, Sorger PK. 2009. Non-genetic origins of cell-to-cell variability in TRAIL-induced apoptosis. *Nature* **459:** 428–432.

Fussenegger M, Bailey JE, Varner J. 2000. A mathematical model of caspase function in apoptosis. *Nat Biotechnol* **18:** 768–774.

An early and fairly successful approach to modeling apoptosis.

Evidence of Bistable Processes in Apoptosis

Pomerening JR. 2008. Uncovering mechanisms of bistability in biological systems. *Curr Opin Biotechnol* **19:** 381–388.

An introduction to bistability.

Yin Z, Qi H, Liu L, Jin, Z. 2017. The optimal regulation mode of Bcl-2 apoptotic switch revealed by bistability analysis. *Biosystems* **162:** 44–52.

This and the following paper model BCL-2 family interactions in apoptosis.

Sun T, Lin X, Wei Y, Xu Y, Shen P. 2010. Evaluating bistability of Bax activation switch. *FEBS Lett* **584:** 954–960.

See above.

Legewie S, Bluthgen N, Herzel H. 2006. Mathematical modeling identifies inhibitors of apoptosis as mediators of positive feedback and bistability. *PLoS Comput Biol* **2:** e120.

Paek AL, Liu JC, Loewer A, Forrester WC, Lahav G. 2016. Cell-to-cell variation in p53 dynamics leads to fractional killing. *Cell* **165:** 631–642.

This and the following paper provide insights into how dynamic regulation of p53 leads to different outcomes.

Hafner A, Stewart-Ornstein J, Purvis JE, Forrester WC, Bulyk ML, Lahav G. 2017. p53 pulses lead to distinct patterns of gene expression albeit similar DNA-binding dynamics. *Nat Struct Mol Biol* **24:** 840–847.

See above.

Ziraldo R, Ma L. 2015. A mathematical model for apoptotic switch in *Drosophila*. *Phys Biol* **12:** 056003.

Making Cells Die

Straathof KC, Pule MA, Yotnda P, Dotti G, Vanin EF, Brenner MK, Heslop HE, Spencer DM, Rooney CM. 2005. An inducible caspase 9 safety switch for T-cell therapy. *Blood* **105:** 4247–4254.

One way of triggering apoptosis, on demand.

Leverson JD, Sampath D, Souers AJ, Rosenberg SH, Fairbrother WJ, Amiot M, Konopleva M, Letai A. 2017. Found in translation: How preclinical research is guiding the clinical development of the BCL2-selective inhibitor venetoclax. *Cancer Discov* **7:** 1376–1393.

The first BH3-mimetic approved for the treatment of human cancer, and how it and others are being investigated.

Oltersdorf T, Elmore SW, Shoemaker AR, Armstrong RC, Augeri DJ, Belli BA, Bruncko M, Deckwerth TL, Dinges J, Hajduk PJ, et al. 2005. An inhibitor of Bcl-2 family proteins induces regression of solid tumours. *Nature* **435:** 677–681.

The first description of ABT-737, and how it was discovered.

Kotschy A, Szlavik Z, Murray J, Davidson J, Maragno AL, Le Toumelin-Braizat G, Chanrion M, Kelly GL, Gong JN, Moujalled DM, et al. 2016. The MCL1 inhibitor S63845 is tolerable and effective in diverse cancer models. *Nature* **538:** 477–482.

A promising, specific MCL1 inhibitor for cancer treatment.

Fulda S. 2017. Smac mimetics to therapeutically target IAP proteins in cancer. *Int Rev Cell Mol Biol* **330:** 157–169.

Brumatti G, Ma C, Lalaoui N, Nguyen NY, Navarro M, Tanzer MC, Richmond J, Ghisi M, Salmon JM, Silke N, et al. 2016. The caspase-8 inhibitor emricasan combines with the SMAC mimetic birinapant to induce necroptosis and treat acute myeloid leukemia. *Sci Transl Med* **8:** 339ra369.

Bykov VJN, Eriksson SE, Bianchi J, Wiman KG. 2017. Targeting mutant p53 for efficient cancer therapy. *Nat Rev Cancer* **18:** 89–102.

Hangauer MJ, Viswanathan VS, Ryan MJ, Bole D, Eaton JK, Matov A, Galeas J, Dhruv HD, Berens ME, Schreiber SL, et al. 2017. Drug-tolerant persister cancer cells are vulnerable to GPX4 inhibition. *Nature* **551:** 247–250.

This paper shows that cancer cells that persist following chemotherapeutic treatment acquire sensitivity to the induction of ferroptosis.

Making Cells Live

Enari M, Hug H, Nagata S. 1995. Involvement of an ICE-like protease in Fas-mediated apoptosis. *Nature* **375:** 78–81.

The first description of the protective effect of caspase (called "ICE-like protease") inhibitors in protection from liver destruction induced by injection of antibodies to CD95 (called "Fas").

Lee H, Shin EA, Lee JH, Ahn D, Kim CG, Kim JH, Kim, SH. 2018. Caspase inhibitors: A review of recently patented compounds (2013–2015). *Expert Opin Ther Pat* **28:** 47–59.

Niu X, Brahmbhatt H, Mergenthaler P, Zhang Z, Sang J, Daude M, Ehlert FGR, Diederich WE, Wong E, Zhu W, et al. 2017. A small-molecule inhibitor of Bax and Bak oligomerization prevents genotoxic cell death and promotes neuroprotection. *Cell Chem Biol* **24:** 493–506.

Small-molecule inhibitors of the mitochondrial pathway of apoptosis.

Degterev A, Hitomi J, Germscheid M, Ch'en IL, Korkina O, Teng X, Abbott D, Cuny GD, Yuan C, Wagner G, et al. 2008. Identification of RIP1 kinase as a specific cellular target of necrostatins. *Nat Chem Biol* **4:** 313–321.

Harris PA, Berger SB, Jeong JU, Nagilla R, Bandyopadhyay D, Campobasso N, Capriotti CA, Cox JA, Dare L, Dong X, et al. 2017. Discovery of a first-in-class Receptor Interacting Protein 1 (RIP1) kinase specific clinical candidate (GSK2982772) for the treatment of inflammatory diseases. *J Med Chem* **60:** 1247–1261.

Angeli JPF, Shah R, Pratt DA, Conrad M. 2017. Ferroptosis inhibition: Mechanisms and opportunities. *Trends Pharmacol Sci* **38:** 489–498.

The Button Experiment

Kauffman S. 1995. *At home in the universe.* Oxford University Press, Oxford.

The first description of the button experiment, a simple idea with intriguing consequences. Many discussions of the button experiment can be found online.

Spierings D, McStay G, Saleh M, Bender C, Chipuk J, Maurer U, Green DR. 2005. Connected to death: The (unexpurgated) mitochondrial pathway of apoptosis. *Science* **310:** 66–67.

An introduction to an extensive online survey of many different proteins that have been described to impact on the mitochondrial pathway of apoptosis.

Buss LW. 1988. *The evolution of individuality.* Princeton Legacy Press, Princeton, NJ.

A classic discussion of the action of evolutionary selection at the cellular level and its implications for development. This book had a major impact on the decision by the author of the book you are holding to explore the processes of cell death.

Index

Page numbers followed by an f denote a figure on the corresponding page.

C

D

N

O

P

R

S

T

U

V

W

X

Z